NEW FACET OF THREE NUCLEON FORCE — 50 YEARS OF FUJITA MIYAZAWA THREE NUCLEON FORCE (FM50)

NEW FACET OF THREE NUCLEON FORCE — 50 YEARS OF FUJITA MIYAZAWA THREE NUCLEON FORCE (FM50)

Proceedings of
The International Symposium

Tokyo, Japan 29 - 31 October 2007

FM50

EDITORS

Hideyuki Sakai
The University of Tokyo
Tokyo, Japan

Kimiko Sekiguchi
RIKEN Nishina Center
Saitama, Japan

Benjamin F. Gibson
Los Alamos National Lab
Los Alamos, New Mexico, U.S.A.

AMERICAN INSTITUTE OF PHYSICS

Melville, New York, 2008
AIP CONFERENCE PROCEEDINGS ■ 1011

Editors

Hideyuki Sakai
Department of Physics
University of Tokyo
Hongo 7-3-1
Bunkyo, Tokyo 113-0033
Japan

E-mail: sakai@phys.s.u-tokyo.ac.jp

Kimiko Sekiguchi
RIKEN Nishina Center
Hirosawa 2-1
Wako, Saitama
351-0198
Japan

E-mail: kimiko@ribf.riken.jp

Benjamin F. Gibson
T-16, MS-B283
Los Alamos National Lab
P.O. Box 1663
Los Alamos, N.M. 87545
U.S.A.

E-mail: bfgibson@lanl.gov

L.C. Catalog Card No. 2008926061

ISBN 978-0-7354-0531-8
ISSN 0094-243X

Printed in the United States of America

CONTENTS

Preface

The International Symposium on New Facets of the Three Nucleon Force – 50 Years of the Fujita-Miyazawa Three Nucleon Force – (FM50) was held in Koshiba Hall on the campus of the University of Tokyo, Japan, October 29–31, 2007. The symposium focused on the topics related to the three nucleon force in nuclear physics. This is because the year 2007 corresponds to the 50 year anniversary of the publication by Fujita and Miyazawa in Progress of Theoretical Physics **17** (1957) 360 of their three nucleon force model based on the two-pion exchange mechanism. At that time Fujita and Miyazawa were young research scientists in their twenties working in the Department of Physics at the University of Tokyo.

It is probably not too much of an exaggeration to say that this model is the origin of quantitative investigation of the three nucleon force in nuclei. Since then many efforts have been made theoretically as well as experimentally to define the essential properties the three nucleon force. Extraordinary progress has been made in theory since the Fujita-Miyazawa publication and the models have approached state-of-the-art. The three nucleon force plays an essential role in the three nucleon continuum Faddeev calculations, *ab initio* calculations describing the nuclear structure of light mass nuclei, in predictions for Λ hypernuclei, *etc*. Experimentally, precise cross sections as well as various spin observables are now routinely measured in nucleon–deuteron elastic scattering plus the breakup and capture processes. Nevertheless, agreement between experiments and theories are sometimes less than satisfactory.

The aim of this symposium was to review the progress in three nucleon force studies, both in theory and in experiment, and to identify achievements made up to now. The purpose was also to discuss future perspectives in three nucleon force studies, to define paths advancing towards a new era. This point is particularly important in Japan because the premier world class radio-active ion beam facility (RIBF) has just initiated operation at RIKEN and the world's most powerful kaon beam will be available in two years from the J-PARC accelerators at Tokai.

The topics covered are:

- The three nucleon force (3NF) study since the Fujita-Miyazawa publication
- 3NF effects in the three nucleon system
- 3NF study in effective field theory
- 3NF effects in hypernuclei
- 3NF effects in the four nucleon system
- 3NF effects in $A > 4$ nuclear structure
- Related nuclear structure and reactions

The symposium consisted of 42 plenary talks including an historical discussion by

Professor Emeritus Hironari Miyazawa. There were 86 participants registered from 12 countries around the world.

The banquet was a memorable event for all participants because Mrs. Kinuko Miyazawa and Mrs. Minako Fujita kindly accepted our invitation to attend.

This symposium was generously supported by the QUESTS (Quantum Extreme System and Their Symmetry) of the Graduate School of Science, University of Tokyo. It was also supported by the RIKEN Nishina Center and the RCNP, Osaka University. It is a pleasure to acknowledge this support, which was essential to make this symposium possible and successful. Last but by no means least, the organizers are grateful to the speakers for their hard work which resulted in a lucid and informative program. We are particularly grateful to Manoel Roberto Robilotta for his beautiful introductory talk.

On behalf of the organizing
committee and the editors,
Hideyuki Sakai
Kimiko Sekiguchi
Benjamin F. Gibson
January 2008

Organizing Committee

- Y. Akaishi, Nihon University / RIKEN Nishina Center
- K. Hatanaka, RCNP, Osaka University
- H. Miyazawa, University of Tokyo (Honorary member)
- T. Motobayashi, RIKEN Nishina Center
- T. Otsuka, University of Tokyo
- K. Sagara, Kyushu University
- H. Sakai, University of Tokyo (Chairman)
- K. Sekiguchi, RIKEN Nishina Center (Scientific Secretary)
- S. Shimoura, CNS, University of Tokyo
- Y. Suzuki, Niigata University
- H. Tamura, Tohoku University

Host :

- the 21st Century COE Program QUESTS (Quantum Extreme Systems and Thier Symmetries) of Graduate School of Science, the University of Tokyo

Co-Hosts :

- Research Center of Nuclear Physics (RCNP), Osaka University
- RIKEN Nishina Center

Koshiba Hall

Three-nucleon interactions: dynamics

M. R. Robilotta

Instituto de Física, Universidade de São Paulo, São Paulo, SP, Brazil

Abstract. A discussion of the dynamics underlying three-body nuclear forces, with emphasis on changes which occurred over several decades, is presented

Keywords: nuclear forces, pion, chiral symmetry.
PACS: 13.75.Cs, 13.75.Gx, 21.30.-x, 21.45.Bc, 21.45.Ff

A WARNING AS INTRODUCTION

In this work, I sketch the development of nuclear three-body forces over a rather large span of time and concentrate narrowly on dynamics. As a consequence, many important subjects are completely omitted, especially those concerning experimental facts and facilities, phenomena and calculational techniques. Even within the restricted domain of dynamics, many relevant contributions cannot be mentioned or discussed, owing to limitations of space. A more comprehensive list of references can be found in a previous related presentation[1].

This work is definitely **not** an historical reconstruction, even if many developments are presented in a rough chronological order. The various sections are organized around themes possessing a kind of internal coherence, which yield something like a research mood. I have tried to convey these moods and to suggest that they change from one period to another. With that purpose in mind, most technicalities are avoided and many original quotations are made, indicated by both quotation marks and italic fonts.

CLASSICAL THREE-BODY FORCES

Classical Electrodynamics, as represented by Maxwell's equations, is **the** paradigm for describing particle interactions mediated by fields. In Electrostatics, the substructure dealing with systems of charges at rest, Coulomb's law holds and the potential energy U is a meaningful concept. In the case of just a pair of particles 1 and 2, the potential energy has the form $U_{12} = KQ_1Q_2/r_{12}$. When several particles are present, one usually invokes the linear superposition principle in order to generalize this result. Accordingly, the potential energy of a system with three particles would be given by

$$U = U_{12} + U_{23} + U_{31} \,. \tag{1}$$

The manifestation of the superpositon principle in this result is the fact that the potential energy of the system is **not** written as

$$U' = U'_{12} + U'_{23} + U'_{31} \,, \tag{2}$$

CP1011, *New Facet of Three Nucleon Force—50 Years of Fujita Miyazawa Three Nucleon Force (FM50), The International Symposium,* edited by H. Sakai, K. Sekiguchi, and B. F. Gibson

with $U'_{ij} \neq U_{ij}$. The superposition principle implements the idea that the interaction between particles 1 and 2 is completely independent of the presence of the third particle. This idea is based on a tacit assumption, namely that the charge distributions of the particles involved are rigid and do not change in the interaction process. In Nature, however, this kind of condition is satisfied only in a few special cases.

Quite generally, if a particle has parts, it must also contain internal forces, responsible for the binding of these parts. As a necessary consequence, when the particle is placed in the presence of sufficiently intense external forces, deformations can be produced. Only elementary particles, which by definition do not have parts, cannot be deformed by external interactions. They also cannot have size, for this would imply the existence of parts. Elementary particles must therefore be point-like, with the word point used in its mathematical sense.

Going back to Electrostatics, one sees that the linear superposition principle can hold only for truly elementary particles. In the case of systems, such as molecules, atoms or nucleons, one indeed has $U'_{ij} \neq U_{ij}$ and eq.(1) corresponds to an approximation. The convenience of this approximation depends on the problem considered and, in many instances, precision demands the inclusion of corrections. These are usually taken into account by means a **three-body potential** W, such that $W \equiv U' - U$. This definition ensures that the full interaction is suitably recovered by the joint use of U and W. Physically, this class of three-body forces is associated with deformations of the interacting objects.

Three-body interactions can also be generated by another kind of mechanism, which depends on the complete Maxwell's picture. When two charged particles 1 and 2 are free to move, both electric and magnetic interactions occur. The force acting on one of the particles depends then on the relative distance r_{12} and on the individual velocities v_1 and v_2. Moreover, as the particles are accelerated, interaction through radiation is also possible. Even when the velocities involved are not very high, the complete force F_1 acting on particle 1 is a complicated function of the form

$$F_1 = F_1(r_{12}, v_1, v_2, a_2)\,. \tag{3}$$

If a third charge is brought into the system, it will give rise to new forces over particles 1 and 2. In particular, the new force over charge 2 will induce a change $a_2 \rightarrow a'_2$ in its acceleration. The inclusion of this effect into eq. (3) promotes an **indirect** influence of the new particle over charge 1, which also corresponds to a three-body force.

EARLY THREE-BODY FORCES

The first work to deal with quantum three-body forces was entitled *"Many-Body Interactions in Atomic and Nuclear Sytems"* and produced in 1939, by Primakoff and Holstein[2]. They begin by considering the second classical mechanism mentioned in the previous section, associated with eq.(3), and then move on to quantum electromagnetic interactions. The conclusion is reached that *"classical many-body potentials"* correspond to the case in which an electron *"simultaneously emits two virtual quanta"*. On the other hand, they note that the process in which *"one electron emits two virtual light quanta in succession"*, gives rise to *"specifically quantum-mechanical many-body potentials"*. The same kind of interactions are considered in the section *"Mesotron Field Theory of Nuclear Interactions"*, with the light quanta replaced by vector Yukawa particles. By estimating the typical sizes and velocities of atomic and nuclear systems, they infer that *"the description of the electromagnetic interaction of electrons in atomic systems, by means of action-at-a-distance two-body potentials is an extraordinarily good approximation [...]"*, whereas *"The usual description of nuclei in terms of two-body potentials cannot [...] be considered satisfactory, except in the case of the deuteron"*.

An extension of this discussion, together with an application to the trinucleon system, were the object of another paper, published in the same year[3].

EARLY STRONG THEORY

The discoveries[4] of the charged pions, in 1947, and of their neutral counterpart, in 1948, were followed by a consistent effort aimed at producing a theory of nuclear forces. This research program relied on the idea of a range expansion, outlined by Taketani, Nakamura and Sasaki[5], in 1951. Soon afterwards, field theory was being employed in nuclear calculations, with great vigour.

In 1952, Lévy published a very influential paper[6], dealing with the description of nuclear forces by means of pion exchanges. The basic πN interaction was assumed to be pseudoscalar, *"since it is only in this case that the intrinsic field theoretical infinities can be separated and re-interpreted consistently"*, and several Feynman diagrams describing the two-body system were calculated. In a subsequent work, which appeared in 1953, Klein[7] corrected some of Lévy's results and considered the role of three-body forces. He also adopted pseudoscalar πN coupling and a pattern that would become important later began to appear, namely that dominant contributions were associated with diagrams containing $N\bar{N}$ intermediate states. Another remark that would have future impact was hidden in a note added in proof: *"Numerical calculations show that the potential obtained in the paper does not agree with experiment. Further work, to be published, indicates moreover, that the perturbation theory doesn't even converge."*

Almost simultaneously, Drell and Huang[8] were investigating the problem of nuclear saturation, which requires a mechanism for short range repulsion. In their abstract, they state: *"[...]The leading term in the n-body potential depends only on the interparticle distance and is repulsive (attractive) for n odd (even)."* This finding is materialized in their interesting fig.6 and the special role of three-body forces is stressed in the concluding section: *"[...]we find that many-body forces, and in particular the three-body repulsion, provide a satisfactory qualitative understanding of nuclear saturation."*

The qualitative findings by both Klein and Drell-Huang were confirmed and summarized by Wentzel[9]: *"In pseudoscalar meson theory (PS coupling), the saturation character of nuclear forces is accounted for by the cooperative action of many-body forces. In particular, the repulsive 3-, 5-, 7-, ... body forces keep the nucleus from collapsing into a very small volume. These nuclear interactions result mainly from matrix elements describing virtual nucleon-pair creation and annihilation processes[...]"*.

These conclusions were disputed, already in 1953, by Brueckner and Watson[10]. In their own words: *"We have seen that it is possible to derive a nucleon-nucleon poten-*

tial, working entirely with a nonrelativistic approximation to the pseudoscalar meson theory. [...] We have depended rather strongly on the suppression of nucleon pair formation by radiative effects; the contributions to the potential from pair formation in high order calculated without taking into account such effects otherwise tend to be so large as to invalidate the power series expansions usually used." This perception underlies their ad hoc calculation procedure, which became known as **pair suppression** and was widely adopted. In another publication along this line, by Brueckner, Levinson and Mahmoud[11], one finds: *"The application of these techniques to a specific problem of considerable interest, namely the two-body potentials of pseudoscalar meson theory (adjusted to fit the low-energy scattering parameters) shows that these potentials also have the necessary characteristics to give an approximately correct description of nuclear saturation."* In the same paper these authors also argue that many-body forces produce negligible effects.

This intense debate about nuclear forces raised many problems regarding the applicability of ordinary perturbation theory to strong interactions. As a consequence, in the mid-fifties, lagrangians had fallen into disgrace and theoretical interest shifted to the analytic structure of the S-matrix, investigated by means of dispersion relation techniques. The rehabilitation of strong lagrangians had to wait for the development of chiral symmetry.

THE FIRST MODERN THREE-BODY FORCE

In the second half of the fifties, the idea of deriving potentials from first principles was put aside and replaced by the more pragmatic attitude of blending theory and empirical information. At that time, the relationship between the off-shell πN amplitude and important components of the potential, as indicated in fig. 1, had already been suggested. Also, many measurements of elastic πN scattering, performed in the same period, had led to a considerable improvement of the empirical information about the on-shell amplitude. If suitably treated by means of dispersion relations, a rather valuable information about the off-shell πN amplitude could be obtained from empirical data.

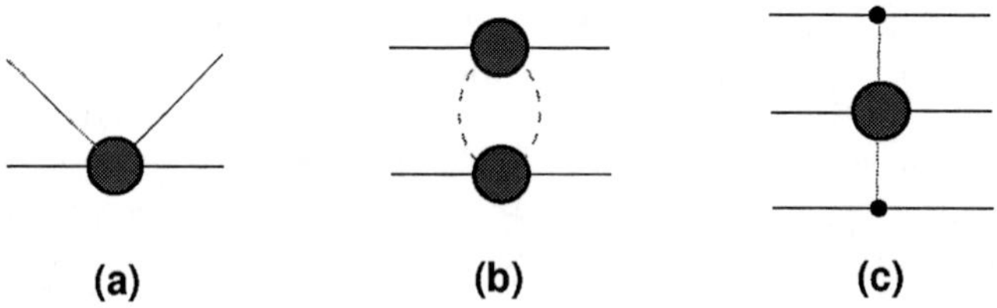

FIGURE 1. (Color online) (a) Free πN amplitude; (b) two-pion exchange two-body potential; (c) two-pion exchange three-body potential.

At the end of 1956, a paper incorporating the new attitude was published by Miyazawa[12]. Its title was *"Interaction of P- and S-Wave Pions with Fixed Nucleons"* and, in the abstract, one finds the clear statement: *"A method is given of replacing pion scattering parts in a Feynman diagram by experimentally observed quantities. [...] Two examples are given: (1) The anomalous magnetic moment of the proton is rigorously expressed in terms of pion-nucleon scattering amplitudes [...]. (2) The internucleon potential is also expressed by means of scattering quantities. In this case the number of virtual pions exchanged between the two nucleons is limited to two, although the number of pions emitted and absorbed by the same nucleon is not limited."*

In the cases of *P*- and *S*-waves, the scattering matrix elements for the process $\pi^i(k)N \to \pi^j(q)N$ were written respectively as

$$\langle j,q|S|i,k\rangle = 2\pi i\delta(k_0-q_0)\left[A(k_0)\,\tau_i\tau_j\,\sigma\cdot k\,\sigma\cdot q + B(k_0)\left(\tau_i\tau_j\,\sigma\cdot q\,\sigma\cdot k + \tau_j\tau_i\,\sigma\cdot k\,\sigma\cdot q\right)\right.$$
$$\left. +\ C(k_0)\,\tau_j\tau_i\,\sigma\cdot q\,\sigma\cdot k\right] e^{i(k-q)\cdot x}\,, \qquad (4)$$

$$\langle j,q|S_s|i,k\rangle = 2\pi i\delta(k_0-q_0)\left[D(k_0)\,\tau_i\tau_j + E(k_0)\,\tau_j\tau_i\right] e^{i(k-q)\cdot x}\,, \qquad (5)$$

and dispersion relations were used to derive the functions *A*, *B* and *C* from empirical total cross sections, whereas *D* and *E* were obtained from scattering lengths. The importance of these results lies in the fact that they hold for both real and virtual pions. The derivation of the two-pion exchange component of the *NN* potential, fig. 1b, was performed with great care, so as to avoid various possibilities of double counting.

In 1957, Fujita and Miyazawa published the work *"Pion Theory of Three-Body Forces"*[13], in which the off-shell πN amplitude also plays an essential role, as in fig. 1c. Their construction of the two-pion exchange three-nucleon potential (*TPE*-3*NP*) is described: *"The process being considered is this. Particle (1) emits a pion which is scattered by particle (2) and then absorbed by particle (3). [...] For the scattering by particle (2), experimental values can be used. Actually, this is a scattering of a virtual pion having zero energy. However, the use of dispersion relations makes it possible to correlate this scattering with real scatterings. In this way all virtual transitions on particle (2) are correctly taken into account."*

This clear understanding of the role played by the off-shell πN amplitude in the *TPE*-3*NP* establishes the work of Fujita and Miyazawa as the first one in the modern tradition. With variations, the essence of their approach is present in the leading terms of potentials constructed ever since.

On the more technical side, an explicit analytic form for the potential was presented in terms of Yukawa functions and numerical values for the constants $A(0)$, $B(0)$ and $D(0)$ were derived from πN scattering data. A little of manipulation allows their original expressions to be recast in the form of *TPE*-3*NP* adopted nowadays, given by

$$V_L(123) = -\frac{\mu}{(4\pi)^2}\left\{\delta_{ab}\left[a\,\mu - b\,\mu^3\,\nabla_{12}\cdot\nabla_{23}\right] + d\,\mu^3\,i\varepsilon_{bac}\tau_c^{(2)}\,i\sigma^{(2)}\cdot\nabla_{12}\times\nabla_{23}\right\}$$
$$\times\left((g_A\mu/2f_\pi)\,\tau_a^{(1)}\,\sigma^{(1)}\cdot\nabla_{12}\right)\left((g_A\mu/2f_\pi)\,\tau_b^{(3)}\,\sigma^{(3)}\cdot\nabla_{23}\right)Y(x_{12})\,Y(x_{23})\,, \qquad (6)$$

where μ is the pion mass and *a*, *b* and *d* are strength constants. The first one is associated with *S*-waves and the other two, with *P*-waves. The interpretation of this result is rather simple. The term within curly brackets describes the off-shell πN scattering on nucleon 2, those within parentheses represent the emission and absorption of a single pion on nucleons 1 and 3, whereas the Yukawa functions *Y* describe pion propagation. To my knowledge, this structure for the potential was first obtained by Fujita and Miyazawa, indicating the power and generality of their approach.

A final comment about their work is in order. It is well known that Δ intermediate states play a dominant role in some low-energy πN cross sections and, by extension, the same happens in the three-body force. On the other hand, the structure of the actual Fujita-Miyazawa potential also incorporates other effects and is rather general. Therefore the widespread identification of the F-M force with Δ intermediate states corresponds, at least, to an undue over-simplification of their results.

The off-shell πN amplitude derived in ref.[12] was also used in the study of more involved components of the three-nucleon force, by Fujita, Kawai and Tanifuji[14], in 1962. In that work, particular attention was paid to both the so called ring diagrams and to processes containing exchanges of one pion with one of the nucleons and two pions with the other, which continue to be object of attention nowadays.

THE AGE OF CHIRAL SYMMETRY

The discovery, in 1956, of parity non-conservation in weak interactions, motivated a great interest about the nature of weak currents. The well known electromagnetic current is represented by a vector (J^μ) and conserves parity. In the case of weak interactions, on the other hand, one finds both vector (V^μ) and axial-vector (A^μ) currents, which respectively conserve and do not conserve parity. The electromagnetic current is conserved ($\partial \cdot J = 0$), but the same does not happen with the weak ones. Nevertheless, in the late fifties, the notion was developed that, in some special limits, weak currents could be considered as being approximately conserved. This was translated into approximate symmetries of the interaction lagrangian, associated with transformations that can be schematically represented as

$$[V,V] \rightarrow V, \qquad [V,A] \rightarrow A, \qquad [A,A] \rightarrow V\,. \tag{7}$$

Loosely speaking, changes in parity occur in images produced by mirrors and one may say that the action of an axial operator is analogous to transforming a right hand into a left one. A vector operator does not change parity and corresponds to transforming a right hand into itself. This kind of analogy led the transformations indicated in eqs.(7) to be called **chiral**, a word derived from **hand** in Greek.

In 1960, **chiral symmetry**, implemented by means of the linear-sigma model[15], has been applied with great success to strong interactions. This model assumes the existence of a scalar-isoscalar particle called σ and proved to have a very rich structure. One of its beautiful features is the picture of a strong vacuum, which is both covariant and not empty.

Another important result was the natural explanation produced for the observed smallness of the πN scattering lengths, which puzzled research in the early fifties. This achievement is crucial for Nuclear Physics. The outer layers of nucleon interactions are due to pion exchanges and many important components of the force depend on the intermediate πN amplitudes shown in fig. 1. In the σ-model, the πN amplitude is given by the three diagrams shown in fig. 2, with pseudoscalar πN coupling. The first two diagrams correspond to the model adopted in the early fifties[6] and explicit calculation of their contributions yields $(a^+_{PS}, a^-_{PS}) = (-1.84, 0.136)\mu^{-1}$, where the superscripts $\pm$ refer to isospin channels. The comparison of these results with the experimental values[16]

$(a^+_{exp}, a^-_{exp}) = (-0.008, 0.092)\mu^{-1}$ shows that the prediction for a^+ is too large by a factor of about 200. The third diagram, containing the exchange of the σ, is the signature of chiral symmetry in this process and produces $(a^+_\sigma, a^-_\sigma) = (1.83, 0)\mu^{-1}$. A large cancellation occurs in $(a^+_{PS}+a^+_\sigma, a^-_{PS}+a^-_\sigma) = (-0.01, 0.136)\mu^{-1}$, turning both sums compatible with empirical orders of magnitude. Related cancellations also happen in the case of nuclear forces and a pedagogical discussion can be found in ref.[17].

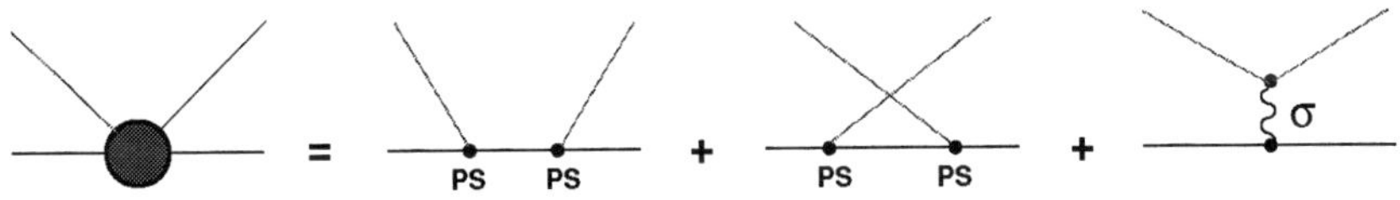

FIGURE 2. (Color online) Structure of the πN amplitude in the linear-sigma model.

Two remarks about the chiral πN amplitude are important at this point. The first one is that the smallness of the full σ-model result is a direct consequence of chiral symmetry. The linear σ-model corresponds to just one among many possibilities for implementing the symmetry in hadronic systems and the order of magnitude of the chiral πN amplitude is independent of the method employed. The second remark is that the symmetry, powerful as it is, cannot predict the full empirical content of the πN interaction. At low-energies, empirical information about this process is usually encoded into a polynomial in the variables ν and t, proposed by Höhler and collaborators[16]. The coefficients of this polynomial are obtained by extrapolating experimental information to the region below threshold, by means of dispersion relations. For this reason, they are called **subthreshold coefficients**. So, in order to construct a πN amplitude suitable to be employed in nuclear interactions, one uses chiral symmetry, supplemented by subthreshold information, as indicated in fig. 3.

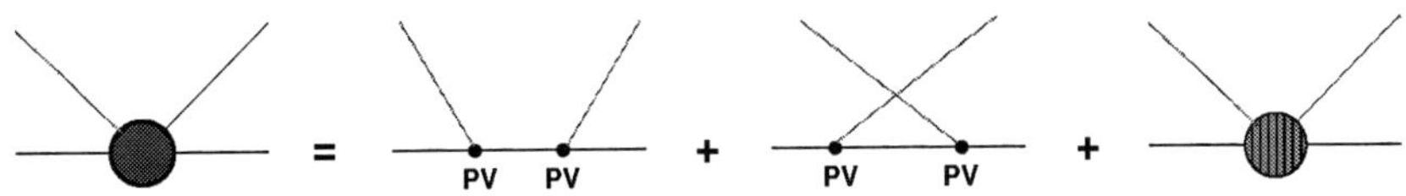

FIGURE 3. (Color online) Chiral tructure of the πN amplitude; the dashed (green) bubble represents a polynomial contribution.

The incorporation of chiral symmetry into hadronic amplitudes is performed rigorously and many results have the status of **theorems**. In the simplest versions, these theorems are exact for low-energy interactions of massless pions. The masses of actual pions, although small, are non-vanishing and it is in this sense that chiral symmetry is approximate. However, the transposition of chiral theorems to the case of massive pions is also performed rigorously and, as a consequence, predictions from the approximate symmetry are both under control and unambiguous.

In general, chiral theorems or amplitudes have the form of power series[1] in a typical scale q, set by either pion four-momenta or nucleon three-momenta, such that $q \ll 1$

[1] This series also includes chiral logarithms an other non-analytic terms.

GeV. As power series[2], these theorems involve both leading order terms and corrections. The former have been evaluated for a variety of problems since the sixties, whereas the latter is th object of the **chiral perturbation theory**, which gained momentum in the last two decades.

THE LEADING ORDER THREE-BODY FORCE

The leading chiral three-body force begins at $\mathcal{O}(q^3)$ and is due to the process shown in fig.1c. It took a long time between the first derivations were performed and a kind of consensus about the uniqueness of the result could be reached by practitioners of the field.

The first application of chiral symmetry to three-body forces appeared in 1968, in a paper by Brown, Green and Gerace[18], in the case of nuclear matter. They state in the abstract: *"[...] it is shown that long-range three-body forces in nuclear matter are essentially zero. This means that they are also small in nuclei."* This conclusion was soon disputed in a letter by McKellar and Rajaraman[19], where one finds: *"[...] we present below two models [...], both of which indicate that three-body forces in nuclei are about as strong as the two-body forces."* A compromise between these extremes was reached in a third paper, by Brown and Green[20], whose abstract begins with: *"An earlier claim about the smallness of three-body effects in nuclear matter is examined in the light of a criticism. Our conclusion is that, whereas the lowest-order three-body effect is small, second order effects are larger,[...]".*

The next step was given in 1974, by Yang[21],who shifted attention to the triton and was the first to use the so called **chiral dynamics**. As he explains in the abstract: *"We have derived a two-pion exchange three-body potential using a chiral invariant Lagrangian for the pion-nucleon interaction. Only the three dominant lowest order processes which give rise to a three-body interaction are included: (a) The second pion is emitted before the first pion is absorbed and the nucleon which emits the second pion goes into a positive-energy state; (b) one nucleon exchanges a ρ meson with a pion; (c) one nucleon is scattered into an $N^*(1236)$."* In the main text, one finds an instructive description of how effective lagrangians were striking back. After emphasizing the role of the off-shell intermediate πN amplitude, he says: *"The PCAC (partially conserved axial-vector current) and the current algebra developed in the late 1960's have been quite successful in describing the soft-pion process, e,g., the Adler's consistency conditions[4] on strong interactions and low-energy parameters[5] of the πN scattering are found to be in good agreement with the experiments. It was noted[6] that for soft-pion processes the predictions obtained with current algebra can also be derived by a different method: Just use the lowest-order graphs generated by a chiral-invariant Lagrangian. We shall apply this method[...]".* His reference [6] is a famous paper by Weinberg[22]. Another passage

---

[2] In a Taylor series such as $\sin x \simeq x - x^3/3! + \cdots$, the r.h.s. is obtained by means of a well defined set of operations over the l.h.s.= $\sin x$. On the other hand, in the case of a chiral amplitude A_χ, one has schematically $A_\chi \simeq A_L q^L + A_{NL} q^{L+1} + \cdots$ and the coefficients A_i must be determined by means of field theory, since one does not know the l.h.s.

shows how chiral symmetry was displacing the old pair-suppression mechanism of the fifties: *"The three-body force arising from a virtual nucleon-antinucleon pair [...] is suppressed, since the soft-pion theory tells us explicitly that* $T^{\rm Born}$ *is computed according to the gradient coupling scheme."* Approximations done in the treatment of process (a), however, does not allow his results to be cast in the form of eq.(6).

Just a little later, in 1975, another paper focused on nuclear matter appeared, by Coon, Scadron and Barrett[23]. Their strategy for dealing with dynamics is presented in the abstract: *"We re-examine the off-shell* πN *amplitude occurring in the two-pion exchange three-body force, subject to all the constraints of current algebra. This amplitude is not dominated by the* $\Delta(1231)$ *isobar; instead, if the* $\sigma-$*term is known, it can be determined from on-shell scattering."* In the third section, *"Off-shell pion-nucleon scattering"*, one finds a detailed description of their approach, as well as a comparison with results form other works. With hindsight, this work can be considered as being a precursor of the Tucson-Melbourne (TM) potential[24], published in 1979. In this later paper, previous results were generalized and its abstract reads: *"We derive the complete three-nucleon potential of the two-pion exchange type, suitable for nuclear structure calculations, by extending away from the forward direction the subthreshold off-pion-mass-shell* πN *amplitude of Coon, Scadron and Barrett. The off-mass-shell extrapolation, subject to current algebra and PCAC constraints, yields approximately model independent amplitudes (in that they depend primarily on* πN *data) in the complete potential. [...]"*. This version of the $3NP$ was restricted to a particular reference frame, but this condition was lifted a little later[25]. In those papers, the notation used in eq.(6) was introduced. The TM force also included another structure, represented by a parameter c within the curly bracket, which was written as $\{\delta_{ab}[a\,\mu - b\,\mu^3\,\nabla_{12}\cdot\nabla_{23} - c\,\mu^3\,(\nabla_{12}^2+\nabla_{23}^2)] + d\,\mu^3\,i\varepsilon_{bac}\tau_c^{(2)}\,i\sigma^{(2)}\cdot\nabla_{12}\times\nabla_{23}\}$. This term proved to be problematic in numerical calculations and has been identified with short range effects[26, 27]. It was eventually removed[3] in a paper by Coon and Han[28], published in 2001, where parameters for a new version, known as TM', were presented.

In 1983, the TPE-$3NP$ known as Brazil potential was produced[4], by Coelho, Das and Robilotta[29]. It was based on effective lagrangians and relied on a chiral model for the πN amplitude derived by Olsson and Osypowski[30] which included, besides the same ρ and Δ intermediate processes considered by Yang[21], a parametrized form for the S-wave interaction. In that model, all the parameters were tuned to empirical subthreshold coefficients and the resulting structure conveyed model independent information. This gave rise to a potential in agreement with eq.(6). In 1986, the possibility that the long range physics in both the TM and Brazil forces could be contaminated by the inclusion of form factors, required by numerical calculations, was discussed[27].

It is worth recalling that chiral expansions are performed by means of well defined theoretical rules and hence their results must be unique. This holds, in particular, for leading terms. In the case of three-body forces, the leading term has the generic structure given by eq.(6), whereas free parameters are determined by subhtreshold coefficients. A

---

[3] For furhter details, the reader is directed to Prof. Coon's contribution to this volume.

[4] The last term of eq.(67) in that paper contains a misprinted sign, which was corrected in ref.[27].

sample of values adopted for potential parameters along five decades is given in table 1. At present, variations in those values reflect just the empirical information used as input and are totally unrelated with chiral symmetry.

TABLE 1. Leading-order parameters for the *TPE-3NP* (dimensionless units).

ref.	year	potential	$a\mu$	$b\mu^3$	$c\mu^3$	$d\mu^3$
[13]	1957	F-M	-0.27	-1.24	0	-0.31
[24]	1979	TM	1.13	-2.58	-1.05	-0.75
[29]	1983	BR	-1.05	-2.29	0	-0.77
[28]	2001	TM'-93	-0.74	-2.53	0	-0.72
[28]	2001	TM'-99	-1.12	-2.80	0	-0.75

CHIRAL PERTURBATION THEORY

Chiral perturbation theory (ChPT) is the art of deriving systematic **corrections** to leading order terms in chiral expansions. In the case of Nuclear Physics, the research program was outlined by Weinberg[31] and one of his papers deals specifically with three-body forces[32]. In ChPT, emphasis is put on a lagrangian which possesses approximate chiral symmetry and is written as a string of terms with different orders in the power-counting parameter q. The coefficients of higher order terms are known as **low-energy constants** (LECs) and their presence is felt in final results for chiral amplitudes. The values of the LECs cannot be fixed by the symmetry and must be determined from experimental information. In the case of nuclear potentials, the basic source of information are the usual πN subthreshold coefficients. For this reason, the result for the **leading** term, which was already unique and given by eq.(6) before ChPT, continues to be so afterwards, in spite possible apparent changes in form due to rephrasing. On the other hand, as stated in the first paper to deal with applications of ChPT, by Ordóñez and van Kolck[33], in 1992, the new method has the clear advantage of stressing the model independence of the results. ChPT is also important for preventing whimsical diagramatics in complex problems.

From an operational point of view, the power of ChPT in nuclear forces is that it allows the generation of new structures within a unified theoretical framework. One possible direction for new developments concerns **short-range** physics. In the case of three-body forces, short-range chiral interactions were considered in 1994 by van Kolck[34] and by Friar, Huber, and van Kolck[35], in 1999. In the later paper[5], the abstract points to the solution of an old problem: *"It is demonstrated that the short-range c term of the Tucson-Melbourne force is unnatural in terms of power counting an should be dropped."*

The second possible direction is related with the production of new mathematical structures, which go beyond eq.(6). At present, next-to-leading order corrections to the three-body force, which involve a rather large number of diagrams, are being calculated

[5] One notes that the values quoted for the F-M and Brazil forces in table II of ref.[35] are in qualitative disagreement with those given in table 1 of the present work.

by Epelbaum, Meissner and collaborators[36]. The reader is directed to the contribution by Prof. Meissner to this volume for a comprehensive account of their results.

A taste of what is about to come can be felt in a recent work by Ishikawa and Robilotta[37], in which a selected subset of corrections to the leading term, associated with the long-range part of the TPE-$3NP$, was considered. The final result has the structure $V(123) = V_L(123) + [V_{\delta L}(123) + \delta V(123)]$, where V_L is the old leading term given by eq.(6), whereas the factor within square brackets includes corrections due to ChPT. The term $V_{\delta L}$ can be obtained directly from V_L, by replacing (a,b,c) with $(\delta a, \delta b, \delta c)$. In other words, this part of the correction corresponds to shifts in the parameters of the leading term, which are numerically smaller than 10%. The factor $\delta V(123)$, on the other hand, represents effects described by new mathematical functions, whose actual forms are too cumbersome to be displayed here. For the present purposes, it suffices to say that they involve both non-local operators and the replacement of the Yukawa functions in eq.(6) by more complicated propagators involving loop integrals. The strengths of these new functions are described by a set of new parameters e_i, which are also typically smaller than 10% of the leading ones. So, the impact of ChPT in table 1 is to produce both small modification in already existing coefficients and the appearance of many new columns. The latter are by far the most interesting ones, since they may contain the explanation for effects such as the A_y puzzle[6].

CONCLUDING REMARKS

Progress in our understanding of the world always involves both conceptual ruptures and continuities. In this work, I have tried to recall that this feature was also present in the development of three-body forces, a highly collective task. If one compares the discussions of strong interactions held in the fifties with those occurring nowadays, one may be tempted to stress discontinuity. On the other hand, reflecting about the work by Fujita and Miyazawa, one notices that their formulation of the problem in terms of dispersion relations corresponds more to a bending of its path than to a complete break with the past. It is therefore best described as a **turning point**. Chiral symmetry came next and has promoted a solid understanding of both meanings and values of the strength parameters of the force. And it is remarkable that chiral perturbation theory, which guides the present wave of research, is not replacing, but rather extending and complementing the results derived fifty years ago by Fujita and Miyazawa.

ACKNOWLEDGMENTS

It was a great pleasure participating in the FM50 Symposium and I would like to thank the organizers for this nice opportunity.

---

[6] Discussed by Prof. Tornow in this volume.

REFERENCES

1. M.R. Robilotta, Few-Body Systems, Suppl. **2**, 35 (1987).
2. H. Primakoff and T. Holstein, Phys. Rev. C **55**, 1218 (1939).
3. L. Jánossy, Proc. Cambridge Phil. Soc. **35**, 616 (1939).
4. C.M.G. Lattes, G.P.S. Occhialini and C.F. Powell, Nature **160**, 453, 486 (1947); E. Gardner and C.M.G. Lattes, Science **107**, 270 (1948).
5. M. Taketani, S. Nakamura, and T. Sasaki, Prog. Theor. Phys. **6**, 581 (1951).
6. M.M. Lévy, Phys. Rev. **88**, 725 (1952).
7. A. Klein, Phys. Rev. **90**, 1101 (1953).
8. S.D. Drell and K. Huang, Phys. Rev. **91**, 1527 (1953).
9. W. Wentzel, Phys. Rev. **91**, 1573 (1953); Helv. Phys. Acta **15**, 111 (1942); **25**, 569 (1952).
10. K.A. Brueckner and K.M. Watson, Phys. Rev. **92**, 1023 (1953).
11. K.A. Brueckner, C.A. Levinson and H.M. Mahmoud, Phys. Rev. **95**, 217 (1954).
12. H. Miyazawa, Phys. Rev. **104**, 1741 (1956).
13. J. Fujita and H. Miyazawa, Progr. Theor. Phys. **17**, 360 (1957).
14. J. Fujita, M. Kawai and M. Tanifuji, Nucl. Phys. **29**, 252 (1962).
15. M. Gell-Mann and M, Lévy, N. Cim. **16**, 705 (1960).
16. G. Höhler, group I, vol.9, subvol.b, part 2 of Landölt-Bornstein Numerical data and Functional Relationships in Science and Technology, ed. H.Schopper, 1983; G. Höhler, H. P. Jacob, and R. Strauss, Nucl. Phys. **B39**, 273 (1972).
17. J-L. Ballot, M. R. Robilotta and C.A. da Rocha, Int. J. Mod. Phys. E **6**, 83 (1997).
18. G.E. Brown, A.M. Green and W.J. Gerace, Nucl. Phys. A **115**, 435 (1968).
19. B.H.J. McKellar and R. Rajaraman, Phys. Rev. Lett. **21**, 450, 1030 (1968).
20. G.E. Brown and A.M. Green, Nucl. Phys. A **137**, 1 (1969).
21. S-N. Yang, Phys. rev. C **10**, 2067 (1974).
22. S. Weinberg, Phys. Rev. Lett. **18**, 507 (1967).
23. S.A. Coon, M.S. Scadron and B.R. Barrett, Nucl. Phys. A **242**, 467 (1975).
24. S.A. Coon, M.D. Scadron, P.C. McNamee, B.R. Barrett, D.W.E. Blatt and B.H.J. McKellar, Nucl. Phys. A **317**, 242 (1979).
25. S. A. Coon and W. Glöckle, Phys. Rev. C **23**, 1790 (1981).
26. M.R. Robilotta, M.P. Isidro Filho, H.T. Coelho and T.K. Das, Phys. Rev. C **31**, 646, (1985); M.R. Robilotta and M.P. Isidro Filho, Nucl. Phys. A **451**, 581 (1986).
27. M.R. Robilotta and H.T. Coelho, Nucl. Phys. A **460**, 645 (1986);
28. S.A. Coon and H.K. Han, Few Body Syst. **30**, 131 (2001).
29. H. T. Coelho, T. K. Das, and M. R. Robilotta, Phys. Rev. C **28**, 1812 (1983).
30. M. G. Olsson and E. T. Osypowski, Nucl. Phys. B **101**,136 (1975); E. T. Osypowski, Nucl. Phys. B **21**, 615 (1970).
31. S. Weinberg, Phys. Lett. B **251**, 288 (1990); Nucl. Phys. B **363**, 3 (1991).
32. S. Weinberg, Phys. Lett. B **295**, 114 (1992).
33. C. Ordóñez and U. van Kolck, Phys. Lett. B **291**, 459 (1992).
34. U. van Kolck, Phys. Rev. C **49**, 2932 (1994).
35. J. L. Friar, D. Huber, and U. van Kolck, Phys. Rev. C **59**, 53 (1999)
36. V. Bernard, E. Epelbaum, H. Krebs and Ulf-G. Meissner, preprint nucl-th/0712.1967.
37. S. Ishikawa and M.R. Robilotta, Phys. Rev. C **76**, 014006 (2007).

Three-nucleon force effects in 3N hadronic and photonic reactions

H. Witała*, R. Skibiński*, J. Golak*, W. Glöckle†, H. Kamada**, A. Nogga‡ and W.N. Polyzou§

*M. Smoluchowski Institute of Physics, Jagiellonian University, PL-30059 Kraków, Poland
†Institut für Theoretische Physik II, Ruhr-Universität Bochum, D-44780 Bochum, Germany
**Department of Physics, Faculty of Engineering, Kyushu Institute of Technology, 1-1 Sensuicho Tobata, Kitakyushu 804-8550, Japan
‡Forschungszentrum Jülich, Institut für Kernphysik (Theorie), D-52425 Jülich, Germany
§Department of Physics and Astronomy, The University of Iowa, Iowa City, Iowa 52242, USA

Abstract. Results on three-nucleon (3N) elastic scattering and breakup below the pion production threshold are discussed with emphasis on the need for a three-nucleon force (3NF). The large discrepancies found between a theory based on numerical solutions of 3N Faddeev equations with modern NN potentials only and data point to the need for 3NF's. Successes and failures of the present 3NF models, mostly of a 2π-exchange nature, to describe high precision 3N data are discussed. Effects due to relativity both in elastic nucleon-deuteron (Nd) scattering and breakup reaction are presented and consequences for a 3NF study pointed out. As an application of 3N bound and scattering states, results for photodisintegration of 3N bound states are shown.

Keywords: nucleon-nucleon potential, three-nucleon force, elastic scattering, breakup, photodisintegration, pd capture
PACS: 21.45.+v; 21.10.-k; 25.10.+s; 25.20.-x

INTRODUCTION

Traditionally in nuclear physics the Hamiltonian has been assumed to be of a nonrelativistic form in which pairwise interactions between nucleons are supplemented by 3NFs for systems with more than two nucleons. The construction of NN potentials guided by a meson picture led to a generation of realistic NN interactions which describe the NN data set with high precision (χ^2/datum ≈ 1) [1, 2, 3]. The 3N system is the first nontrivial case where those realistic potentials can be tested. In that system also the first time 3NFs come into play making it a valuable source of information on 3NF properties and their significance in the nuclear Hamiltonian.

The need for 3NFs was established when three- and four-nucleon bound states were solved exactly using standard integration and differentiation methods [4, 5]. Later, using stochastic techniques low energy states for nuclei up to A=8 were calculated [6]. It turned out that in all cases studied realistic NN forces alone provided clear underbinding, which for ^{3}H and ^{3}He amounts to $\approx 0.5 - 0.9$ MeV and for ^{4}He to $\approx 2-4$ MeV.

It was natural to look for an explanation of this underbinding introducing 3NFs in the nucleonic Hamiltonian. An important example for a process which leads to a 3NF is a $\pi - \pi$ exchange between three nucleons with an intermediate Δ excitation of one nucleon, considered by Fujita-Miyazawa [7]. Later, this process was incorporated

CP1011, *New Facet of Three Nucleon Force—50 Years of Fujita Miyazawa Three Nucleon Force (FM50), The International Symposium,* edited by H. Sakai, K. Sekiguchi, and B. F. Gibson

into the Urbana IX 3NF, where it is supplemented by a phenomenological short-range spin- and isospin-independent part [8]. The Tucson-Melbourne (TM) model [9] was constructed using a more general πN amplitude. It also contains a strong form factor parametrization with a cut-off parameter Λ which can be adjusted to the ^{3}H or ^{3}He binding energy when this 3NF is combined with a particular NN interaction. For the combination of the AV18 NN potential and the Urbana IX 3NF calculations are available for nuclei up to $A = 8$ which show that one can reach a reasonable description of the low lying bound states in those nuclei [6]. However, the results show, for instance, an insufficient spin-orbit splitting and an insufficient charge dependence, which indicate defects of this 3NF. Taking additional 3NF ring-diagrams into account the situation could be improved [10].

A rich set of spin observables for elastic Nd scattering and the breakup reaction is a source of valuable information on the spin and momenta structure of 3NFs. In the next section we briefly review the 3N scattering formalism and give some examples where data are compared to various NN potential predictions alone or combined with different 3NFs. These examples are chosen to show the importance of the 3NF in the 3N system. The importance of 3NFs seems to increase with increasing energy of the 3N system. Therefore we discuss the importance of relativistic effects in these reactions and their significance in the study of 3NF effects. As an application of rigorous 3N bound and scattering states we discuss next the photodisintegration of 3N bound states and the proton-deuteron (pd) capture process. In the last section we give the summary.

REACTIONS IN THE 3N CONTINUUM

All observables for elastic Nd scattering and the breakup reaction can be obtained from an amplitude $T|\phi\rangle$ which fulfills the 3N Faddeev equation [11]

$$\begin{aligned} T|\phi\rangle &= tP|\phi\rangle + (1+tG_0)V_4^{(1)}(1+P)|\phi\rangle + tPG_0T|\phi\rangle \\ &+ (1+tG_0)V_4^{(1)}(1+P)T|\phi\rangle \, . \end{aligned} \tag{1}$$

The initial channel state $|\phi\rangle$ is composed of a deuteron and a momentum eigenstate of the projectile nucleon. On top of 2N forces with their off-the-energy shell t-matrix t for nucleons 2 and 3, also a 3NF is included and $V_4^{(1)}$ is the part of it which is symmetrical under exchange of nucleons 2 and 3. The permutation operator P takes into account the identity of the nucleons and G_0 is the free 3N propagator.

Using realistic NN forces: AV18 [1], CD Bonn [2], Nijm I, II, and 93 [3], one obtains, in general, predictions for 3N scattering observables which agree well with data at energies below $\approx$ 30 MeV. A fairly complete overview of those theoretical predictions in comparison to data is presented in [12, 13]. At higher energies discrepancies develop. They are exemplified for the total neutron-deuteron (nd) cross section in Fig. 1 and for the elastic scattering cross section in Fig. 2. The large discrepancy in the total cross section obtained with NN forces only seen for energies above $\approx$ 60 MeV is removed for energies below $\approx$ 140 MeV when 3NFs which reproduce the experimental triton binding energy are included. Also, the large discrepancy in the minimum of the elastic scattering cross section using NN forces only is removed when these 3NFs are included [19, 13].

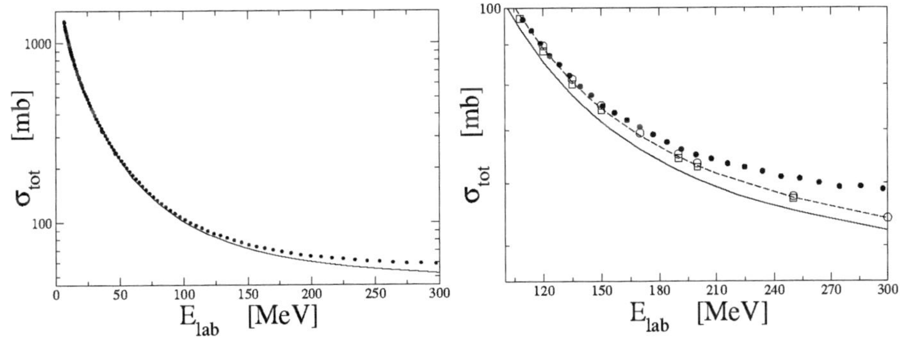

FIGURE 1. The total cross section for nd scattering as a function of the incoming neutron lab energy E_{lab}. Experimental data (solid dots) are from [14]. The solid and dashed lines are the CD Bonn and CD Bonn+TM 3NF predictions. The open squares and circles are the results of AV18+Urbana IX 3NF and CD Bonn+TM99 [15] 3NF predictions.

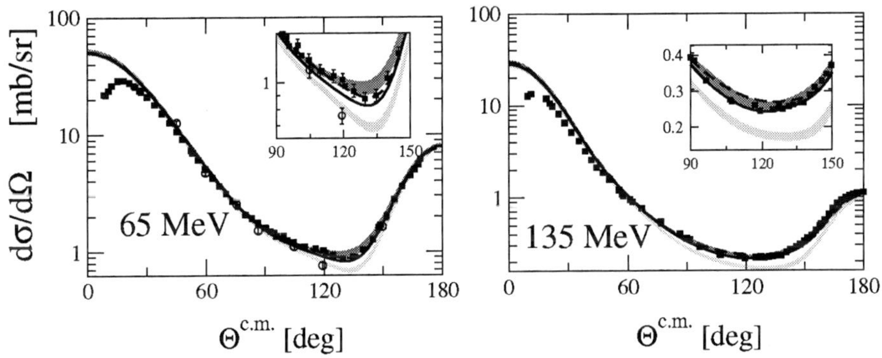

FIGURE 2. The angular distribution for elastic Nd scattering. The light shaded band contains theoretical predictions of the AV18, CD Bonn, Nijm I, II, and Nijm 93 potentials. The dark band contains the predictions when these potentials are combined with the TM 3NF. The solid and dashed lines are the AV18+Urbana IX and CD Bonn+TM99 predictions. Open circles are 65 MeV nd data from [16]. Full squares are 65 MeV pd data from [17] and 135 MeV pd data are from [18].

A similar behavior shows up for the high energy deuteron vector analyzing power $A_y(d)$ [13, 20, 18] (see Fig. 3). But there are many spin observables for which large 3NF effects are predicted and where the TM and the Urbana IX do not reproduce the data [13]. This is the case *e.g.* for the nucleon analyzing power A_y [13, 22] (see Fig. 3). This also happens for the tensor analyzing powers [13] (see Fig. 4). In both cases the data cannot be reproduced by pure 2N force predictions either.

The Nd breakup reaction in specific kinematically complete configurations also promises to provide information on the nuclear Hamiltonian. Cross sections and some spin observables exhibit very large 3NF effects at higher energies, which moreover are different for the TM and Urbana IX models [23]. Also large discrepancies between high

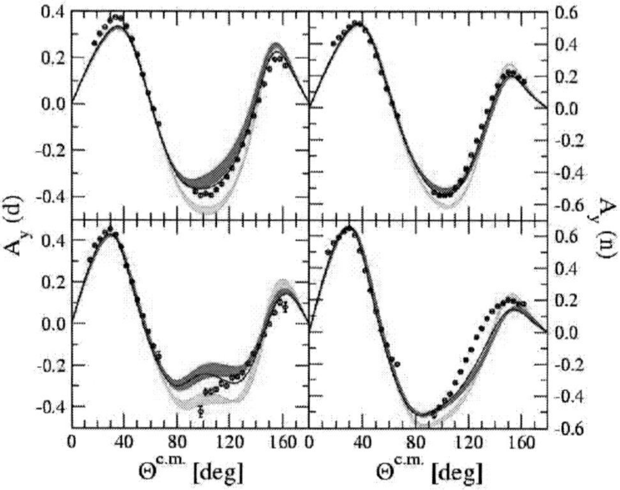

FIGURE 3. The deuteron $A_y(d)$ and neutron $A_y(n)$ vector analyzing powers in elastic nd scattering at incoming neutron lab energy $E = 135$ MeV (first row) and $E = 190$ MeV (second row). Circles are the pd data from [21]. The light bands contain the NN force predictions: AV18, CD Bonn, Nijm I and II. The dark bands contain the combinations of the NN+TM99 3NF predictions. The solid curve is the AV18+Urbana IX 3NF prediction.

energy breakup cross section data and theory based on NN forces only are found (see Fig. 5) which cannot be removed by adding TM or Urbana IX 3NFs [23]. Therefore, observables of elastic Nd scattering and the breakup process can be identified, which are sensitive to the 3NF structure. With precise data on such spin observables it should be possible to reveal the proper spin structure of the 3NF. It should be pointed out that the present breakup data base is more restricted than the data base for elastic scattering.

RELATIVISTIC EFFECTS IN THE 3N CONTINUUM

There are large discrepacies at higher energies between data and theory, both in elastic Nd scattering and breakup reaction, which cannot be removed by adding standard 3NFs (see Fig. 1, Fig. 5, and Fig. 6). They require one to study the magnitude of relativistic effects. We used an instant form relativistic approach which encompasses relativistic kinematics, boost corrections, and Wigner spin rotations [27, 28]. The boost effects turned out to be the most significant for the elastic scattering cross section at higher energies as shown in Fig. 6. They diminish the transition matrix elements at higher energies and lead, in spite of the increased relativistic phase-space factor as compared to the nonrelativistic one, to rather small effects in the cross section, mostly restricted to the backward angles [27].

At energies below ≈ 20 MeV boost and Wigner spin rotation lower the maximum of the nucleon analyzing power A_y increasing the discrepancy between theory and

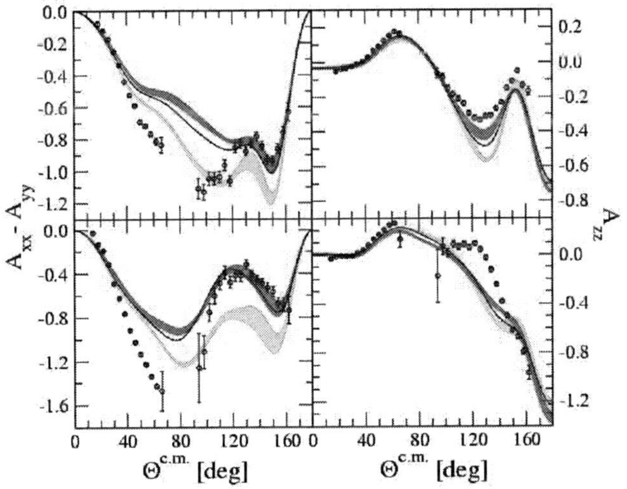

FIGURE 4. The same as in Fig.3 but for the tensor analyzing powers $A_{xx} - A_{yy}$ and A_{zz}.

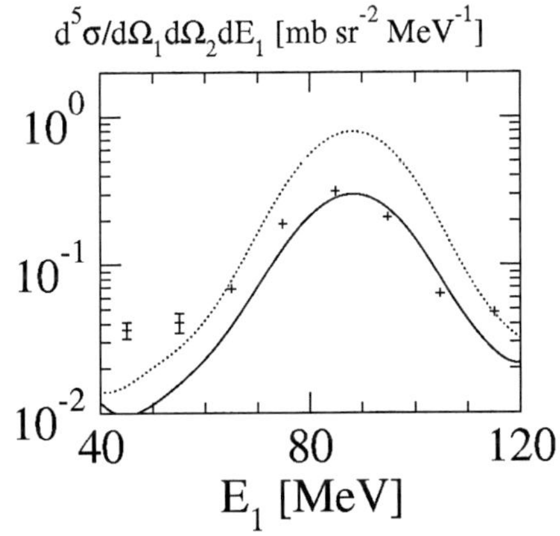

FIGURE 5. The five-fold cross section $\frac{d^5\sigma}{d\Omega_1 d\Omega_2 dE_1}$ for the $d(n,np)n$ breakup reaction at incoming neutron lab energy $E = 200$ MeV. The angles of detected nucleons are: $\theta_1 = 52^o$, $\theta_2 = 45^o$, $\phi_{12} = 180^o$. The $d(p,pn)p$ data are from [24]. The nonrelativistic and relativistic cross sections are shown by dotted and solid lines, respectively.

data [28] (see Fig. 7). This calls for even larger 3NF effects to explain the low energy A_y puzzle. Higher energy elastic scattering spin observables are only slightly modified by relativity. The selectivity of breakup singles out this reaction as a tool to look for localized effects which, when averaged, are difficult to see in elastic scattering. At higher energies this selectivity of breakup makes it possible to find the configurations with large relativistic and/or 3NF effects for the breakup cross section. The configurations with

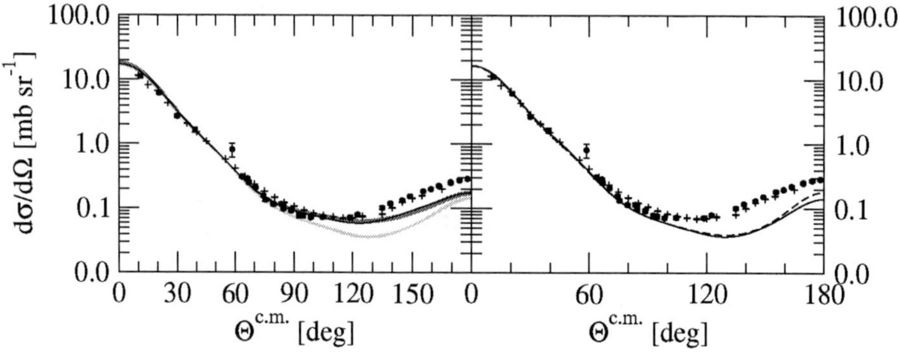

FIGURE 6. The Nd elastic scattering angular distribution at the incoming nucleon lab energy $E =$ 250 MeV. On the left side the light shaded band contains theoretical predictions of the AV18, CD Bonn, Nijm I and II potentials. The dark band contains predictions when these potentials are combined with TM99 3NF. The solid line is the prediction obtained with AV18+Urbana IX 3NF. On the right side the solid line is the prediction of the CD Bonn potential and the dashed line is the relativistic result with boost effects included. Crosses are pd data from [25] and solid circles are nd data from [26].

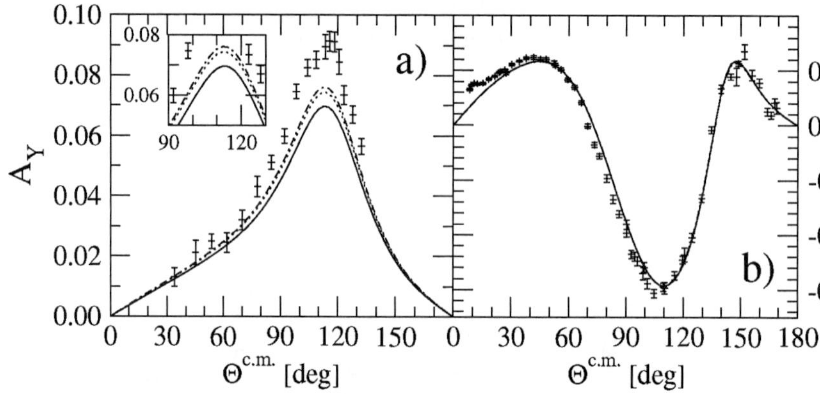

FIGURE 7. The nucleon analyzing power A_y in elastic Nd scattering at $E_{lab}^{N} = 5$ MeV (a) and 65 MeV (b). The dashed-dotted line is nonrelativistic result obtained with CD Bonn potential. The corresponding relativistic predictions are given by dotted and solid lines for the cases without and with Wigner rotations, respectively. The nd data in a) are from [33] and pd data in b) from [17].

large increasing and decreasing relativistic effects are localized in specific regions of the breakup phase-space [29, 30] and two cases are exemplified in Fig. 8. Some breakup data seem to support these relativistic effects [30] (see Fig. 5).

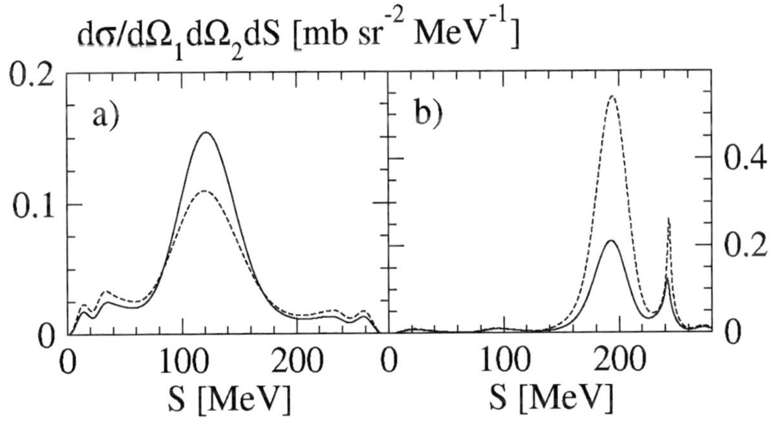

FIGURE 8. The cross section for exclusive d(n,nn)p breakup at $E^n_{lab} = 200$ MeV at fixed angle $\theta_2 = 37.5^o$, $\phi_{12} = 180.0^o$, and varying θ_1: a) $\theta_1 = 32.5^o$, b) $\theta_1 = 62.5^o$. The nonrelativistic and relativistic cross sections are shown by dashed and solid lines, respectively. The calculations are based on the CD Bonn potential.

PHOTODISINTEGRATION OF 3N BOUND STATES AND THE PD CAPTURE REACTION

All observables for these electromagnetic processes can be calculated from the nuclear matrix element

$$N_\tau = \langle \Psi_f^{(-)} | j_\tau(\vec{Q}) | \Psi_b M_b \rangle \ , \tag{2}$$

which contains the ^{3}He (^{3}H) bound state $|\Psi_b M_b\rangle$ with spin projection M_b, the spherical components $j_\tau(\vec{Q})$ $(\tau = \pm 1)$ of the nuclear current operator defined in relation to the photon momentum $\vec{Q} \parallel \hat{z}$, and the final interacting state $\langle \Psi_f^{(-)}|$. Using the Faddeev decomposition of $\langle \Psi_f^{(-)}|$, N_τ can be calculated using numerical methods standard in the treatment of the 3N continuum [31].

We use two different forms of the electromagnetic current operator. We either take the single nucleon current augmented by explicit π- and ρ-like two-body meson exchange currents or employ Siegert's theorem, which induces some many-body contributions to the current operator. For a fairly complete overview of theoretical predictions in comparison to data we refer to [31, 32].

Cross sections for the two-body photodisintegration of ^{3}He reveal a clear signature of the 3NF. Including 3NFs increases this cross section with increasing energy of the incoming photons (see Fig. 9). This increase is supported by some existing data as shown in Fig. 10

At lower energies the angular distribution for photodisintegration and pd capture are rather well described [31]. However, the effects of final (initial) state interactions are very large. For proton analyzing powers there are some discrepancies at low and higher

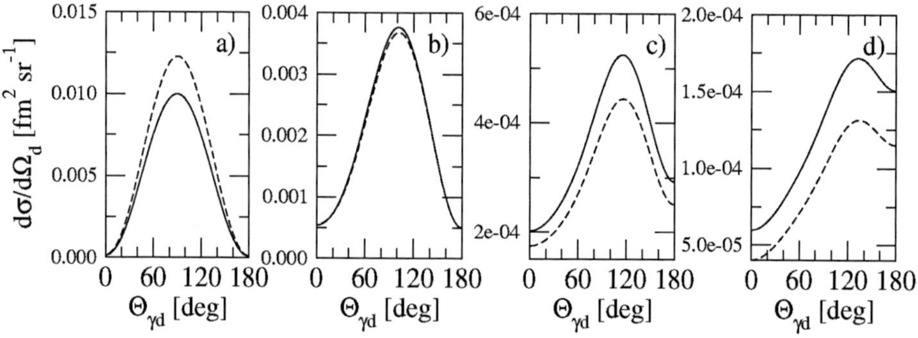

FIGURE 9. The angular distributions of deuterons from two-body photodisintegration of ^{3}He at four photon energies: a) $E_\gamma = 10$ MeV, b) $E_\gamma = 28$ MeV, c) $E_\gamma = 80$ MeV, d) $E_\gamma = 140$ MeV. $\Theta_{\gamma d}$ is the lab angle of the outgoing deuteron. The dashed (solid) lines show the result of AV18 (AV18+Urbana IX) Siegert calculations.

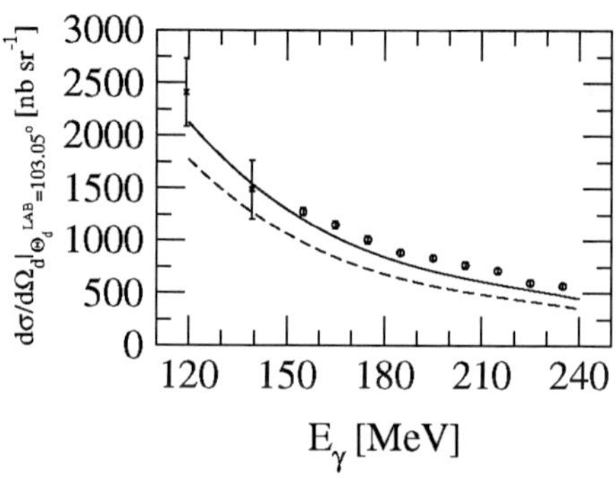

FIGURE 10. Deuteron angular distribution for the process ^{3}He(γ,d)p at $\theta_d^{lab} = 103.5^o$ as a function of the photon energy E_γ. Lines show results of calculations with the AV18 alone (dashed) and with the AV18 + Urbana IX (solid). Explicit π- and ρ-like MEC's are included in the current operator. Data are from [34] (x-es) and [35] (circles).

energies [31].

At higher energies the descrepancies between the cross section data and theory resembles those for elastic Nd scattering. Including a 3NF improves the description of the data. However, the theoretical predictions depend on the form of the current operator used (see Fig. 11). For the proton analyzing power the influence of the 3NF is smaller but results for our two choices of the current operator are quite different (Fig. 12).

SUMMARY

Solving 3N scattering exactly in a numerical sense up to energies below the pion production threshold allows one to test 3N Hamiltonian based on modern NN potentials

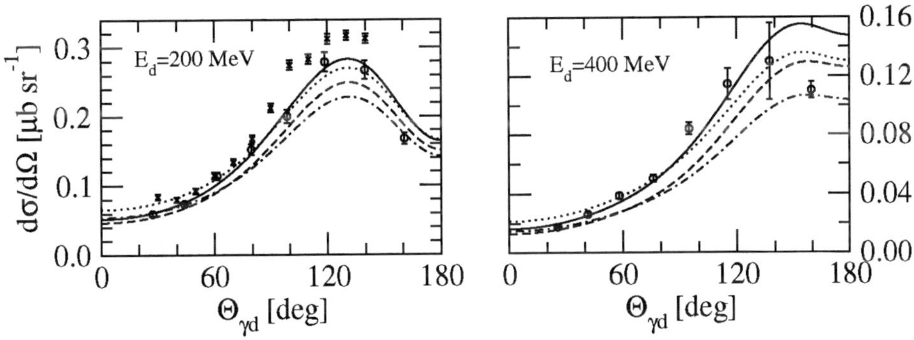

FIGURE 11. The c.m. pd capture cross sections at two deuteron lab energies calculated with four different dynamical inputs: MEC + AV18 (dashed line), Siegert + AV18 (dot-dashed line), MEC + AV18 + Urbana IX (solid line), Siegert + AV18 + Urbana IX (dotted line). Data at 200 MeV are from [36] (circles) and [37] (x-es). At 400 MeV the data are from [36].

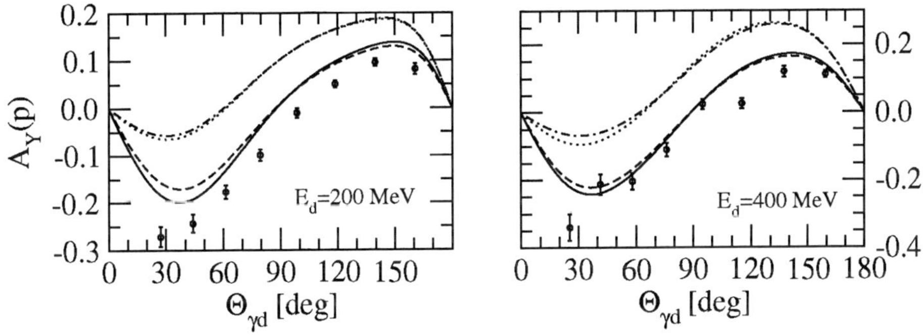

FIGURE 12. The proton anlyzing power A_y in pd capture at a) 200 MeV and b) 400 MeV deuteron lab energies. Lines and data as in Fig. 11.

and 3NFs. At the higher energies for some observables large 3NF effects are predicted when using present day models such as TM and Urbana IX. Some Nd elastic scattering cross sections and polarization data support these predictions. In some other cases, however, defects of the present day 3NFs are demonstrated. It can be expected that a precise and rich data basis, comprising elastic scattering and breakup data, will suffice to establish the proper spin structure of 3N forces.

Relativistic effects are found to be small for elastic scattering cross section. In the case of breakup reactions their importance depends on the phase-space region studied. The existing high energy discrepancies which remain even when Urbana IX or TM 3NF's are included indicate the importance of short-range contributions to a 3NF.

A well founded 3N Hamiltonian is a prerequisite for a theoretical analysis of electroweak processes in 3N systems. Exact 3N bound and scattering states are mandatory in such an analysis to account for the final state interactions. Results depend on the form of the electromagnetic current operator used.

Chiral perturbation theory, providing consistent NN and 3N forces [38] together with consistent electromagnetic current operator, will play an important role in these studies.

ACKNOWLEDGMENTS

This work was partially supported by the Helmholtz Association through funds provided to the virtual institute "Spin and strong QCD"(VH-VI-231). The numerical calculations were performed on the IBM Regatta p690+ of the NIC in Jülich, Germany.

REFERENCES

1. R. B. Wiringa, V. G. J. Stoks, R. Schiavilla, Phys. Rev. C**51**, 38 (1995).
2. R. Machleidt, Phys. Rev. C**63**, 024001 (2001).
3. V. G. J. Stoks et al., Phys. Rev. C**49**, 2950 (1994).
4. A. Nogga, H. Kamada, W. Glöckle, Phys. Rev. Lett. **85**, 944 (2000).
5. M. Viviani, Nucl. Phys. **A631**, 111c (1998).
6. R. B. Wiringa et al., Phys. Rev. C**62**, 014001 (2000).
7. J. Fujita, H. Miyazawa, Prog. Theor. Phys. **17**, 360 (1957).
8. B. S. Pudliner et al., Phys. Rev. C**56**, 1720 (1997).
9. S. A. Coon et al., Nucl. Phys. **A317**, 242 (1979); S. A. Coon, W. Glöckle, Phys. Rev. C**23**, 1970 (1981).
10. S. C. Pieper et al., Phys. Rev. C**70**, 054325 (2004).
11. D. Hüber et al., Acta Phys. Polonica **B28**, 167 (1997).
12. W. Glöckle et al., Phys. Rep.**274**, 107 (1996).
13. H. Witała et al., Phys. Rev. C**63**, 024007 (2001).
14. W. Abfalterer et al., Phys. Rev. Lett. **81**, 57 (1998).
15. S. A. Coon and H. K. Han, Few-Body Systems **30**, 131 (2001).
16. H. Rühl et al., Nucl. Phys. **A524**, 377 (1991).
17. H. Shimizu et al., Nucl. Phys. **A382**, 242 (1982).
18. H. Sakai et al., Phys. Rev. Lett. **84**, 5288 (2000).
19. H. Witała et al., Phys. Rev. Lett. **81**, 1183 (1998).
20. R. V. Cadman et al., Phys. Rev. Lett. **86**, 967 (2001).
21. B. v. Przewoski et al., Phys. Rev. C **74**, 064003 (2006).
22. R. Bieber et al., Phys. Rev. Lett. **84**, 606 (2000).
23. J. Kuroś-Żołnierczuk et al., Phys. Rev. C**66**, 024003 (2002); Phys. Rev. C**66**, 024004 (2002).
24. W Pairsuwan et al., Phys. Rev. C **52**, 2552 (1995).
25. K. Hatanaka et al., Phys. Rev. C **66**, 044002 (2002).
26. Y. Maeda et al., Phys. Rev. C **76**, 014004 (2007).
27. H. Witała et al., Phys. Rev. C **71**, 054001 (2005).
28. H. Witała et al., in preparation.
29. H. Witała et al., Phys. Lett. **B 634**, 374 (2006).
30. R. Skibiński et al., Eur. Phys. J. A**30**, 369 (2006).
31. J. Golak et al., Phys. Rep. **415**, 89 (2005).
32. R. Skibiński, Phys. Rev. C**67**, 054001 (2003); Phys. Rev. C**67**, 054002 (2003).
33. W. Tornow et al., Phys. Lett. **B 257**, 273 (1991).
34. D. I. Sober et al., Phys. Rev. C**28**, 2234 (1983).
35. N. M. O'Fallon, L. Koster Jr., J. Smith, Phys. Rev. C**5**, 1926 (1972).
36. M. A. Pickar et al., Phys. Rev. C**35**, 37 (1987).
37. T. Yagita et al., Mod. Phys. Lett. A**18**, 322 (2003).
38. E. Epelbaum, Prog. Part. Nucl. Phys. **57**, 654 (2006).

Nd Scattering at Intermediate Energies : An Epoch-making Experimental Study of the Three Nucleon Force

Kimiko Sekiguchi

RIKEN Nishina Center, Saitama 351-0198, Japan

Abstract. Recent progress in our three nucleon force study with three nucleon scattering at intermediate energies ($E/A \approx 100$ MeV) is presented, especially focusing on experimental deuteron–proton elastic scattering and breakup reactions at RIKEN. One signature of three nucleon forces is identified in the cross section minimum for deuteron–proton elastic scattering by direct comparison between precision data and state–of–the–art Faddeev calculations based on realistic nucleon–nucleon forces plus 2π–exchange three nucleon forces. However, the polarization observables are not always well described by adding the model three nucleon forces, indicating that components other than a 2π–exchange three nucleon force are required to describe the data.

Keywords: ^{1}H$(\vec{d},d)^1$H, ^{1}H$(\vec{d},p)^1$H,^{2}H$(\vec{d},\vec{p}p)n$, Analyzing powers, Polarization transfer coefficients
PACS: 21.30.-x,21.45.+v,25.10.+s,25.45.-z,25.60.Bx,25.60.Gc

INTRODUCTION

Experimentally, one must utilize systems with more than two nucleons ($A \geq 3$) to investigate properties of three nucleon forces (3NFs). The 3NFs arise naturally in the standard meson exchange picture in which the main ingredient is considered to be a 2π–exchange between three nucleons along with the Δ–isobar excitation, initially proposed by Fujita and Miyazawa a half century ago [1]. Further augmentations have led to the Tucson–Melbourne (TM) [2], the Urbana IX 3NF [3], *etc*. However, 3NFs are relatively small compared to the nucleon–nucleon (NN) forces and their effects are easily masked. Therefore it is hard to find evidence for them experimentally.

Three nucleon (3N) scattering has been studied for a long time as one of the most promising tools to explore the properties of 3NFs, because this system provides a rich set of energy dependent spin observables and differential cross sections. At lower energies ($E/A \leq 20$ MeV), very high precision measurements were carried out in proton–deuteron (*pd*) and neutron–deuteron (*nd*) scattering, including elastic and breakup reactions. However, theoretically predicted 3NF effects are rather small and a generally good description for nucleon–deuteron (*Nd*) elastic scattering data is obtained by exact solutions of 3N Faddeev equations employing only NN forces [4, 5] [1]. Study of the 3NF has changed since the end of 1990's. The following advances have made it possible to explore the 3NF effects contained in 3N scattering.

[1] Exceptions are the vector analyzing powers A_y and iT_{11} for *pd* elastic scattering.

CP1011, *New Facet of Three Nucleon Force—50 Years of Fujita Miyazawa Three Nucleon Force (FM50), The International Symposium*, edited by H. Sakai, K. Sekiguchi, and B. F. Gibson

(i) Generation of the so–called realistic NN forces (*e.g.*, AV18 [6], CDBonn [7], Nijmegen I, II and 93 [8]) which reproduce a rich set of experimental NN data for laboratory energy up to 350 MeV with an accuracy of $\chi^2 \sim 1$.

(ii) Achievement of rigorous numerical Faddeev calculations based on the realistic NN potentials below the π–threshold energy (the incident nucleon energy $E/A \leq$ 215 MeV) [4].

(iii) Development of experimental techniques to obtain precision data for 3N scattering at intermediate energies ($E/A \approx 100$ MeV).

In 1996 Sakamoto *et al.* reported measurement at RIKEN of the cross section and the deuteron analyzing powers A_y^d, A_{yy}, A_{xx}, A_{xz} for elastic dp scattering at 135 MeV/A in the angular range $\theta_{\rm c.m.} = 57^\circ$–$138^\circ$ [9]. This was the first precision measurement of both the cross sections and the deuteron analyzing powers for this reaction at intermediate energies. The data were compared with Faddeev calculations using the AV14 NN force[10]. The results of the comparison were rather surprising. The agreement for the analyzing powers was generally good, while the calculated cross sections were smaller by 30% than the observed values at the angles where the cross sections exhibit minima. In the same year a discrepancy of similar size in the cross section minimum was found for results calculated with the realistic NN forces [4]. In 1998 two theory groups incorporated 3NFs in elastic Nd scattering at intermediate energies, and they suggested that the difference found in the cross section minima is the signature of 3NF effects [11, 12]. Since then, experimental studies of intermediate–energy pd and nd elastic scattering have been performed by groups at RIKEN, RCNP, KVI, IUCF and Uppsala, providing precision data for cross sections and a variety of spin observables [13, 14, 15, 16, 17, 18, 19, 20, 21]. Complete dp breakup ($d+p \rightarrow p+p+n$) reactions would be more interesting because they cover different kinematic conditions. By selecting a particular kinematic configuration, one hopes to enhance the effects which are sensitive to specific components of the 3NFs. Thus, the study of dp breakup reactions has been the second step in investigating 3NF dynamics.

The importance of 3NFs has also been noted in other instances; *e.g.*, descriptions of the binding of light mass nuclei and the empirical saturation point of symmetric nuclear matter. Microscopic calculations, such as Green's Function Monte Carlo [22, 23] and *ab initio* no-core shell model [24] , of light mass nuclei $A \leq 12$ with realistic two– and three–nucleon forces highlight the necessity of including 3NFs to explain the binding energies of these nuclei. As for the density of symmetric nuclear matter, it has been reported that all NN potentials provide larger saturation points, and the 3NF is one possibility to shift the theoretical results to the empirical point [25].

It is clear that the testing of 3NF models has just begun. In the following sections, the experimental challenges in dp scattering at RIKEN [13, 14, 15, 16] are presented.

EXPERIMENT

dp Elastic Scattering

The dp elastic scattering experiments were performed at the RIKEN Accelerator Research Facility (RARF). The observables we obtained are cross sections and the complete deuteron analyzing powers (A_y^d, A_{yy}, A_{xx}, A_{xz}) in the angular range $\theta_{\rm c.m.} = 10^\circ - 180^\circ$ at 70 – 135 MeV/A (corresponding to deuteron energies of $E_d = 140 - 270$ MeV). Later, we extended the measurement to the deuteron–to–proton polarization transfer coefficients ($K_y^{y'}$, $K_{xx}^{y'}$–$K_{yy}^{y'}$, and $K_{xz}^{y'}$) and the proton induced polarization $P^{y'}$ for the angles $\theta_{\rm c.m.} = 90^\circ - 180^\circ$ at 135 MeV/A. Note that $P^{y'} = -A_y^p$ assuming time-reversal invariance, where the A_y^p denotes the proton analyzing power for *pd* elastic scattering.

The vector and tensor polarized deuteron beams [26] were accelerated by the AVF and Ring cyclotron. They bombarded a liquid hydrogen or polyethylene (CH_2) target. Either the scattered deuteron or the recoil proton was momentum analyzed by the magnetic spectrograph SMART (Swinger and Magnetic Analyzer with Rotator and Twister) [27] depending on the scattering angle and detected at the focal plane (see Fig. 1). For the polarization transfer measurement, the polarization of the elastically scattered protons from the hydrogen target was measured with the focal-plane polarimeter [28]. One characteristic feature of the RIKEN polarized deuteron beams is that we could obtain the beams the axis of which was controlled in an arbitrary direction on the target, making it possible for us to obtain all the deuteron analyzing powers A_y^d, A_{yy}, A_{xx}, A_{xz}. The polarization axis of the deuteron beams was controlled by the spin rotator Wien Filter prior to acceleration [29]. Due to the single-turn extraction feature of the RARF cyclotrons the polarization amplitudes were maintained during acceleration. The beam polarizations were monitored with the beam line polarimeter by using the analyzing powers for dp elastic scattering. To obtain the absolute values of the deuteron beam polarizations, the analyzing powers for dp elastic scattering were calibrated by using the $^{12}C(d,\alpha)^{10}B^*\,[2^+]$ reaction, the $A_{yy}\,(0^\circ)$ of which is exactly $-1/2$ because of parity conservation [30]. In all the measurements the actual magnitudes of the polarizations were 60–80% of the theoretical maximum values.

Determination of Absolute Values of the Cross Section. It is essential to obtain precise absolute values of the cross section to compare with the state–of–the–art Faddeev calculations. However, it is usually difficult to know experimentally the systematic uncertainty. We performed the cross section measurements with three different experimental techniques and tried to estimate the systematic uncertainties. First, we made a measurement at RIKEN with the proton beam at 135 MeV and a CD_2–CH_2 sandwiched solid target at the angles where the *pp* and *pd* elastic scattering were simultaneously measured with the magnetic spectrograph SMART. Using the well–known elastic *pp* cross sections we can estimate the overall systematic uncertainty for the *pd* cross section. Secondly, to confirm the angular distribution we measured with 135 MeV/A deuterons, a CH_2 solid target, and the SMART system. In the measurement we tried to check the fluctuations of the target thickness during the experiment by measuring the dp scattering at the fixed angle $\theta_{\rm c.m.} = 69.7^\circ$, where the scattered deuterons and recoil protons were detected in coincidence in the scattering chamber. The cross section at $\theta_{\rm c.m.} = 165.1^\circ$ was measured for that same purpose with the SMART system several times during the experiment. We also measured the carbon background events. Finally, we performed a totally independent measurement at the Research Center for Nuclear Physics (RCNP) of Osaka University, using a 135 MeV proton beam and deuterated polyethylene target.

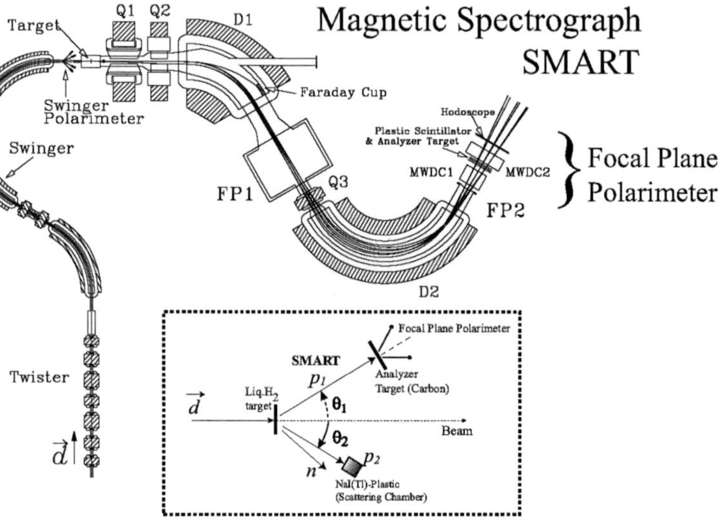

FIGURE 1. Schematic view of the experimental setup for the measurements of *dp* elastic and breakup processes at RIKEN.

The absolute normalization of the cross sections has been performed by taking data with a D_2 gas target and the double slit system for which the RCNP group has already established the procedure to obtain the absolute *pd* cross section [17]. The data taken at RCNP are shown as black solid diamonds. The open squares are the data measured with the proton beam at RIKEN and the open circles are the data taken with the deuteron beam at the same laboratory. Only statistical errors are presented. The very good agreement between the independent measurements allows us to conclude that the systematic uncertainty due to the detection setup is small [13, 14, 16].

dp Breakup Reaction

The *dp* breakup experiment was also performed with the SMART spectrograph using polarized deuteron beams at 135 MeV/A. The measurement was focused on the deuteron–to–proton polarization transfer coefficient $K_{yy}^{y'}$ in the particular coplanar configurations in which different results are seen between the TM′99 [32] (the TM′99 is a version of the Tucson–Melbourne 3NF which is more consistent with chiral symmetry [33, 34]) and the Urbana IX 3NFs. In addition we measured all deuteron analyzing powers A_y^d, A_{yy}, A_{xx}, A_{xz} where strong sensitives to the current 3NFs are predicted. The measured angles were $\theta_1 = 27°$–$33°$ for one of the emerging protons p_1 and $\theta_2 = 31°$ for the other proton p_2. The relative azimuthal angle between p_1 and p_2 ($\phi_{12} \equiv \phi_1 - \phi_2$) was 180°. These configurations include the final state interaction between the scattered proton p_2 and the neutron. The two emerging protons were detected in coincidence to identify the *dp* breakup events. The proton p_1 was momentum analyzed by the magnetic spectrograph SMART and its polarizations were measured with the focal plane polarimeter. The proton p_2 was detected by E–dE counters consisting of NaI(Tl) and plastic scintillators installed in a scattering chamber (see Fig. 1). The kinetic energies were covered from 164 MeV–180 MeV for p_1 and 30 MeV–75 MeV for p_2, respectively.

RESULTS AND DISCUSSION

Cross Section in elastic dp scattering. Let us compare the cross section data for elastic dp scattering with the theoretical predictions. The data at 135 MeV/A are shown with open circles in Fig. 2 together with the Faddeev calculations by the Bochum–Cracow–KIT (BCK) group. The dark shaded bands in the figure are the results of the calculations based on the high precision NN potentials, AV18, CDBonn, Nijmegen I, II only. The light shaded bands in the figure contain the predictions of the four NN forces with the TM′99 3NF. The solid lines are the theoretical predictions obtained using the AV18 potential combined with the Urbana IX 3NF. Based on the comparison between the data and the NN force predictions, clear discrepancies are found in the angular range where the cross sections reach a minimum at the measured incident energies 70 and 135 MeV/A (the data at 70 MeV/A are not shown here). They become larger as an incident energy increases. The discrepancies are explained by taking into account the 2π exchange type 3NF models (TM′99, and Urbana IX). All 2π–exchange 3NF potentials considered here provide 3NF effects for the cross section which are comparable in magnitude and sign.

An alternative theoretical prediction has been reported by the Hannover–Lisbon group in which the Δ–isobar excitations are explicitly included in the coupled-channel approach [35]. The contributions are based on all meson exchanges, i.e. π, ρ, σ, and ω exchanges contained in the coupled-channel potential. Thus the π–π, π–ρ, ρ–ρ 3NFs with Δ–isobar excitations as well as the Illinois type [23] many–π rings are effectively incorporated. The CDBonn potential is taken as the NN interaction. Recently this group has succeeded in incorporating the Coulomb force via the screening and renormalization approach within a momentum–space framework [36]. The results are shown in Fig. 2. The dashed curves are the coupled–channel approach obtained with the Δ–isobar excitations and the dotted curves are the predictions in which the Coulomb interactions are additionally included. The predictions with only NN forces are not shown in the figures because they provide results similar to those of the CDBonn potential by the BCK group. Δ–isobar effects improve the agreement in the angular range where the cross sections reach a minimum, similar to the results by the BCK group. It should also be noted that the dashed curve shows very good agreement at forward angles, which indicates that the Coulomb interaction is significant at these angles.

Polarization Observables in Elastic dp Scattering. The polarization observables in dp elastic scattering at 135 MeV/A are shown as open circles in Figs. 3 and 4. The deuteron analyzing powers are expressed in the spherical frame (iT_{11},T_{20},T_{21},T_{22}). Only statistical errors are shown in the figure and their values are less than 0.03 for all the measured observables. Generally, the discrepancies between the data and the pure NN

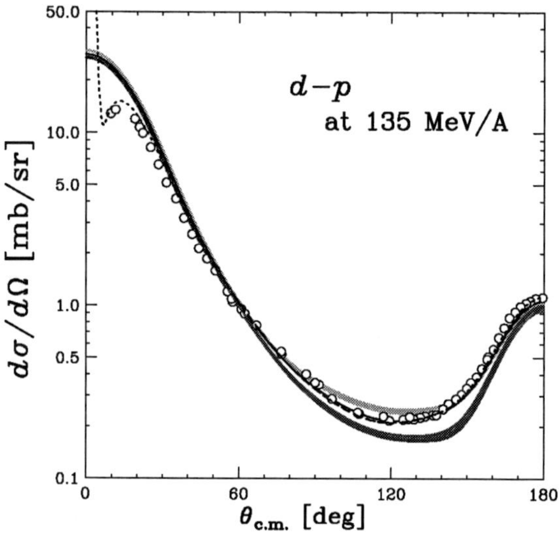

FIGURE 2. Cross section for dp elastic scattering at 135 MeV/A. For the description of the theoretical prediction, see text.

force predictions (dark shaded bands) are clearly seen at the angles where the cross sections reach a minimum. The 2π–exchange 3NF potentials, TM′99 and Urbana IX, provide large and almost identical 3NF effects for all the measured observables. As is the case for cross sections, all the predictions with the contemporary 3NFs are in good agreement with the data for the proton analyzing power A_y^p, the deuteron vector analyzing power iT_{11}, and the polarization transfer coefficients $K_{xx}^{y'} - K_{yy}^{y'}$. However, no calculation is superior for the other spin observables, *i.e.* the tensor analyzing powers T_{20}, T_{21}, T_{22} and some polarization transfer coefficients $K_y^{y'}$, $K_{xz}^{y'}$ (the $K_y^{y'}$, $K_{xz}^{y'}$ data are not shown here). As for the predictions by the coupled channel approach (dashed curves), the Δ-isobar effects are similar to those of the TM′99 and Urbana IX 3NFs for almost all the observables.

Polarization Observables in dp Breakup Reactions. In Fig. 5 selected experimental results are shown as open circles along with the kinematic curves of the two emerging protons (S-curve). Only statistical errors are shown, and they are less than 0.03 for all the measured observables. The data were obtained with an *S*–curve energy bin of 8 MeV for the analyzing powers and 12 MeV for the polarization transfer coefficients. The predictions shown here are averaged over the S–curve energy bin of 8 and/or 12 MeV for direct comparison between the data and the calculations.

For the polarization transfer $K_{yy}^{y'}$, the shapes of the predictions change drastically with inclusion of the 3NFs. The TM′99 and Urbana IX 3NFs provide different results. The data rather support the Urbana IX 3NF. For the deuteron analyzing powers A_y^d and A_{xx}, the two 3NFs provide large but model independent effects. The tensor analyzing power

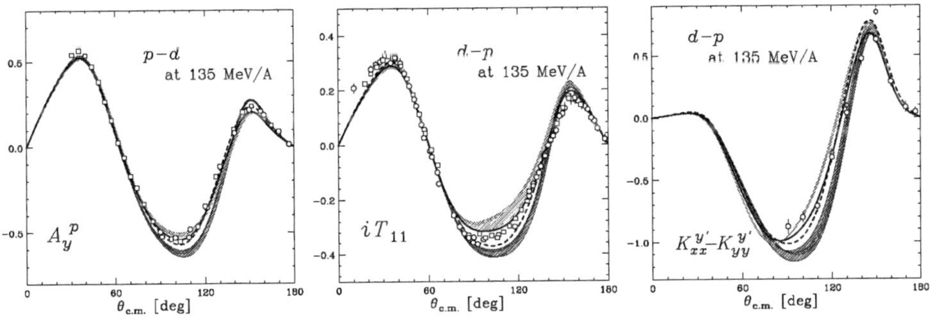

FIGURE 3. Proton and deuteron vector analyzing powers A_y^p,iT_{11}, and the polarization transfer coefficients $K_{xx}^{y'} - K_{yy}^{y'}$ for dp elastic scattering at 135 MeV/A. For the description of the theoretical predictions, see text.

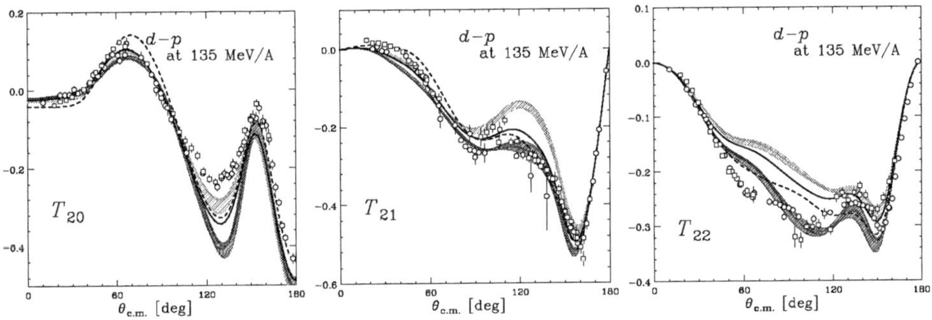

FIGURE 4. Deuteron tensor analyzing powers T_{20}, T_{21}, T_{22} for dp elastic scattering at 135 MeV/A. For the description of the theoretical predictions, see text.

A_{xx} data are well explained by inclusion of 3NFs; however, the vector analyzing power A_y^d is reproduced by the predictions with only NN forces and the inclusion of 3NFs worsens the agreement. Again, the measured kinematic configuration includes the final state interaction of the proton p_2 and the neutron, that means this configuration is close to the elastic scattering. It is interesting to note that the results of the comparison are opposite to that seen in elastic dp scattering.

The results of the cross section for elastic dp scattering have opened up a new possibility to explore 3NFs via 3N scattering at intermediate energies. However, the spin observables for dp elastic scattering are not properly described by contemporary 3NFs, which is also the case for the spin observables of the measured coplanar configurations in dp breakup. These results clearly indicate that effects other than 2π–exchange 3NFs should be included. Also the results imply that the observables where the large 3NF effects are predicted are useful to test the current theory and study the detailed properties of 3NFs. It would be interesting to see how well the theoretical approaches, *e.g.* inclusion of 3NFs other than 2π–exchange types, relativistic treatment [37], and the potentials

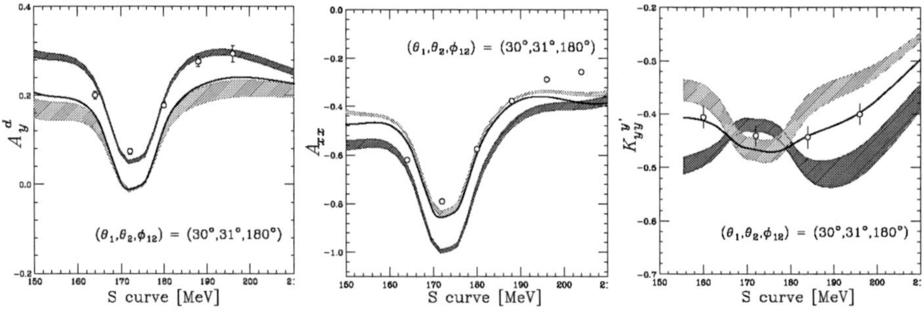

FIGURE 5. Analyzing powers (A_y^d, A_{xx}) and deuteron to proton polarization transfer coefficient $K_{yy}^{y'}$ for the dp breakup reaction at 135 MeV/A. For the description of the theoretical predictions, see text.

based on chiral effective field theory [38], describe these data. Experimentally, other kinematic configurations of the dp breakup reactions at higher energies would provide a valuable source of information on 3NFs, and such reactions should be measured.

Recently, relativistic effects have been estimated using Lorentz boosted NN potentials [41]. The predicted effects are small and only visible at the very backward angles for dp elastic scattering [16, 21, 37], while large relativistic effects are predicted for the particular kinematic configurations of the breakup reactions [37].

Developments of chiral perturbation theory (χPT) lead to NN potentials, where the one–pion and multi–pion exchange are treated unambiguously according to chiral symmetry (for example, see Ref. [38]). The χPT approach allows us to construct three– and many–nucleon forces consistent with the NN interactions [39, 40]. The first non-vanishing 3NFs appear at the next–next–leading order (NNLO) in χPT. So far theoretical predictions for the 3N scattering are available below a laboratory energy 100 MeV/A.

SUMMARY

The 3NFs are now accepted as key elements in understanding various nuclear phenomena, such as the binding of light mass nuclei and the empirical saturation point of nuclear matter density. The Nd scattering data provide rich sources with which to explore the properties of 3NFs such as momentum and spin dependence. Since the end of the 1990's extensive experimental measurements of proton–deuteron and neutron–deuteron elastic scattering at intermediate energies ($E/A \approx 100$ MeV) have been performed at several facilities. Study of Nd breakup reactions has followed as the second step. In this talk, the experimental work on dp elastic scattering and dp breakup reactions at RIKEN was presented. Comparing the data with state–of–the–art Faddeev calculations, the cross section data for elastic dp scattering are explained by incorporating 2π–exchange 3NFs; that is the first clear signature of 3NF effects in three nucleon scattering. However, polarization observables are not always properly described by adding the model 3NFs, indicating that additional elements are required to describe the data. To understand these spin observables, theoretical approaches incorporating 3NFs other than 2π–exchange types, relativistic treatment, Coulomb effects, and completely new approaches based on chiral effective field theory are now in progress. Experimentally, other channels of the dp breakup reactions at higher energies would provide a valuable source of information on 3NFs, and such reactions should be measured.

ACKNOWLEDGMENTS

The experiments were performed under the RIKEN SMART collaboration; H. Sakai, H. Okamura, N. Sakamoto, T. Uesaka, A. Tamii, T. Wakasa, Y. Satou, T. Ohnishi, T. Kawabata, K. Yako, K. Suda, S. Sakoda, H. Kato, Y. Maeda, M. Hatano, T. Saito, N. Uchigashima, H. Kuboki, K. Itoh, M. Sasano, S. Sakaguchi, Y. Sasamoto. The author would like to acknowledge the collaboration on the polarization transfer measurement at RIKEN by N. Kalantar-Nayestanaki and K. Ermisch. Finally, the author would like to acknowledge the outstanding work of the RIKEN Accelerator group for delivering superb polarized deuteron beams.

REFERENCES

1. J. Fujita, and H. Miyazawa, Prog. Theor. Phys. **17**,360 (1957).
2. S. A. Coon and W. Glöckle, Phys. Rev. C**23**, 1790 (1981).
3. B. S. Pudliner *et al.*, Phys. Rev. C**56**, 1720 (1997).
4. W. Glöckle, H. Witała, D. Hüber, H. Kamada and J. Golak, Phys. Rep. **274**, 107 (1996).
5. A. Kievsky, M. Viviani, and S. Rosati, Phys. Rev. C **64**, 024002 (2001).
6. R. B. Wiringa *et al.*, Phys. Rev. C **51**, 38 (1995).
7. R. Machleidt, Phys. Rev. C **63**, 024001 (2001).
8. V. G. J. Stoks *et al.*, Phys. Rev. C **49**, 2950 (1994).
9. N. Sakamoto *et al.*, Phys. Lett. B **367**, 60 (1996).
10. R. B. Wiringa, R. A. Smith, and T. L. Ainsworth, Phys. Rev. C **29**, 1207 (1984).
11. H. Witała *et al.*, Phys. Rev. Lett. **81**, 1183 (1998).
12. S. Nemoto *et al.*, Phys. Rev. C **58**, 2599 (1998).
13. H. Sakai *et al.*, Phys. Rev. Lett. **84**, 5288 (2000);
14. K. Sekiguchi *et al.*, Phys. Rev. C **65**, 034003 (2002);
15. K. Sekiguchi *et al.*, Phys. Rev. C **70**, 014001 (2004);
16. K. Sekiguchi *et al.*, Phys. Rev. Lett **95**, 162301 (2005).
17. K. Hatanaka *et al.*, Phys. Rev. C **66**, 044002 (2002).
18. R. Bieber *et al.*, Phys. Rev. Lett. **84**, 606 (2000); K. Ermisch *et al.*, Phys. Rev. Lett. **86**, 5862 (2001); K. Ermisch *et al.*, Phys. Rev. C **68**, 051001 (2003); K. Ermisch *et al.*, Phys. Rev. C **71**, 064004 (2005); H. R. Amir-Ahmadi *et al.*, Phys. Rev. C **75**, 041001 (2007); H. Mardanpour *et al.*, Eur. Phys. J. A **31**, 383 (2007).
19. E. J. Stephenson *et al.*, Phys. Rev. C **60**, 061001 (1999); R. V. Cadman *et al.*, Phys. Rev. Lett. **86**, 967 (2001); B. v. Przewoski *et al.*, Phys. Rev. C **74**, 064003 (2006).
20. P. Mermod *et al.*, Phys. Rev. C **72**, 061002 (2005).
21. Y. Maeda *et al.*, Phys. Rev. C **76**, 014004 (2007).
22. R. B. Wiringa *et al.*,Phys. Rev. C **62**, 014001 (2000).
23. S. C. Pieper *et al.*, Phys. Rev. C. **64**, 014001 (2001).
24. P. Navrátil and W. E. Ormand, Phys. Rev. C **68**, 034305 (2003).
25. see for example, A. Akmal *et al*,, Phys. Rec. C **58**, 1804 (1998).
26. H. Okamura *et al.*, AIP Conf. Proc. **293**, 84 (1994).
27. T. Ichihara *et al.*, Nucl. Phys. **A569**, 287c (1994).
28. S. Ishida *et al.*, AIP Conf. Proc. **343**, 182 (1995).
29. H. Okamura *et al.*, AIP Conf. Proc. **343**, 123 (1995).
30. K. Suda *et al.*, Nucl. Instrum. Methods Phys. Res. A **572**, 745 (2007).

31. R. A. Arndt and L. D. Roper, *Scattering Analysis Program* (SAID), Virginia Polytechnic Institute and State University (unpublished), see also Phys. Rev. C **56**, 3005 (1997) and references therein.
32. S. A. Coon and H. K. Han, Few Body Syst., **30**, 131 (2001).
33. J. L. Friar *et al.*, Phys. Rev. C **59**, 53 (1999).
34. D. Hüber *et al.*, Few-Body Syst. **30**, 95 (2001).
35. A. Deltuva *et al.*, Phys. Rev. C **68**, 024005 (2003).
36. A. Deltuva *et al.*, Phys. Rev. C **71**, 054005 (2005).
37. H. Witała, *et al.*, Phys. Lett. B **634**, 374 (2006).
38. E. Epelbaum Prog. Part. Nucl. Phys. **57**, 654 (2006).
39. U. van Kolck, Phys. Rev. C **49**, 2932 (1994).
40. E. Epelbaum, *et al.*, Nucl. Phys. **A747**, 362 (1998).
41. H. Kamada *et al.*, Phys. Rev. C **66**, 044010 (2002).

What have we learned about three-nucleon systems at intermediate energies?

N. Kalantar-Nayestanaki

KVI, University of Groningen, Zernikelaan 25, 9747 AA, Groningen, The Netherlands

Abstract. In the past few years, a program has been set up at KVI to study the effects of three-nucleon forces in various reactions. Several observables in elastic and break-up channels in proton-deuteron scattering have been measured and compared with state-of-the-art Faddeev calculations using modern two-body forces and a number of three-body forces. For the elastic scattering process, the data-base is now large enough to perform a partial wave analysis to provide an unambiguous description of nuclear forces. For the break-up reaction, data are emerging at intermediate energies revealing interesting aspects of nuclear forces.

Keywords: nuclear forces, nucleon-nucleon interaction, three-nucleon forces, proton-deuteron elastic scattering, spin-transfer coefficients, scaling
PACS: 13.75.Cs, 21.30.-x, 21.45.+v, 25.10.+s

INTRODUCTION

In the last decade, several semi-phenomenological two-nucleon models, namely CD-Bonn, Argonne-V18 (AV18), Nijmegen-I, Nijmegen-II and Reid93 [1, 2, 3, 4] have become available, which describe two-nucleon scattering observables accurately. Furthermore, due to the availability of fast computers, Faddeev-type three-nucleon calculations are now-a-days routinely performed. The use of modern two-nucleon potentials, to describe three-nucleon scattering observables, leads to various degrees of agreement between the calculations and the experimental data, depending on the observable being studied and the incident beam energy. At low incident beam energies, up to ≈ 30 MeV, the differential cross sections in nucleon-deuteron scattering are described generally well using solely two-nucleon potentials, with small changes when the three-nucleon force is included. In contrast, efforts to describe the analyzing power have failed and the inclusion of three-nucleon forces (3NFs) into the calculations has not remedied the discrepancy, leading to the well-known A_y-puzzle. Also at low energies, tensor-analyzing powers and spin-transfer coefficients are rather well described using solely NN forces [5], whereas at intermediate and higher energies, calculations based on two-nucleon forces generally fail to describe many experimental observables [6, 7, 8, 9, 10, 11, 12, 13, 14].

In order to have a comprehensive model of the three-body forces, one should not only study spin-averaged cross section but also observables which involve spin such as analyzing power, spin-transfer coefficients, and spin-correlation coefficients for a large energy and angular range. Only with such a reservoir of data, which has now emerged from various laboratories for the elastic channel, one can understand in detail the interplay between the two and three-body forces. The understanding of the nature of forces is strongly coupled to the theoretical efforts which attempt to generate nuclear

CP1011, *New Facet of Three Nucleon Force—50 Years of Fujita Miyazawa Three Nucleon Force (FM50), The International Symposium*, edited by H. Sakai, K. Sekiguchi, and B. F. Gibson

forces [15, 16, 17] in a consistent manner such as those in the effective field theory [18, 19, 20].

The Dutch-Polish-Japanese collaboration has set up a program to study many of the observables for various reactions of interest systematically. These include elastic proton-deuteron scattering [8, 9, 10, 11, 12], proton-deuteron break-up reaction [21, 22, 23] and proton-deuteron capture reaction [24]. For more details on the break-up channel, one is referred to the contribution by St. Kistryn elsewhere in these proceedings. In this article, the results of the elastic-scattering cross sections which have been measured in the past are revisited from a different point of view. In addition, results of a more recent experiment to measure spin-transfer coefficients, also in the elastic scattering process, are discussed. Finally, preliminary results for a sample of data from the proton-deuteron break-up reaction will be presented.

RESULTS AND DISCUSSION

Proton-deuteron elastic scattering

About a decade ago, it was proposed by Witała et al. and Nemoto et al. to search for the effects of three-body forces in the minimum of the proton-deuteron elastic scattering cross section [25, 26]. At KVI, a number of measurements were carried out to study this effect as a function of the beam energy. Measurements were done with six incident energies, namely 108, 120, 135, 150, 170 and 190 MeV. The results of these measurements have already been published in Refs. [9, 10, 11]. At 135 MeV, a discrepancy has arisen between the KVI data and those obtained at RIKEN [27]. To resolve this disagreement, a new measurement has been performed at KVI with a different detection system. The results of the new measurement which have been submitted for publication show that both experiments performed in the past, and in particular the RIKEN data, had an absolute normalization problem. In addition, the angular distributions of the cross sections turned out to be different in shape in both experiments. In order to do a global analysis for the elastic channel in proton-deuteron scattering, a partial-wave analysis (PWA) is, therefore, called for. Within the framework of the PWA, problems with the shapes and normalization factors of various experiments could be examined more thoroughly. Presently, the task of performing a PWA seems to be not feasible, though. Instead, we have carried out a systematic study of elastic cross sections as a function of momentum transfer.

The cross section of the elastic proton-deuteron scattering process as a function of angle shows that scattering at small angles (low momentum transfers) is dominated by the direct term in the two-nucleon potential which should then be rather well described by the calculations using only two-nucleon forces. At the backward angles (large momentum transfers), the exchange term in the two-nucleon potential is responsible for the increase of the cross sections. In these two regions, the effect of the three-body forces are predicted to be small. Finally, in the intermediate angles where the cross section has a minimum, one would expect the effects of three-nucleon forces to be largest as was shown in Refs. [25, 26]. The cross section can be expressed as a multiplication of the square of the invariant matrix element with a phase space factor which has a large vari-

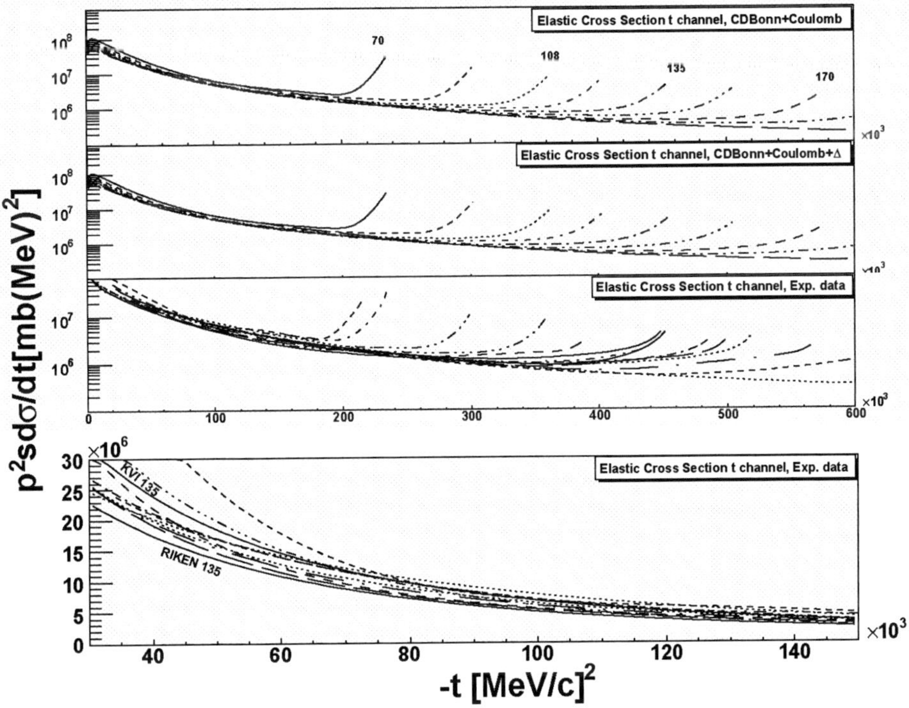

FIGURE 1. The elastic scattering cross sections divided by the phase-space factor is shown as a function of t, for a number of incident beam energies between 70 and 250 MeV. In the top panel the results of the calculations using the CD-Bonn potential are shown, while in the second panel from the top, the effect of three-nucleon force has also been added. All calculations include the Coulomb force. The third panel shows the same but for the experimental data. The lowest panel zooms into a small area at low momentum transfers.

ation with the incident beam energy. The dynamical part of the cross section, i.e. the square of the invariant matrix element, can be obtained and studied by taking out the trivial phase-space dependence by multiplying the cross section with $p^2 s$ where p is the momentum of the incident beam and s is the total energy available in the center of mass. This is done in Fig. 1 for a large number of bombarding energies between 70 and 250 MeV where the square of the invariant matrix element is plotted against the Mandelstam variable, $t = |p_1 - p_3|^2$, where p_1 and p_3 are the four momenta of the incoming and the outgoing particles. The variable t corresponds to the relativistic momentum transfer in the reaction. The calculations are all done by the Hannover-Lisabon group and include the Coulomb effect [28, 29, 30].

There are several features which are very clear: at low values of t, one observes scaling for the square of the invariant matrix element at all incident energies shown in the figure. This implies that the dynamical part of the cross section predominantly depends on momentum transfer and has a very small energy dependence. It is exactly in this region where one could expect all the experiments to agree with each other (modulo a small energy dependence). For the intermediate values of t, one enters the region of the cross section minimum. In this region, the results of the calculations including the three-nucleon force show indeed that their effect is as large as 50% in the minimum. However, the energy dependence of the effect is again very small in the energy range shown (see the second panel from the top).

The third panel from the top in Fig. 1 shows the experimentally-obtained square of the invariant matrix element. Each line corresponds to a fit through the data for a given energy. The spread in the band at lower values of t represents the uncertainty in the normalization of various data sets. The width of this band is generally explained by the systematic uncertainties quoted by various experiments. Incidentally, the two recent data sets at 135 MeV which disagree with each other reside on the two sides of the band (the KVI data above and the RIKEN data below the band). For more details, see the lowest panel in Fig. 1.

Similarly, one can study the dependence on the Mandelstam variable, $u = |p_1 - p_4|^2$ for which p_4 is the momentum of the recoil particle. This variable represents the relativistic exchange momentum transfer. The conclusions are again the same: there is the same sort of scaling as in Fig. 1 but then at backward angles. Even though the region of scaling is smaller than for the t variable, since the exchange term is more sharply peaked at backward angles than the direct term at forward angles, one can still use this region to judge the absolute normalization of the various data sets. One should note that this simple "scaling" breaks down for energies below 100 MeV (see the curve for 70 MeV in Fig. 1). The origin of this may be due to the fact that the de Broglie wavelength is getting so large that the coherence effects cannot be ignored anymore.

Inspired by the observed discrepancies in the elastic cross section and analyzing power data, we have studied other spin observables as well. Many experiments have been performed in the past at KVI and RIKEN to study the analyzing powers in the three-body system. More recently, experiments have been performed at RIKEN and KVI to study spin-transfer coefficients. Figure 2 shows the results of an experiment performed recently at KVI in which a polarized deuteron beam of 180 MeV scatters off a hydrogen target to study all possible observables in the elastic channel [31]. In addition to cross section, vector and tensor analyzing powers, spin-transfer coefficients were studied by measuring the polarization of the outgoing protons and correlating them with the incident beam polarization. These measurements require a secondary scattering from a Carbon target positioned in the focal-plane of the Big-Bite Spectrometer [32, 33, 34]. The Carbon "analyzer" had to be calibrated in a separate experiment to provide the effective analyzing powers for the energy range of the outgoing protons in the present experiment [35]. The observables obtained for this reaction were cross section ($d\sigma/d\Omega$), vector (A_y) and tensor (A_{yy}) analyzing powers, vector ($K_y^{y'}$) and tensor ($K_{yy}^{y'}$) spin transfer coefficients and the induced polarization (P_y). The last observable is the same as the analyzing power in the inverse reaction with an opposite sign. We used the information from our earlier measurements on the analyzing power to fix the overall small asymmetry of the detection system. It should, therefore, be noted that for P_y, only the shape of the observable should be compared with the results of the calculations.

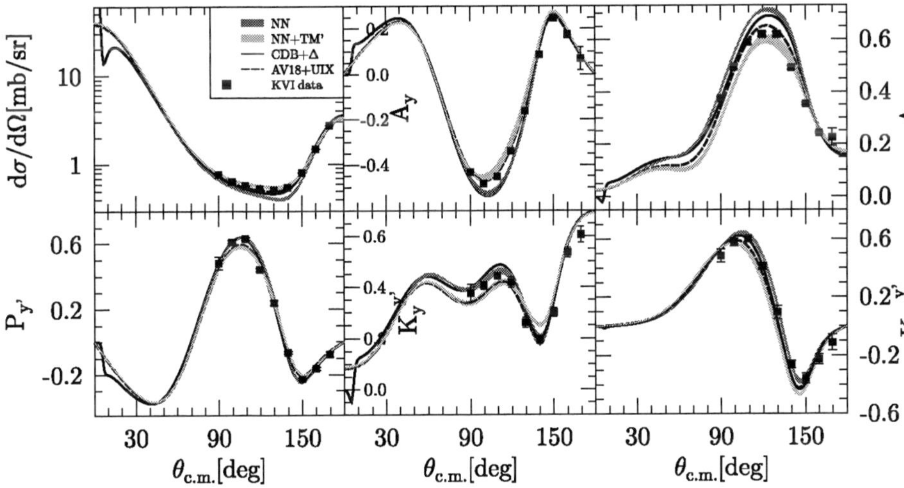

FIGURE 2. Data for the cross sections, $d\sigma/d\Omega$, the vector, A_y, and tensor, A_{yy}, analyzing powers, the induced polarizations, $P_{y'}$, and the vector, $K_y^{y'}$, and tensor, $K_{yy}^{y'}$ spin-transfer coefficients in the elastic deuteron-proton scattering. Theoretical predictions based on NN potential alone are shown by the dark grey bands, while those with NN+TM′ are presented by the light grey bands. Dashed (solid) lines show the predictions using AV_{18}+UIX (CDB+Δ). For A_y, the solid line is on the lower edge of the dark grey band.

The error bars are for some data points smaller than the symbol size of the point. This error accounts for the statistical uncertainties and a very small point-to-point (PTP) uncertainty. The statistical uncertainties come from the spin-dependent cross sections and the statistical uncertainty in determining the incoming-beam polarization with the in-beam polarimeter [36]. The PTP error accounts for a very small instability of the experimental apparatus over long periods of time and background subtraction. In addition to the statistical and the PTP errors there are other types of systematic errors. These errors originate from the target thickness measurement, the estimation of the opening angle of the detector, the total collected charge in the Faraday cup, and the systematic error of the incoming beam polarization. The resultant systematic error is $\sim 5\%$ for the cross sections and $\leq 3\%$ for all other spin-dependent observables.

For the cross sections, it is clear that the inclusion of the 3NF is needed and the magnitude of the correction due to 3NF is in agreement with the trend seen in Fig. 1 for these energies. For the vector and tensor analyzing powers, calculations including 3NF seem to describe the data better than those based only on two-nucleon forces, with the exception of CDB+Δ. This picture is completely opposite for the spin-transfer coefficients where the calculations based on 2NF agree with those from CDB+Δ and are also in better agreement with the experimental data. The data show that the full spin structure of 3NF is not understood at this energy either and needs further attention. The shape of the induced polarization is in good agreement with the results of all calculations.

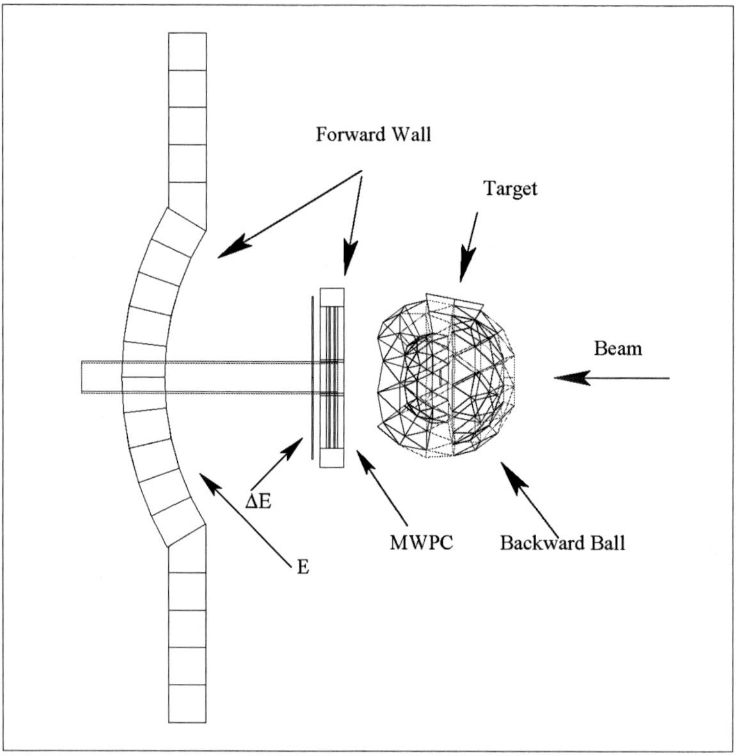

FIGURE 3. The side view of the new detection system, BINA with all its components marked in the figure.

Proton-deuteron break-up channel

The break-up channel offers a very rich kinematics as it involves three particles in the final state. Kinematically, one would then need to measure five out of the nine available kinematical variables to define an event completely. For this purpose, a detection system was designed which has an almost complete coverage of the phase space of the break-up reaction. This detector which is called Big Instrument for Nuclear-polarization Analysis (BINA) has been operational since 2005. BINA is a setup with a nearly 4π geometrical acceptance and has been used in various few-nucleon scattering experiments to measure the scattering angles and energies of protons and deuterons with the possibility for particle identification. The detector is composed of a forward and a backward part. A side view of BINA has been sketched in Fig. 3. The forward part consists of a Multi-Wire Proportional Chamber (MWPC) and a segmented hodoscope of vertically-placed thin scintillators with a thickness of 2 mm followed by ten horizontally-placed scintillators with a thickness of 12 cm each. The thick scintillators were mounted in a cylindrical shape, with center of the arc at the target position. The MWPC of BINA consists of X, Y, and U planes [37]. These planes are parallel arrays of equally spaced anode wires to read out the position of the scattered protons and deuterons. The X plane is composed of 96 vertically-oriented wires, the Y plane of 96 horizontally-oriented wires, and the U plane has 128 wires placed at an angle of 45° with respect to the X and Y planes. The MWPC has a central hole for a safe passage of the incident beam through the beam pipe. The detector covers scattering angles from 10° up to 32° with a full azimuthal coverage and up to 37° with a limited azimuthal acceptance. The detection efficiency of the MWPC is above 90% for most energies and particles in our experiments. The backward part has 149 phoswich scintillators covering polar angles between 40° to 165° with a nearly-full azimuthal coverage. These scintillators are glued together to operate as the scattering chamber. Each element consists of a pyramid-shaped fast plastic scintillator and, glued to its face surface, a 1 mm thin slow scintillator. The target used for these studies is liquid hydrogen or deuterium [38].

For the proton-deuteron break-up reaction, three measurements have been performed at beam energies of 100 (deuteron), 135 (proton) and 190 (proton) MeV with BINA. For the results of the measurements, with another detection system at KVI, called SALAD [39], and in particular the effects of Coulomb interaction at an incident deuteron beam energy of 130 MeV, refer to the contribution of Kistryn elsewhere in these proceedings. Many kinematical configurations have been measured and analyzed. Here, only a very small sample of the preliminary analyzing-power results are shown for the experiment done at an incident proton beam energy of 190 MeV. The results shown in Fig. 4 are for configurations where the two protons scatter to forward angles of $(25°,25°)$ (left panels) and $(25°,20°)$ (right panels). The azimuthal opening angles between the two protons vary between 180° (top panels) to 20° bottom panels. The results of various calculations are also shown. As can be seen from the top panels, various models agree with each other and also with experimental data for co-planar geometries. However, as one decreases the co-planarity angle from 180° to 20°, serious disagreements between various calculations set in and more importantly, they all disagree with the experimental data. Adding a 3NF to two-nucleon potential increases the disagreement even further. Even the effect of relativity [40] goes in the wrong direction. This problem needs to be further investigated.

CONCLUSIONS

To understand the properties of the three-nucleon forces, a large data-base including spectra of light nuclei and scattering data in three-body systems with all possible observables measured is mandatory. For the cross section of the elastic proton-deuteron scattering, a large data-base has been obtained which asks for a partial wave analysis for this channel. The results show unambiguously that two-nucleon potentials are not enough to describe the bulk of data. However, none of the three-nucleon potentials available in the literature is at a stage of describing all the data presented here and elsewhere consistently. The self-consistent treatment of the Δ in the Hanover approach which does a relatively good job in describing the analyzing powers in the elastic proton-deuteron scattering at energies between 100 and 200 MeV, performs less satisfactorily for vector and tensor analyzing powers in case of polarized beam of deuterons. The advantage of

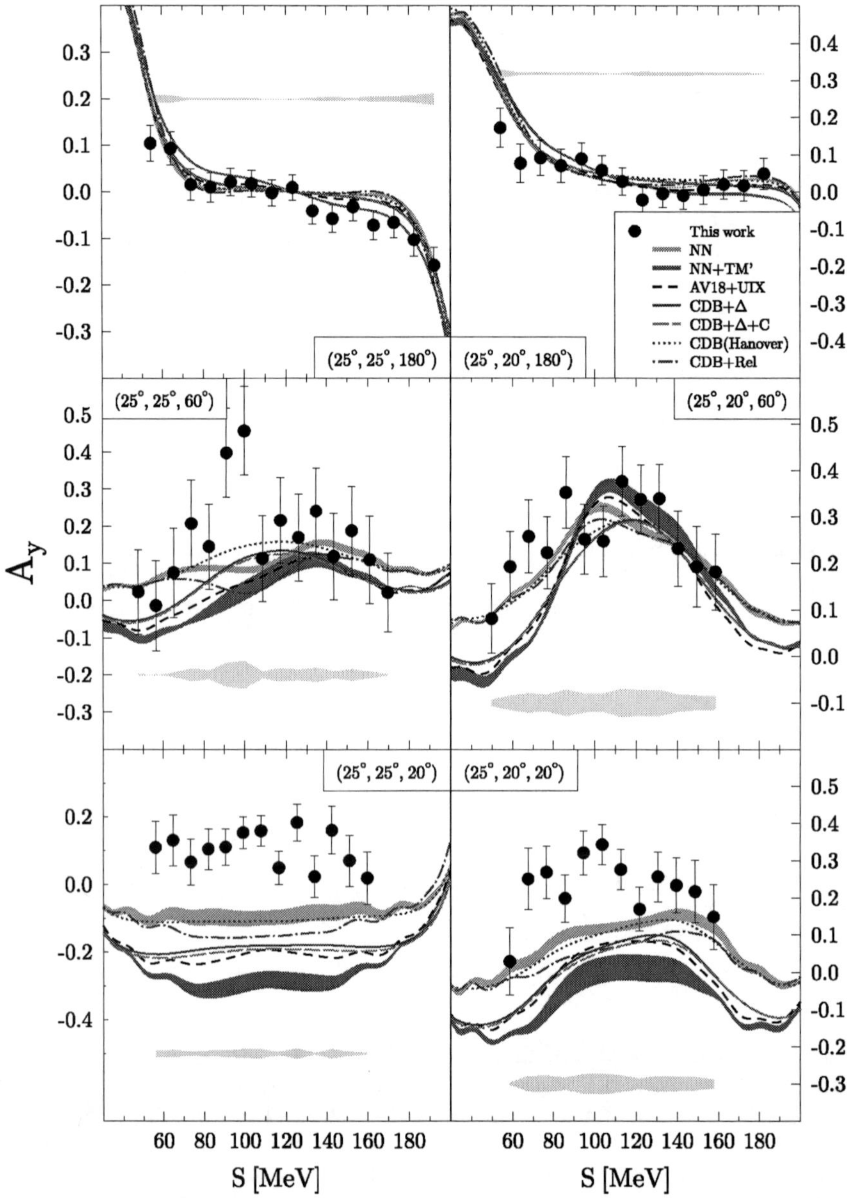

FIGURE 4. The preliminary results for analyzing powers of proton-deuteron break-up reaction at $E_p = 190$ MeV as a function of the kinematical variable, S, for various polar-angle combinations of the two outgoing protons and the opening azimuthal angle between them as marked in the figure by $(\theta_1, \theta_2, \phi_{12})$. The calculations are from the Hannover-Lisabon group (CDB, CDB+Δ, and CBD+Δ+Coulomb) and Bochum-Cracow group (NN, NN+TM',AV18+UIX, and CDB+Relativity). The systematic errors are shown by the light gray band in the panels.

this calculation is the inclusion of Coulomb force for the first time which is shown to be rather small in the elastic channel except at small scattering angles. Note that the Coulomb effects is rather sizable for some configurations in the break-up channel (see the contribution of St. Kistryn elsewhere in these proceedings). For the break-up channel, results show clearly that the nucleon-nucleon interaction alone is not enough to describe the data obtained at many configuations. However, the existing three-nucleon forces and also the model which includes some relativistic effects are far from being able to describe the analyzing-power data, specially at small co-planarity angles. For these kinematics, they even make the disagreement larger. Preliminary cross-section data also show large disagreements between the data and the calculations but then for large co-planarity angles.

It is clear that one needs to go beyond the nucleon-nucleon interaction to understand the dynamics of the three-body system. However, the characteristics of the three-nucleon forces, such as the central part and the spin part of the force, are not yet under control to be able to make high-precision predictions for scattering observables. The illusive nature of three-body forces might be exposed completely when the results of calculations based on effective field theory become available for these intermediate energies.

ACKNOWLEDGMENTS

I would like to thank my graduate students Karsten Ermisch and Hamid Amir-Ahmadi whose Ph.D. thesis work on the elastic channel were presented in this presentation. A. Ramazani-Moghadam-Arani should also be thanked for working on the scaling issue. A. Deltuva from the Hannover-Lisabon group and H. Witała from the Bochum-Cracow group should be thanked for providing the calculations shown in the figures.

REFERENCES

1. R. Machleidt, Phys. Rev. C63, 024001 (2001).
2. R.B. Wiringa et al., Phys. Rev. C51, 38 (1995).
3. V.G.J. Stoks et al., Phys. Rev. C48, 792 (1993).
4. V.G.J. Stoks et al., Phys. Rev. C49, 2950 (1994).
5. W. Glöckle et al., Phys. Rep. 274, 107 (1996).
6. N. Sakamoto et al., Phys. Lett. B367, 60 (1996).
7. H. Sakai et al., Phys. Rev. Lett. 84, 5288 (2000).
8. R. Bieber et al., Phys. Rev. Lett. 84, 606 (2000).
9. K. Ermisch et al., Phys. Rev. Lett. 86, 5862 (2001).
10. K. Ermisch et al., Phys. Rev. C68, 051001 (2003).
11. K. Ermisch et al., Phys. Rev. C71, 064004 (2005).
12. H. Mardanpour et al., Eur. Phys. J. **A31**, 383 (2007).
13. K. Sekiguchi et al., Phys. Rev. C65, 034003 (2002).
14. K. Sekiguchi et al., Phys. Rev. C70, 014001 (2004).
15. S.A. Coon et al., Phys. Rev. C48, 2559 (1993); Few Body Syst. 30, 131 (2001).
16. B.S. Pudliner et al., Phys. Rev. C56, 1720 (1997).
17. S.C. Pieper et al., Phys. Rev. C64, 014001 (2001).
18. U. van Kolck, Phys. Rev. C49, 2932 (1994).
19. E. Epelbaum et al., Phys. Rev. C66, 064001 (2002).

20. N. Kalantar-Nayestanaki and E. Epelbaum, Nuclear Physics News Inter. 17, No. 3, 22 (2007) and nucl-th/0703089.
21. St. Kistryn et al., Phys. Rev. C68, 054004 (2003).
22. St. Kistryn et al., Phys. Rev. C72, 044006 (2005).
23. St. Kistryn et al., Phys. Lett. B641, 23 (2006).
24. A. Mehmandoost-Khajeh-Dad et al., Phys. Lett. B617, 18 (2005).
25. H. Witała et al., Phys. Rev. Lett. 81, 1183 (1998).
26. S. Nemoto et al., Phys. Rev. C58, 2599 (1998).
27. K. Sekiguchi et al., Phys. Rev. Lett. **95**, 162301 (2005).
28. A. Deltuva et al., Phys. Rev. C68, 024005 (2003).
29. A. Deltuva et al., Phys. Rev. C71, 054005 (2005).
30. A. Deltuva et al., Phys. Rev. C72, 054004 (2005).
31. H.R. Amir-Ahmadi et al., Phys. Rev. **C75**, R041001 (2007).
32. A.M. van den Berg, Nucl. Instr. and Meth. B99, 637 (1995).
33. H.J. Wörtche, Nucl. Phys. A687, 321c (2001).
34. V. Hannen et al., Nucl. Inst. and Meth. in Phys. Res. A500, 68 (2003).
35. H.R. Amir-Ahmadi et al., Nucl. Inst. and Meth. in Phys. Res. A562, 338 (2006).
36. R. Bieber et al., Nucl. Inst. and Meth. in Phys. Res. A457, 12 (2001).
37. M. Volkerts et al., Nucl. Inst. and Meth. in Phys. Res. A428, 432 (1999).
38. N. Kalantar-Nayestanaki et al., Nucl. Inst. and Meth. in Phys. Res. A417, 215 (1998).
39. N. Kalantar-Nayestanaki et al., Nucl. Inst. and Meth. A444, 591 (2000).
40. H. Witała et al., Phys. Lett. B634, 374 (2006).

Model study of short-range three-nucleon forces

S. Ishikawa

Science Research Center, Hosei University, 2-17-1 Fujimi, Chiyoda, Tokyo 102-8160, Japan

Abstract. A phenomenological three-nucleon force is introduced to get rid of discrepancies between experimental data and calculations with a realistic nucleon-nucleon force plus the two-pion exchange three-nucleon force in low-energy three-nucleon scattering observables. Effects of this force, which consists of scalar, vector, and tensor components in spin space, on observables at higher energies are examined.

Keywords: Three-nucleon forces, Nucleon-deuteron polarization observables
PACS: 21.45.Ff, 25.10.+s

INTRODUCTION

As is well known, calculations with modern realistic two-nucleon forces (2NFs) give smaller values of the triton binding energy, B(^{3}H), than the empirical value, and the missing attraction is successfully explained by introducing a three-nucleon force (3NF) arising from the exchange of two pions among three nucleons (2πE-3NF). The introduction of the 2πE-3NF is also known to contribute to explaining differential cross sections in nucleon-deuteron (N-d) elastic scattering. This kind of 3NF originated from the work of Fujita and Miyazawa [1], and nowadays is considered as the long-range part of the 3NF [2, 3].

The next question might be whether any evidence for a short range part of the 3NF exists or not. Here we point out that, to reproduce B(^{3}H), we need to introduce a rather small value of cutoff mass parameter ($\Lambda_\pi \sim 700$ MeV) for the pion-Nucleon-Nucleon vertex form factor. This indicates that the 2πE-3NF in nature produces a large amount of attraction in three-nucleon (3N) systems, and the small value of Λ_π may simulate a repulsive effect caused from another kind of 3NF, which may be shorter range than the 2πE-3NF. In addition, it is remarked that the 2πE-3NF gives effects that are too small or contradictory in explaining experimental data for some polarization observables in N-d scattering (as shown later).

In general, nuclear forces are decomposed into scalar, vector, and tensor components in spin space. While the success of the 2πE-3NF is considered to be because of its spin scalar components, the failure in N-d polarization observables may be due to vector and tensor components in the 2πE-3NF [4, 5]. Thus we may expect additional 3NFs, which should have a repulsive character and should have vector and tensor components that contribute to N-d polarization observables.

Since there is no consensus for possible mechanism of such 3NFs, in this paper we take a phenomenological approach, in which we will introduce a (toy) model 3NF to reproduce experimental data, and study effects of such 3NF in 3N observables. A model 3NF has a form that has components typical of 2NFs, i.e., central spin-independent,

CP1011, *New Facet of Three Nucleon Force—50 Years of Fujita Miyazawa Three Nucleon Force (FM50), The International Symposium,* edited by H. Sakai, K. Sekiguchi, and B. F. Gibson

tensor, and spin-orbit components, which are modified in the presence of third nucleon [6].

THREE-BODY CALCULATIONS

Calculations in the present work are based on the coordinate space Faddeev approach, whose basic formulation is given in Refs. [7, 8]. In calculations at energies below the 3N breakup threshold, the long-range Coulomb interactions between two protons can be fully taken into account [7].

We use the Argonne V_{18} (AV18) model [9] as a typical 2NF, which gives 7.63 MeV for $B(^3H)$ compared to the empirical value of 8.48 MeV. The Brazil model [3] is used for the 2πE-3NF, in which the cutoff mass is chosen to be 680 MeV (BR_{680}) to reproduce $B(^3H)$.

In Fig. 1, experimental data for the vector analyzing power $A_y(\theta)$ in neutron-deuteron (*n-d*) elastic scattering below 30 MeV [10, 11, 12, 13, 14] is compared with calculations. In the figure, calculations with AV18 are plotted as dotted curves and those with AV18 + BR_{680} by dashed curves. (The meaning of the solid curves in the figure will be shown in the next section.) As Fig. 1 shows, effects of the 2πE-3NF are too small to remedy the discrepancy between the AV18 calculations and the experimental data for $A_y(\theta)$.

In Fig. 2, experimental data of the tensor analyzing power $T_{21}(\theta)$ in proton-deuteron (*p-d*) scattering at $E_p = 3.0$ MeV [15, 16] and $E_p = 28.0$ MeV [17] are compared with results of the calculations. The meaning of the curves in this figure is the same as in Fig. 1. The figure shows that the 2πE-3NF provides wrong effects in $T_{21}(\theta)$ compared with the experimental data [5].

MODEL 3NF

As shown in the previous section, the 2πE-3NF is not sufficient to explain all of the *N-d* scattering data. Here we consider a phenomenological 3NF, V^{phe}, of the following form [4, 6, 18],

$$V^{phe} = \sum_{i<j} e^{-(\frac{r_{ik}}{r_G})^2-(\frac{r_{jk}}{r_G})^2} \{V_0 + V_T S_T(ij)\hat{P}_{11}(ij)\} + V_{ls} e^{-\alpha\rho} \sum_{i<j} [\boldsymbol{l}_{ij} \cdot (\boldsymbol{S}_i + \boldsymbol{S}_j)]\hat{P}_{11}(ij), \tag{1}$$

where $S_T(ij)$ is the tensor operator acting between nucleon pair (i,j), $\hat{P}_{11}(ij)$ the projection operator to the spin and isospin triplet state of the (i,j) pair, and $\rho^2 = \frac{2}{3}(r_{12}^2 + r_{23}^2 + r_{31}^2)$. The range parameter r_G is taken to be 1.0 fm as in Refs. [5, 18], and α to be 1.5 fm^{-1}, which is the shortest one in Ref. [4].

The parameters for the strength of the central spin-independent component (C) V_0, the tensor component (T) V_T, and the spin-orbit component (SO) V_{ls} are determined as follows: First, the Brazil model is chosen with $\Lambda_\pi = 800$ MeV (BR_{800}), which gives overbinding for the triton by about 1 MeV. In Ref. [18], it is reported that one can simulate the 2πE-3NF by Eq. (1) with a choice of $(V_0, V_T) = (-38 \text{ MeV}, +20 \text{ MeV})$, and then get a remarkable improvement in $T_{21}(\theta)$ at $E_p = 3.0$ MeV by changing the

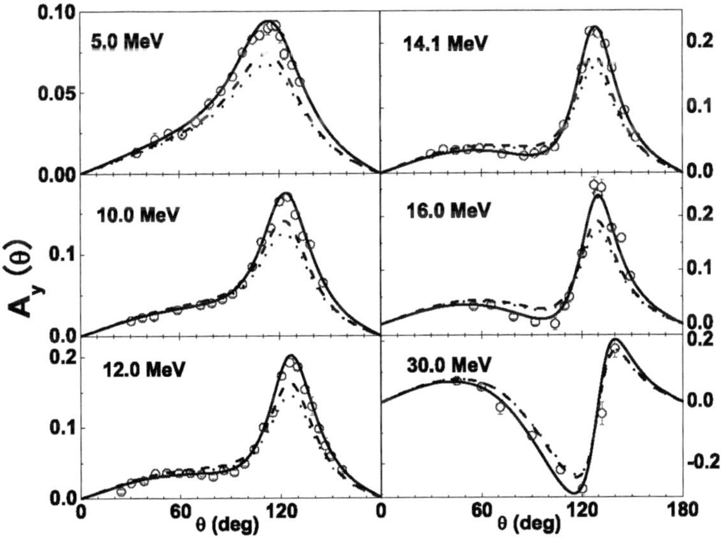

FIGURE 1. Neutron vector analyzing power $A_y(\theta)$ in *n-d* scattering. Dotted curves denote the AV18 calculations; dashed curves AV18+BR$_{680}$ calculations; solid curves AV18+BR$_{800}$+V^{phe}. Experimental data are from Ref. [10] for 5.0 MeV; from Ref. [11] for 10.0 MeV; from Ref. [12] for 14.1 MeV; from Ref. [13] for 16.0 MeV; and from Ref. [14] for 30.0 MeV.

sign of V_T and readjusting V_0 to fit the triton binding energy. Thus, in the present case, we choose the variation $-20-(+20) = -40$ MeV for the value of V_T. Hereafter, we call this procedure as *tensor inversion*. The strength of the spin-orbit component V_{ls} is determined to be -16 MeV to reproduce $A_y(\theta)$ at $E_p = 3.0$ MeV. Finally, the strength of the spin-independent component V_0 is determined to reproduce the triton binding energy. The values obtained are $(V_0, V_T, V_{ls}) = (+25\text{ MeV}, -40\text{ MeV}, -16\text{ MeV})$. It is noted that the C-, T-, and SO-3NFs are responsible for explaining B(^{3}H), $T_{21}(\theta)$ at $E_p = 3.0$ MeV, and $A_y(\theta)$ at $E_p = 3.0$ MeV, respectively.

Results of the combination of AV18, BR$_{800}$, and this model 3NF (AV18+BR$_{800}$+V^{phe}) at higher energies are displayed as the solid curves in Figs. 1 and 2. Figure 1 shows that the SO-3NF is successful in explaining $A_y(\theta)$ even at higher energies. However, the T-3NF is not enough to reproduce $T_{21}(\theta)$ at $E_p = 28.0$ MeV although it almost cancels out the effect of the 2πE-3NF.

In Fig. 3, calculations of the nucleon vector analyzing power at $E_N = 65$ MeV is compared with experimental data [19]. It is rather surprising that V^{phe} is still effective in improving the results of AV18 or AV18 + BR$_{680}$ calculations for $A_y(\theta \sim 90^\circ)$ at this energy.

Here, we study the energy dependence of $A_y(\theta)$ at two typical values of the angle θ: one is $\theta = 90^\circ$ and the other is the angle where $A_y(\theta)$ reaches a maximum, which varies depending on the energy. In Fig. 4, we plot the energy dependence of $A_y(\theta = 90^\circ)$ and $A_y(\theta = \theta_{max})$. In this figure, in addition to the results for AV18 (dotted curves), AV18

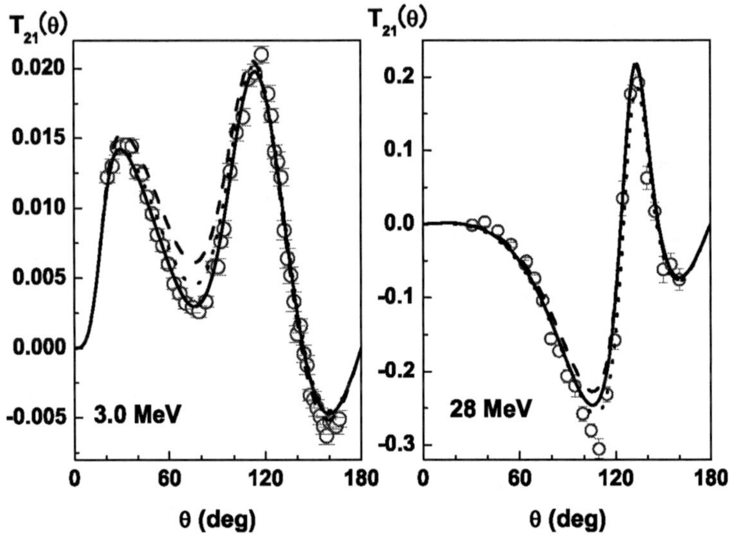

FIGURE 2. Deuteron tensor analyzing power $T_{21}(\theta)$ in *N*-*d* scattering at $E_N = 3.0$ and 28.0 MeV (or $E_d = 6.0$ and 56.0 MeV). Calculations are for *p*-*d* for 3.0 MeV and *n*-*d* for 28.0 MeV: dotted curves denote the AV18 calculation; dashed curves AV18+BR_{680}; solid curves AV18+BR_{800}+V^{phe}. Experimental data for 3.0 MeV are from Refs. [15, 16] and those for 28.0 MeV from Ref. [17].

+ BR_{680} (dashed curves), and AV18 + BR_{800} + V^{phe} (solid curves), we plot calculations with AV18 + BR_{680}-2NF plus the SO-3NF by dotted-dashed curves to distinguish effects of the spin-orbit component and the tensor component. The coincidence of the solid and the dotted-dashed curves for $A_y(\theta = \theta_{max})$ indicates that the effect of the SO-3NF is significant in reproducing the vector analyzing power at the maximum. On the other hand, the deviation of the dotted-dashed curve from the solid curve for $A_y(\theta = 90°)$, which seems to reproduce the data well, indicates that the effect of the tensor inversion becomes significant at $\theta = 90°$ as the energy increases.

SUMMARY

We introduced a model 3NF, V^{phe}-3NF, consisting of spin-independent, spin-orbit, and tensor components, which are essential in reproducing the triton binding energy, the proton vector analyzing power $A_y(\theta)$, and the deuteron tensor analyzing power $T_{21}(\theta)$, in *p*-*d* scattering at 3 MeV, respectively.

Effects of the V^{phe}-3NF on some *N*-*d* polarization observables at higher energies are examined.

The spin-orbit component plays a significant role in reproducing $A_y(\theta)$. The tensor component has non-negligible effects in $T_{21}(\theta \sim 90°)$, which partially cancels the undesirable effect of 2πE-3NF, and begins to play an important role in $A_y(\theta \sim 90°)$ for $E_N > 30$ MeV.

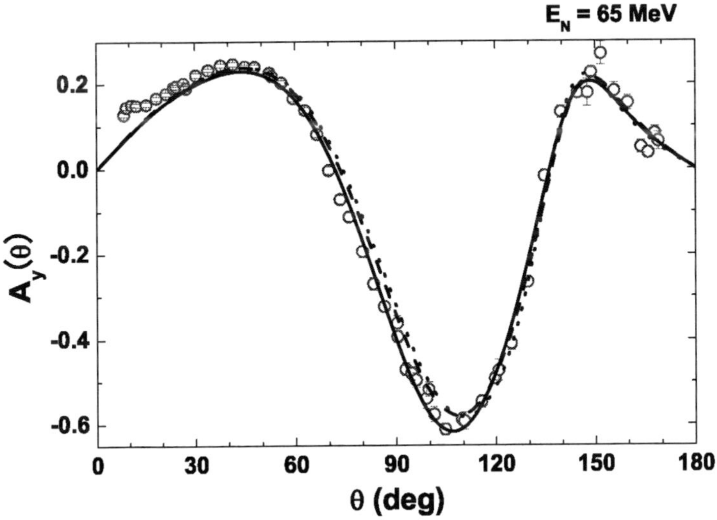

FIGURE 3. Neutron vector analyzing power $A_y(\theta)$ in n-d scattering at $E_N = 65$ MeV. Dotted curve denotes the AV18 calculation; dashed curve AV18+BR$_{680}$; solid curve AV18+BR$_{800}$+V^{phe}. Experimental data are from Ref. [19].

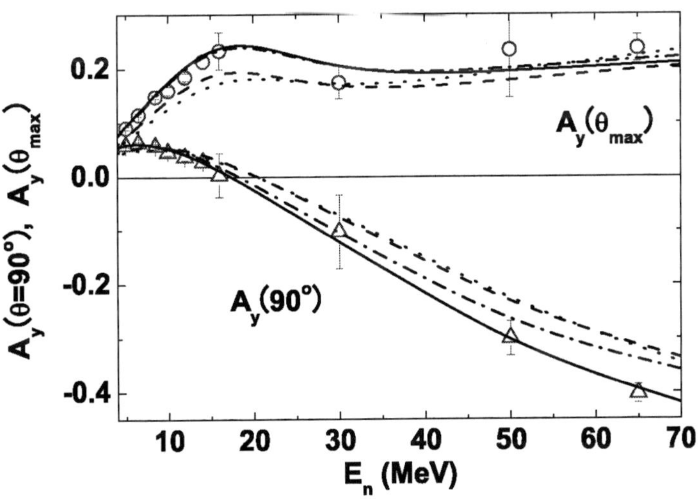

FIGURE 4. Energy dependence of $A_y(\theta)$ at $\theta = 90°$ and $\theta = \theta_{max}$. Dotted curves denote the AV18 calculation; dashed curves AV18+BR$_{680}$; solid curves AV18+BR$_{800}$+V^{phe}; dotted-dashed curves AV18+BR$_{680}$+SO. Data points are extracted from Refs. [10, 11, 12, 13, 14, 20, 21, 19].

If the model 3NF presented in the present work were really successful in explaining all of experimental data, it would be interesting to see what kind of realistic 3NFs are simulated by this. In Ref. [22], 3NFs arising from the exchange of a pion and a scalar object, such as σ or a scalar part of two-pion exchange, are examined. It is shown that these types of 3NFs give qualitatively similar effects to tensor inversion, but not quantitatively. Of course, this is not conclusive due to a large number of uncertainties in the description of such processes, and further study is required.

ACKNOWLEDGMENTS

The numerical calculations were supported by the Research Center for Computing and Multimedia Studies, Hosei University, under Project No. lab0003.

REFERENCES

1. J. Fujita and H. Miyazawa, Prog. Theor. Phys. **17**, 360 (1957).
2. S. A. Coon, M. D. Scadron, P. C. McNamee, B. R. Barrett, D. W. E. Blatt, and B. H. J. McKellar, Nucl. Phys. **A317**, 242 (1979).
3. H. T. Coelho, T. K. Das, and M. R. Robilotta, Phys. Rev. C **28**, 1812 (1983).
4. A. Kievsky, Phys. Rev. C **60**, 034001 (1999).
5. S. Ishikawa, M. Tanifuji, and Y. Iseri, Phys. Rev. C **67**, 061001-R (2003).
6. S. Ishikawa, Phys. Rev. C **75**, 061002(R) (2007).
7. S. Ishikawa, Few Body Syst. **32**, 229 (2003).
8. S. Ishikawa, Few Body Syst. **40**, 145 (2007).
9. R. B. Wiringa, V. G. J. Stoks, and R. Schiavilla, Phys. Rev. C **51**, 38 (1995).
10. W. Tornow, C. R. Howell, M. Alohali, Z. P. Chen, P. D. Felsher, J. M. Hanly, R. L. Walter, G. Weisel, G. Mertens, I. Slaus, H. Witała, and W. Glöckle, Phys. Lett. **B257**, 273 (1991).
11. W. Tornow, C. R. Howell, R. C. Byrd, R. S. Pedroni, and R. L. Walter, Phys. Rev. Lett. **49**, 312 (1982).
12. W. Tornow, R. C. Byrd, C. R. Howell, R. S. Pedroni, and R. L. Walter, Phys. Rev. C **27**, 2439 (1983).
13. K. Fujita, Ph.D Thesis (1999), Kyushu University; K. Sagara (private communication).
14. H. Dobiasch, R. Fischer, B. Haesner, H. O. Klages, P. Schwarz, B. Zeitnitz, R. Maschuw, K. Sinram, and K. Wick, Phys. Lett. **76B**, 195 (1978).
15. K. Sagara, H. Oguri, S. Shimizu, K. Maeda, H. Nakamura, T. Nakashima, and S. Morinobu, Phys. Rev. C **50**, 576 (1994).
16. S. Shimizu, K. Sagara, H. Nakamura, K. Maeda, T. Miwa, N. Nishimori, S. Ueno, T. Nakashima, and S. Morinobu, Phys. Rev. C **52**, 1193 (1995).
17. K. Hatanaka, N. Matsuoka, H. Sakai, T. Saito, K. Hosono, Y. Koike, M. Kondo, K. Imai, H. Shimizu, T. Ichihara, K. Nisimura, and A. Okihana, Nucl. Phys. **A426**, 77 (1984).
18. S. Ishikawa, M. Tanifuji, and Y. Iseri, Proc. of the Seventeenth International IUPAP Conference on Few-Body Problems in Physics (Few-Body 17). Durham, North Carolina, USA, June 5-10, 2003, edited by W. Glöckle, W. Tornow, (Elsevier, 2004) S61.
19. H. Shimizu, K. Imai, N. Tamura, K. Nisimura, K. Hatanaka, T. Saito, T. Koike, and Y. Taniguchi, Nucl. Phys., **A382**, 242 (1982).
20. J. L. Romero, J. L. Ullmann, F. P. Brady, J. D. Carlson, D. H. Fitzgerald, A. L. Sagle, T. S. Subramanian, C. I. Zanelli, N. S. P. King, M. W. Mc Naughton, and B. E. Bonner, Phys. Rev. C **25**, 2214 (1982).
21. J. W. Watson, R. Garrett, F. P. Brady, D. H. Fitzgerald, J. L. Romero, J. L. Ullman, C. I. Zanelli, Phys. Rev. C **25**, 2219 (1982).
22. S. Ishikawa, Nucl. Phys. **A790**, 389c (2007).

The Fujita-Miyazawa force in the light of effective field theory

Ulf-G. Meißner

HISKP, Universität Bonn, D-53115 Bonn, Germany
IKP-3, Forschungszentrum Jülich, D-52425 Jülich, Germany

Abstract. I discuss the Fujita-Miyazawa three-nucleon force in the light of chiral effective field theory.

Keywords: Chiral Lagrangians, effective field theory, nuclear forces
PACS: 13.75.Cs, 21.30.-x

THE FUJITA-MIYAZAWA FORCE

In their seminal paper, Fujita and Miyazawa (FM) [1] derived a three-nucleon force (3NF) from meson field theory, extending the idea of Yukawa beyond systems of two nucleons. More precisely, they sandwiched the pion-nucleon scattering amplitude between nucleon lines, thus generating the 3NF of longest range $\sim 1/M_\pi$, with M_π the pion mass. It reads (see the diagrammatic representation in Fig. 1)

$$\begin{aligned} V^{(3)} &= -\Big[\mathscr{A}\Big\{(\tau^1\tau^2)(\tau^2\tau^3)+\frac{3}{5}(\tau^3\tau^2)(\tau^3\tau^1)\Big\}(\sigma^1\nabla^1)(\sigma^2\nabla^1)(\sigma^2\nabla^3)(\sigma^3\nabla^3) \\ &+ \text{perm.}+\text{S}-\text{wave}\Big]\frac{e^{-x_{12}}}{x_{12}}\frac{e^{-x_{23}}}{x_{23}}\ , \quad \mathscr{A}=\frac{f^2}{4\pi}\frac{5}{72\pi^2}\int\frac{\sigma_{33}}{\omega_p^3}dp\ , \end{aligned} \tag{1}$$

with f the pseudoscalar pion-nucleon coupling constant, the indices $1,2,3$ refer to the nucleons, x_{ij} is the distance between nucleons i and j and the two Yukawa functions e^{-x}/x are responsible for the range of this force. FM realized the dominant contribution from the P_{33} partial wave, i.e. the excitation of the $\Delta(1232)$, as encoded in the partial cross section σ_{33}. At that time, the S-wave corrections were believed to be small, but could not be quantified precisely. That changed with the work of Tomozawa [2] and

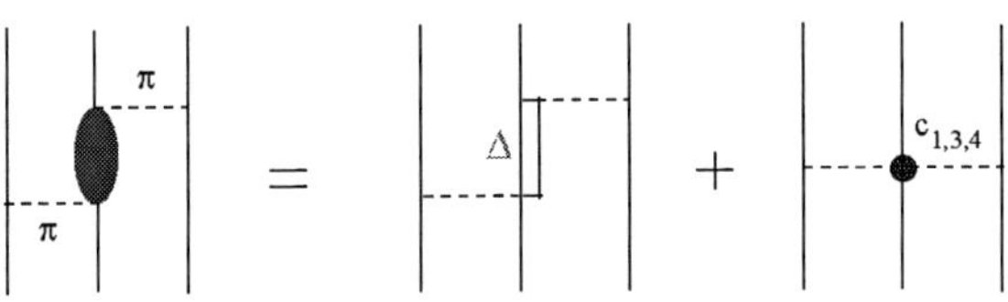

FIGURE 1. The Fujita-Miyazawa force. Its dominant contribution comes from the Δ resonance in the P_{33} partial wave of the pion-nucleon scattering amplitude (as depicted by the filled blob in the left-most diagram). The smaller S-wave contributions shown on the very right will be discussed later.

CP1011, *New Facet of Three Nucleon Force—50 Years of Fujita Miyazawa Three Nucleon Force (FM50), The International Symposium,* edited by H. Sakai, K. Sekiguchi, and B. F. Gibson

Weinberg [3], who used current algebra to show that S-wave scattering was expected to be small due to chiral symmetry. In fact, the work of FM has been the seed for many meson-theoretical approaches to the three-nucleon force like the families of Tucson-Melbourne [4, 5], Brazilian [6] or Urbana-Illinois [7, 8] 3NFs, as discussed by various speakers at this symposium. Despite all these modifications and extensions, the FM force turned out to be a very good approximation, and only with the advent of chiral nuclear effective field theory this could be understood, as I will explain in what follows.

FROM QCD TO NUCLEAR FORCES

The spontaneously, explicitely and anomalously broken chiral symmetry of QCD determines the low-energy interactions of pions with themselves and other hadrons. All this is captured by a suitably tailored EFT with hadronic degrees of freedom - chiral perturbation theory (CHPT). For nuclear systems, a straightforward application of this scheme is not possible due the very large S-wave nucleon-nucleon scattering lengths or the appearance of shallow nuclear bound states. As Weinberg [9] showed, this problem can be overcome by using CHPT to construct systematically the effective (irreducible) nuclear potential and iterate this in a Schrödinger or Lippmann-Schwinger equation. Such a scheme naturally leads to an explanation of the phenomenological observation that NN forces are much bigger than 3N forces, which are then also bigger than 4N forces [10]. By now the theory based on pions and nucleons as active degrees of freedom has reached a precision for NN forces that is comparable to the one of the so-called high precision phenomenological potentials, but in the EFT scheme many-nucleon forces and the couplings to electroweak currents can be worked out consistently with the NN forces and calculations can be improved systematically. Space forbids a more detailed discussion of these exciting developments, I refer the reader to the comprehensive review by Epelbaum [11]. Note that in the EFT with pions and nucleons the effect of the heavier states like from the delta or the rho-meson are hidden in the values of the pertinent low-energy constants (LECs), as analyzed for the pion-nucleon system in Ref. [12]. This is referred to as resonance saturation. The corresponding corrections from the subleading effective Lagrangian proportional to the LECs c_i to the pion-nucleon scattering amplitude are also shown in Fig. 1. In fact, in the pure pion-nucleon theory this diagram includes the effects of the delta and gives the leading contribution of the long-range part of the 3NF. Thus, to better understand the role of the FM force in the light of these developments, one needs to include the delta resonance in the EFT [13]. A systematic framework to extend the power counting is the so-called small scale expansion (SSE) of Ref. [14], in which the $N\Delta$ splitting is counted as an additional small parameter. Since this splitting does not vanish in the chiral limit, the connection to QCD requires the fulfillment of the decoupling theorem, for details I refer to [15]. Now let me come back to the FM force. We first discuss the underlying power counting. In the Weinberg scheme, irreducible diagrams with two or more nucleons which give rise to the nuclear forces are ordered according to the power ν of the expansion parameter Q/Λ, where Q collectively denotes small pion four-momenta, the pion mass, baryon three-momenta and the nucleon-delta mass splitting and Λ is the pertinent hard scale. For an irreducible N-nucleon diagram, the power ν is given by:

$$\nu = -2 + 2N + 2(L - C) + \sum_i V_i \Delta_i \,. \tag{2}$$

Here, L, C and V_i refer to the number of loops, separately connected pieces and vertices of type i, respectively. Further, the vertex dimension Δ_i is given by

$$\Delta_i = d_i + \frac{1}{2} b_i - 2 \,, \tag{3}$$

where b_i is the number of baryon field operators and d_i is the number of derivative and/or insertions of M_π.

As a first step in the analysis of the nuclear forces beyond leading order, one needs to analyze pion-nucleon scattering in the EFT with explicit deltas. To the order we are working, the analysis of $\pi N \to \pi N$ needs to be carried out up-to-and-including next-to-leading order effects [16]. This determines the four dimension-two LECs $c_{1,2,3,4}$ together with the dimension-three LEC $b_3 + b_8$. The underlying effective chiral Lagrangian reads

$$\begin{aligned}
\mathscr{L} &= \mathscr{L}^{(1)} + \mathscr{L}^{(2)} + \ldots \\
\mathscr{L}^{(1)} &= \bar{N} \mathscr{A}^{(1)}_{\pi N} N + \bar{T} \mathscr{A}^{(1)}_{\pi\Delta} T + \left(\bar{T} \mathscr{A}^{(1)}_{\pi N\Delta} N + \text{h.c.} \right) \,, \\
\mathscr{L}^{(2)} &= \bar{N} \mathscr{A}^{(2)}_{\pi N} N + \left(\bar{T} \mathscr{A}^{(2)}_{\pi N\Delta} N + \text{h.c.} \right) \,,
\end{aligned} \tag{4}$$

with

$$\begin{aligned}
\mathscr{A}^{(1)}_{\pi N} &= i v \cdot D + g_A u \cdot S \,, \\
\mathscr{A}^{(2)}_{\pi N} &= c_1 \langle \chi_+ \rangle + c_2 (v \cdot u)^2 + c_3 u \cdot u + c_4 [S^\mu, S^\nu] u_\mu u_\nu \,, \\
\mathscr{A}^{(1)}_{\pi\Delta} &= -\left(i v \cdot D^{ij} - (m_\Delta - m_N) \delta^{ij} + \ldots \right) g^{\mu\nu} \,, \\
\mathscr{A}^{(1)}_{\pi N\Delta} &= h_A P^{\mu\nu} w^i_\alpha \,, \\
\mathscr{A}^{(2)}_{\pi N\Delta} &= P^{\mu\alpha} i (b_3 + b_8) w^i_{\alpha\beta} v^\beta \,,
\end{aligned} \tag{5}$$

where N denotes the large component of the nucleon field, T is an abbreviation for the large component of the delta field, $T \equiv T^i_\mu$, with i an isospin and μ a Lorentz index. We use standard notation: $U(x) = u^2(x)$ collects the pion fields, $u_\mu = i(u^\dagger \partial_\mu u - u \partial_\mu u^\dagger)$, $\chi_+ = u^\dagger \chi u^\dagger + u \chi^\dagger u$ includes the explicit chiral symmetry breaking, $\langle \ldots \rangle$ denotes a trace in flavor space and $D_\mu (D^{ij}_\mu)$ is the chiral covariant derivative for the nucleon (delta) fields. Furthermore, $P_{\mu\nu}$ is the standard projector on the 3/2-components, $P_{\mu\nu} = g_{\mu\nu} - v_\mu v_\nu - 4 S_\mu S_\nu / (1-d)$, with v_μ the four-velocity, S_μ the covariant spin vector and d the number of space-time dimensions. We also have $w^i_\alpha = \langle \tau^i u_\alpha \rangle / 2$ and $w^i_{\alpha\beta} = \langle \tau^i [\partial_\alpha, u_\beta] \rangle / 2$. In the following, we also use the notation $\Delta \equiv m_\Delta - m_N$ for the $N\Delta$ mass splitting (which can not be confused with the same symbol denoting the delta field). The pertinent LECs at leading order are the nucleon axial-vector coupling g_A and the $\pi N\Delta$

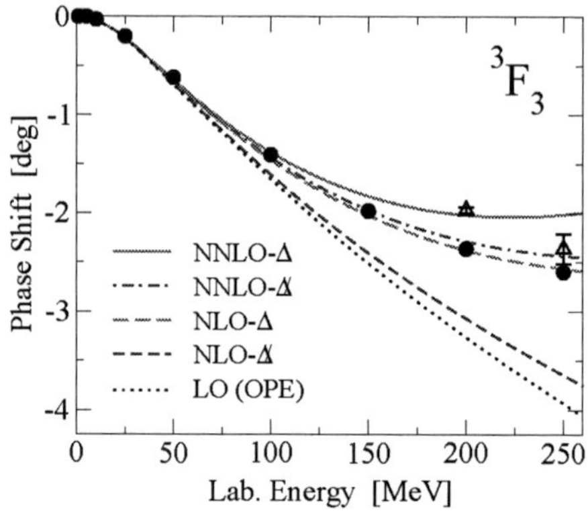

FIGURE 2. $^3F_3$ NN phase shift in EFT with and without explicit deltas calculated in first Born approximation using the spectral-function cutoff 700 MeV. The filled circles (open triangles) depict the results from the Nijmegen (Virginia Tech) PWA.

axial coupling h_A. At NLO, we have the four dimension-two πN LECs c_i $(i = 1,2,3,4)$ and the combination of $\pi N\Delta$ LECs $b_3 + b_8$. Strictly speaking, all couplings and masses appearing in the effective Lagrangian should be taken at their chiral limit values, but to the accuracy we are working, we can use their pertinent physical values. The values of the LECs c_i are different in the delta-less and delta-full theories. At subleading order, the determination of c_i from the πN S- and P-wave threshold coefficients yields in the delta-less theory: $c_1 = -0.57, c_2 = 2.84, c_3 = -3.87, c_4 = 2.89$, where only central values are given and the units are GeV^{-1}. Including the Δ-contributions and utilizing the large-N_c relation between the $\pi N\Delta$ and πNN axial-vector coupling, $h_A = 3g_A/(2\sqrt{2})$, leads to $b_3 + b_8 = 1.40\ \mathrm{GeV}^{-1}$ and $c_1 = -0.57, c_2 = -0.25, c_3 = -0.79, c_4 = 1.33$. As expected, the LECs $c_{2,3,4}$ are strongly reduced in magnitude. Note, however, that c_1 is entirely given by scalar-isoscalar two-pion exchange and c_4 receives a sizeable contribution from the ρ, see Ref. [12] (for a more detailed discussion of these LECs, see Ref. [17]). A more complete analysis of πN scattering at one-loop order in the SSE is given in Ref. [18] (see also Ref. [19]).

Armed with this information, we can now study the two-pion-exchange (TPE) potential between two nucleons, which receives first contributions due to intermediate deltas at NLO. These where worked out long ago [20, 21]. At NNLO, further corrections including the LECs c_i enter [16]. Their reduced values in the delta-full theory lead to a much more natural convergence pattern for the TPE potential. Contrary to the delta-less theory, the numerically dominant contributions to the central, tensor and spin-spin components of the TPE potential are now generated at NLO with the NNLO contributions yielding typically only modest corrections to the NLO result. The improved convergence is also clearly visible in peripheral NN partial waves which were calculated using the Born approximation and, at this order, are only sensitive to the OPE and TPE potentials, see Fig. 2 for a representative example. On the other hand, the NNLO TPE potential in the delta-less theory is found to provide a surprisingly good approximation to the potential resulting at the same order in the delta-full theory. This indicates that the saturation of the LECs $c_{3,4}$ is the most important effect of the Δ-isobar at the order considered.

EMERGENCE OF THE FUJITA-MIYAZAWA FORCE

We now analyze the three-nucleon forces in the framework of a chiral EFT with pions, nucleons and deltas, based on Weinbergs power counting. For the calculation of the 3NF up to next-to-next-to-leading order (NNLO), we use the effective Lagrangian given in Eq. (4). The only additional terms that need to be considered for the analysis of the 3NF are the leading-order $NN \to N\Delta$ and $N\Delta \to NN$ contact interactions

$$\bar{T}_i^{\mu} N \bar{N} S_{\mu} \tau^i N + \text{h. c.}. \tag{6}$$

The contact interactions in Eq. (6) were already considered by van Kolck [10] who worked out the corresponding contributions to the 3NF. As pointed out in Ref. [11], matrix elements of the resulting 3NFs between antisymmetrized $|NNN\rangle$ states vanish, see also [22] for a related discussion. The reason for vanishing of these 3NF contributions can be understood already at the level of the effective Lagrangian. To that aim, we rewrite Eq. (6) including the spin- and isospin-3/2 projectors $P^{3/2}_{\mu\nu}$ and $\xi^{3/2}_{ij}$ explicitely:

$$\bar{T}_i^{\mu} N \bar{N} S_{\mu} \tau^i N = \bar{T}_j^{\nu} \xi^{3/2}_{ji} P^{3/2}_{\nu\mu} N \bar{N} S^{\mu} \tau^i N, \tag{7}$$

with

$$\xi^{3/2}_{ij} = \frac{2}{3}\delta_{ij} - \frac{i}{3}\varepsilon_{ijk}\tau^k, \quad P^{3/2}_{\mu\nu} = g_{\mu\nu} - v_{\mu}v_{\nu} - \frac{4}{1-d} S_{\mu} S_{\nu}\,. \tag{8}$$

Here, v_{μ} denotes the baryon four-velocity and d the number of space-time dimensions. In four dimensions with $v = (1,0,0,0)$, the spin-3/2 projector reduces to

$$P^{3/2}_{ij} = -\left[\frac{2}{3}\delta_{ij} - \frac{i}{3}\varepsilon_{ijk}\sigma^k\right], \tag{9}$$

with σ^k denoting the Pauli spin matrices. It is straightforward to see that the antisymmetrized Feynman rules for the particular contact interactions vanish:

$$\xi^{3/2}_{ij}(1) P^{3/2}_{kl}(1) \tau_2^j \sigma_2^l \mathscr{A}_{12} = \xi^{3/2}_{ij}(2) P^{3/2}_{kl}(2) \tau_1^j \sigma_1^l \mathscr{A}_{12} = 0\,, \tag{10}$$

where the antisymmetrization operator for momentum-independent interactions $\mathscr{A}_{12}$ has the form

$$\mathscr{A}_{12} = \frac{1-P_{12}}{2}, \quad \text{with} \quad P_{12} = \frac{1+\vec{\sigma}_1\cdot\vec{\sigma}_2}{2}\frac{1+\vec{\tau}_1\cdot\vec{\tau}_2}{2}\,. \tag{11}$$

The labeled projection operators $\xi^{3/2}_{ij}(X)$ and $P^{3/2}_{ij}(X)$ in Eq. (10) are defined according to

$$\xi^{3/2}_{ij}(X) = \frac{2}{3}\delta_{ij} - \frac{i}{3}\varepsilon_{ijk}\tau_X^k, \quad P^{3/2}_{ij}(X) = -\left[\frac{2}{3}\delta_{ij} - \frac{i}{3}\varepsilon_{ijk}\sigma_X^k\right], \quad X \in \{1,2\}\,. \tag{12}$$

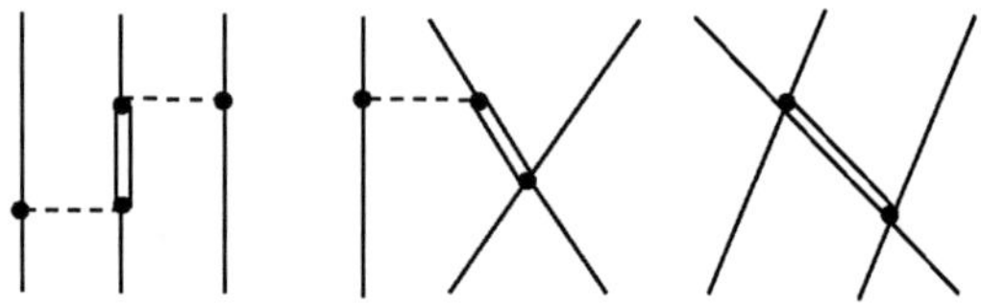

a) b) c)

FIGURE 3. Leading contributions to the 3NF at NLO in the delta-full theory. Solid, double and dashed lines denote nucleons, deltas and pions, in order. The circles represent leading vertices.

d) e) f) g) h)

FIGURE 4. Subleading contributions to the 3NF at NNLO in the delta-full theory. Solid, double and dashed lines denote nucleons, deltas and pions, in order. The circles (square) are leading (subleading) vertices. Graphs g) and h) do not contribute at this order.

Here and in what follows, the subscripts of the Pauli spin and isospin matrices refer to the nucleon labels.

We are now in the position to discuss the 3NF contributions due to intermediate Δ-excitations. The leading 3NF appears at NLO and this given by the topologies a), b) and c) in Fig. 3. Due to vanishing of the lowest-order contact interaction as just discussed, the complete effect due the Δ is given by a single two-pion (2π) exchange diagram a) in Fig. 3 – which is nothing but the delta contribution to the pion-nucleon scattering amplitude considered by Fujita and Miyazawa (see Fig. 1). It is well known that the corresponding 3NF contribution is exactly reproduced at NNLO in EFT without explicit Δ-fields via resonance saturation of the low-energy constants $c_{3,4}$ accompanying the subleading $\pi\pi NN$ vertices: $c_3 = -2c_4 = -4h_A^2/(9\Delta)$.

Let us now analyze the NNLO contributions, i.e. the first corrections to the FM force. First, we emphasize that there are still no diagrams involving contact interactions at this order. In fact, since $NNN\Delta$ contact interactions contain at least two derivatives, they start to contribute to the 3NF only at N^{3}LO. Therefore, the only diagrams which need to be considered are the ones which result from graph a) in Fig. 3 by substituting one of the leading $\pi N\Delta$ vertices by the subleading one which is proportional to the combination of LECs $(b_3 + b_8)$, see Refs. [18, 16].[1] These contributions are represented in Fig. 4 by graph g). Since this vertex involves a time derivative, the corresponding 3NF

[1] Due to parity invariance, the subleading πNN vertices contain two derivatives or one insertion of M_π^2 more than the leading one and thus need not be considered at NNLO.

contribution is shifted to higher orders due to the $1/m_N$-suppression. We, therefore, end up with the conclusion that the 2π-exchange 3NF from diagram a) in Fig. 4 represents the only additional contribution which arises in EFT with explicit deltas up to NNLO. At this order, there are non-delta contributions from graphs d), e) and f). The two-pion exchange contribution d) is given in terms of the LECs $c_{1,3,4}$ with their reduced values given above. The LECs D and E related to the shorter-ranged graphs e) and f) can not be determined in the pion-nucleon system. While E can only be determined in systems with at least three nucleons, the LEC D can also be pinned down e.g. from the process $NN \to NN\pi$, as first pointed out in Ref. [23].

Let me summarize the central findings of this chapter. In chiral EFT with explicit delta degrees of freedom, the 3NF appears first at NLO and is thus expected to be smaller than the two-nucleon force – that is what is seen in nature. Also, at this order, the only graph that contributes to the 3NF is the two-pion exchange graph with an intermediate delta. This is nothing but what has frequently been called the FM force. Corrections to it arise at NNLO, but these do not involve effects of the delta resonance. There are two-pion exchange corrections proportional to the reduced LECs c_i. These parameterize the physics of higher baryon resonances and, even more important, of the non-Goldstone mesons like the ρ (c_4) or the σ (c_1). In addition, there are corrections of shorter distance involving 4- and 6-nucleon contact operators - only within EFT one can systematically generate all these different contribution.

ISOSPIN-VIOLATING THREE-NUCLEON FORCES

So far, we have considered an ideal world of strong interactions only and with equal quark masses $m_u = m_d$. In such a world, isospin symmetry is exact. In nature, however, the strong and the electromagnetic interactions lead to isospin violation (IV), and thus one has to consider the IV corrections to the FM force. This can be done consistently within EFT. As it turns out, the leading IV effects can all be expressed in terms of the splittings in the pertinent hadron multiplets, i.e. the nucleon doublet, the pion triplet and the delta quartet, for details see Ref. [24]. The new ingredient in this investigation is the first quantitative analysis of the delta quartet splitting (chiral extrapolation functions for various delta masses were already considered in [25]). The corresponding Lagrangian that parameterizes the splittings within the quartet contains two independent operators

$$\mathscr{L}_{\Delta,\mathrm{mass,IV}} = -\bar{T}_i^\mu \left[-\delta m_\Delta^1 \frac{1}{2} \tau^3 \delta_{ij} - \delta m_\Delta^2 \frac{3}{4} \delta_{i3} \delta_{j3} \right] g_{\mu\nu} T_j^\nu , \tag{13}$$

where the LECs $\delta m_\Delta^{1,2}$ are combinations of strong and electromagnetic dimension-two LECs, see [24] for details. Switching from the Rarita-Schwinger representation to physical delta fields $(\Delta^{++}, \Delta^+, \Delta^0, \Delta^-)$ one can express the various delta masses as follows:

$$\begin{aligned} m_{\Delta^{++}} &= \tilde{m}_\Delta + \frac{1}{2} \delta m_\Delta^1 , \quad m_{\Delta^+} = \tilde{m}_\Delta + \frac{1}{6} \delta m_\Delta^1 + \frac{1}{2} \delta m_\Delta^2 , \\ m_{\Delta^0} &= \tilde{m}_\Delta - \frac{1}{6} \delta m_\Delta^1 + \frac{1}{2} \delta m_\Delta^2 , \quad m_{\Delta^-} = \tilde{m}_\Delta - \frac{1}{2} \delta m_\Delta^1 , \end{aligned} \tag{14}$$

where the mass $\tilde{m}_\Delta$ contains an isospin-invariant delta mass shift δm_Δ, defined as $\tilde{m}_\Delta = \overset{\circ}{m}_\Delta + \delta m_\Delta$, with $\overset{\circ}{m}_\Delta$ the delta mass in the chiral limit. Notice also that in the absence of electromagnetic corrections (no δm_Δ^2-term), there is an equal spacing between the members of the quartet. In that case, one has $\bar{m}_\Delta = \tilde{m}_\Delta$ (with $\bar{m}_\Delta = (m_{\Delta^{++}} + m_{\Delta^+} + m_{\Delta^0} + m_{\Delta^-})/4$ the average delta mass) and all splittings are equal to $\delta m_\Delta^1/3$. We also recover the SU(6) results $m_{\Delta^{++}} - m_{\Delta^-} = 3(m_{\Delta^+} - m_{\Delta^0})$ independent of the strength of the electromagnetic splitting, see e.g. [26]. To pin down the values for $\tilde{m}$, δm_Δ^1 and δm_Δ^2, we use as input the available data on $m_{\Delta^{++}}$, m_{Δ^0} and the average delta mass $\bar{m}_\Delta$. For the latter, we adopt the value $\bar{m}_\Delta = 1233$ MeV which is consistent with the estimation of the PDG, $\bar{m}_\Delta = 1231\ldots1233$ MeV, as well as with the most recent determination from Ref. [27], $\bar{m}_\Delta = 1233.4 \pm 0.4$ MeV. This leads to

$$\tilde{m}_\Delta = 1233.4 \pm 0.7\,\text{MeV}\,, \qquad \delta m_\Delta^1 = -5.3 \pm 1.9\,\text{MeV}\,, \qquad \delta m_\Delta^2 = -1.7 \pm 2.7\,\text{MeV}\,, \tag{15}$$

where the error bars result from using different input values for $m_{\Delta^{++}}$ and m_{Δ^0}. Notice that the obtained results for δm_Δ^1 and δm_Δ^2 are of a natural size. Indeed, based on the naive dimensional analysis, one expects $|\delta m_\Delta^1| \sim |\varepsilon M_\pi^2/M_\rho| = 8.2$ MeV and $|\delta m_\Delta^2| \sim e^2 M_\rho/(4\pi)^2 = 1.5$ MeV, with $\varepsilon = (m_d - m_u)/(m_d + m_u) \approx 0.3$, $e^2/(4\pi) \approx 1/137$ and $M_\rho = 770\,\text{MeV}$ (for another determination of $\tilde{m}_\Delta$, δm_Δ^1 and δm_Δ^2 using the quark model relation, see [24]).

The calculation of the IV violating 3NF in the delta-full theory can now be done straightforwardly. A technical simplification arises if one performs field redefinitions to eliminate the nucleon mass shift from the effective Lagrangian. This allows to get rid of diagrams which involve reducible topologies and, therefore, enables one to use the Feynman graph technique to derive the 3NF. As already demonstrated in Ref. [28] for the EFT without explicit Δ resonances, the neutron-proton mass difference can be eliminated from $\mathscr{L}_{\text{eff}}$ via a suitable redefinition of the pion and nucleon fields (or, equivalently, via a suitable *local* $SU(2)_V$ transformation) in favour of new vertices proportional to δm_N. This approach can be generalized to include the Δ as an explicit degree of freedom, as detailed in Ref. [24]. A straightforward evaluation of the Feynman diagrams in Fig. 5 yields charge-symmetry conserving (CSC) and charge-symmetry breaking (CSB) contributions to the 3NF, the latter one is given by

$$\begin{aligned} V_{3N}^{\text{CSB}} &= -\sum_{i\neq j\neq k} \frac{g_A^2 h_A^2 (\delta m_\Delta^1 - 3\delta m_N)}{216 F_\pi^4 \Delta^2} \frac{\vec{q}_i \cdot \vec{\sigma}_i \vec{q}_k \cdot \vec{\sigma}_k}{[q_i^2 + M_\pi^2][q_k^2 + M_\pi^2]} \Big(5\,[\vec{\tau}_i \times \vec{\tau}_k]^3\,[\vec{q}_i \times \vec{q}_k] \cdot \vec{\sigma}_j \\ &\quad + 4(\vec{\tau}_j \cdot \vec{\tau}_k \tau_i^3 - 2\vec{\tau}_i \cdot \vec{\tau}_k \tau_j^3)\,\vec{q}_i \cdot \vec{q}_k \Big)\,. \end{aligned} \tag{16}$$

This can be transformed into configuration space:

$$V_{3N}^{\text{CSB}} = -\sum_{i\neq j\neq k} \frac{g_A^2 h_A^2 (\delta m_\Delta^1 - 3\delta m_N) M_\pi^6}{3456 \pi^2 F_\pi^4 \Delta^2} \vec{\sigma}_i \cdot \vec{\nabla}_{ij}\, \vec{\sigma}_k \cdot \vec{\nabla}_{kj} \Big[5[\vec{\tau}_i \times \vec{\tau}_k]^3\,[\vec{\nabla}_{ij} \times \vec{\nabla}_{kj}] \cdot \vec{\sigma}_j$$

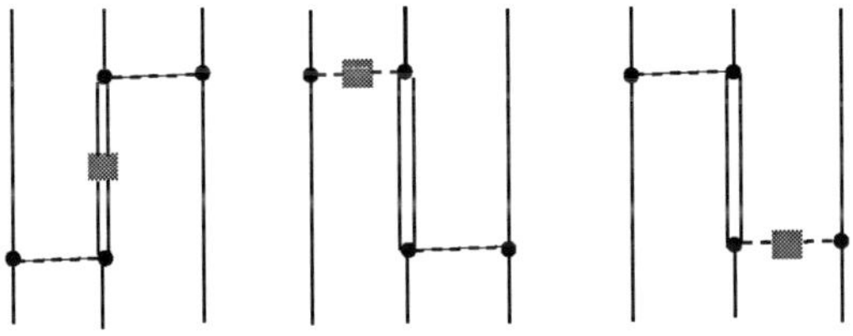

FIGURE 5. Leading isospin-breaking contributions to the 3NF with a Δ-excitation. Solid, dashed and double lines represent nucleons, pions and deltas, respectively. Solid dots denote the leading isospin-invariant vertices, filled squares are the isospin-breaking vertices from the $SU(2)_V$-rotated effective Lagrangian as discussed in the text.

$$+4(\vec{\tau}_j\cdot\vec{\tau}_k\tau_i^3 - 2\vec{\tau}_i\cdot\vec{\tau}_k\tau_j^3)\,\vec{\nabla}_{ij}\cdot\vec{\nabla}_{kj}\Big]\frac{e^{-x_{ij}}}{x_{ij}}\frac{e^{-x_{kj}}}{x_{kj}}\,, \tag{17}$$

where $\vec{r}_{ij} \equiv \vec{r}_i - \vec{r}_j$ is the distance between the nucleons i and j, $\vec{x}_i \equiv M_\pi \vec{r}_i$, $\vec{\nabla}_i$ act on $\vec{x}_i$ and $x_{ij} \equiv |\vec{x}_{ij}|$. To get a rough idea about the size of the isospin breaking 3NF contributions to e.g. the 3N binding energy, one can look at the strength of the corresponding r-space potentials in Eq. (17). The estimated size of the CSB 3NF using $\delta m_\Delta^1 = -5.3\,\mathrm{MeV}$ is

$$|\delta m_\Delta^1 - 3\delta m_N|\,\frac{g_A^2 h_A^2 M_\pi^6}{432\pi^2 F_\pi^4 \Delta^2} \sim 3\ \mathrm{keV}, \tag{18}$$

which is comparable to the typical size of the leading CSB 3NF obtained in Ref. [29] and based on EFT without explicit deltas, $\delta m_N g_A^4 M_\pi^4/(256\pi^2 F_\pi^4) \sim 7$ keV. We further emphasize that the CSB 3NF in Eq. (16) vanishes exactly if one adopts the quark model relation $m_{\Delta^{++}} - m_{\Delta^-} = 3(m_{\Delta^+} - m_{\Delta^0})$. The CSC 3NF generated by these diagrams is discussed in Ref. [24], where it is also shown how these IV 3NF contributions are reproduced in the EFT without explicit deltas.

SUMMARY

As I have shown in this contribution, only 50 years after the seminal paper of Fujita and Miyazawa [1] one could obtain a deeper understanding of their work. Of course, phenomenological arguments could always be brought forward to justify their ansatz, but only with the rise of chiral EFT a systematic explanation was possible. This is truly a pleasure. Now it is time is ripe to go further - hopefully progress is made on shorter time scales.

ACKNOWLEDGMENTS

Most of the work presented here has been done in collaboration with Evgeny Epelbaum and Hermann Krebs. I also would like to thank the organizers for making such an interesting symposium possible. I especially enjoyed discussions with Professor Miyazawa. This work was supported in part by the DFG (SFB/TR 16 "Subnuclear Structure of Matter"), by the EU Integrated Infrastructure Initiative Hadron Physics Project under contract number RII3-CT-2004-506078 and by funds provided from the Helmholtz Association to the virtual institute "Spin and strong QCD" (grant VH-VI-231).

REFERENCES

1. J.-I. Fujita and H. Miyazawa, Prog. Theo. Phys. **17**, 360 (1957).
2. Y. Tomozawa, Nuovo Cim. **46A**, 707 (1966).
3. S. Weinberg, Phys. Rev. Lett. **17**, 616 (1966).
4. B. H. J. McKellar and R. Rajaraman, Phys. Rev. Lett. **21**, 450 (1968).
5. S. A. Coon, M. D. Scadron and B. R. Barrett, Nucl. Phys. A **242**, 467 (1975).
6. H. T. Coelho *et al.*, Phys. Rev. C **28**, 1812 (1983).
7. B. S. Pudliner *et al.*, Phys. Rev. C **56**, 1720 (1997).
8. S. C. Pieper *et al.*, Phys. Rev. C **64**, 014001 (2001).
9. S. Weinberg, Nucl. Phys. B **363**, 3 (1991).
10. U. van Kolck, Phys. Rev. C **49**, 2932 (1994).
11. E. Epelbaum, Prog. Part. Nucl. Phys. **57**, 654 (2006), [arXiv:nucl-th/0509032].
12. V. Bernard, N. Kaiser and U.-G. Meißner, Nucl. Phys. A **615** (1997) 483 [arXiv:hep-ph/9611253].
13. E. E. Jenkins and A. V. Manohar, Phys. Lett. B **259**, 353 (1991).
14. T. R. Hemmert, B. R. Holstein and J. Kambor, J. Phys. G **24**, 1831 (1998) [arXiv:hep-ph/9712496].
15. J. Gasser and A. Zepeda, Nucl. Phys. B **174**, 445 (1980).
16. H. Krebs, E. Epelbaum and U.-G. Meißner, Eur. Phys. J. A **32**, 127 (2007) [arXiv:nucl-th/0703087].
17. V. Bernard, Prog. Part. Nucl. Phys. **60**, 82 (2008) [arXiv:0706.0312 [hep-ph]].
18. N. Fettes and U.-G. Meißner, Nucl. Phys. A **679**, 629 (2001) [arXiv:hep-ph/0006299].
19. P. J. Ellis and H. B. Tang, Phys. Rev. C **57**, 3356 (1998) [arXiv:hep-ph/9709354].
20. C. Ordóñez, L. Ray, and U. van Kolck, Phys. Rev. C **53**, 2086 (1996), hep-ph/9511380.
21. N. Kaiser, S. Gerstendorfer, and W. Weise, Nucl. Phys. A **637**, 395 (1998), nucl-th/9802071.
22. E. Epelbaum, A. Nogga, W. Glöckle, H. Kamada, U.-G. Meißner and H. Witala, Phys. Rev. C **66**, 064001 (2002) [arXiv:nucl-th/0208023].
23. C. Hanhart, U. van Kolck and G. A. Miller, Phys. Rev. Lett. **85**, 2905 (2000) [arXiv:nucl-th/0004033].
24. E. Epelbaum, H. Krebs and U.-G. Meißner, arXiv:0712.1969 [nucl-th].
25. B. C. Tiburzi and A. Walker-Loud, Nucl. Phys. A **764**, 274 (2006) [arXiv:hep-lat/0501018].
26. H. R. Rubinstein, F. Scheck, and R. H. Sokolov, Phys. Rev. **154**, 1608 (1967).
27. R. A. Arndt, W. J. Briscoe, I. I. Strakovsky, and R. L. Workman, Phys. Rev. C **74**, 045205 (2006), [arXiv:nucl-th/0605082].
28. J. L. Friar, U. van Kolck, M. C. M. Rentmeester, and R. G. E. Timmermans, Phys. Rev. C **70**, 044001 (2004), [arXiv:nucl-th/0406026].
29. E. Epelbaum, U.-G. Meißner, and J. E. Palomar, Phys. Rev. C **71**, 024001 (2005), [arXiv:nucl-th/0407037].

Partial wave decomposition of 2π-1π exchange three-nucleon force in chiral effective field theory

H. Kamada*, E. Epelbaum†,**, A. Nogga†, Ulf-G. Meißner†,**, H. Witała‡, J. Golak‡, R. Skibiński‡ and W. Glöckle§

*Department of Physics, Faculty of Engineering, Kyushu Institute of Technology, 1-1 Sensuicho Tobata, Kitakyushu 804-8550, Japan
†Forschungszentrum Jülich, Institut für Kernphysik(Theorie), D-52425 Jülich, Germany
**Universität Bonn, Helmholtz-Institut für Strahlen- und Kernphysik (Theorie), D-53115 Bonn, Germany
‡M. Smoluchowski Institute of Physics, Jagiellonian University, PL-30059 Kraków, Poland
§Institut für Theoretische Physik II, Ruhr-Universität Bochum, D-44780 Bochum, Germany

Abstract. The 3-nucleon force at N^3LO in chiral effective field theory is briefly reviewed. The two-pion-one-pion-exchange 3-nucleon force is decomposed into partial waves.

Keywords: nucleon-nucleon potential, three-body forces, chiral effective field theory
PACS: 21.45.-v,21.45.Bc,21.45.Ff

INTRODUCTION

Three-nucleon forces (3NFs) are present in systems with more than three nucleons. Since Fujita and Miyazawa [1] first introduced a 3NF, various 3NF models have been developed, for example, Tucson-Melbourne [2, 3], Brazil [4, 5], Urbana [6], Texas [7], Ruhr-Pot [8] etc. These meson-theoretical models are based on the ideas of the Delta-isobar excitation, axial current algebra, chiral perturbation theory.

Recently not only for the 3-nucleon bound state but also for 3-nucleon scattering we have been able to solve the Faddeev equations with these 3NFs and compare with precision experimental data [9, 10, 11, 12, 13]. In the low energy region where we do not expect a large contribution from a 3NF we succeeded [14, 15] in explaining the data except for some observables. The typical exceptions are A_y [16, 17] for Nd elastic scattering at very low energy and the differential cross section of the space-star [18, 19] break-up configuration. In order to remove the discrepancies between our theory and expermental data, we solved the Faddeev equations with a 3NF of the 2-π exchange type (Fujita-Miyazawa, Tucson-Melbourne and Urbana), however, without success. Still, since the Sagara discrepancy [20, 21, 22] was to a large extent explained by the inclusion of a 3NF, it has become clear that this ingredient of the 3N Hamiltonian might play an important role. Namely, the minimum of the Nd differential cross section at intermediate energies was explained by incorporating the 2-π exchange Tucson-Melbourne 3NF. However, at intermediate energy many observables [23, 24, 25, 26] emerge where the theoretical prediction including only a 2-π exchange 3NF is poor. The effects of relativity have been investigated [13] but it was shown that relativity has only a tiny effect in elastic Nd scattering. For a relativistic calculation of Nd scattering, see

CP1011, *New Facet of Three Nucleon Force—50 Years of Fujita Miyazawa Three Nucleon Force (FM50), The International Symposium,* edited by H. Sakai, K. Sekiguchi, and B. F. Gibson

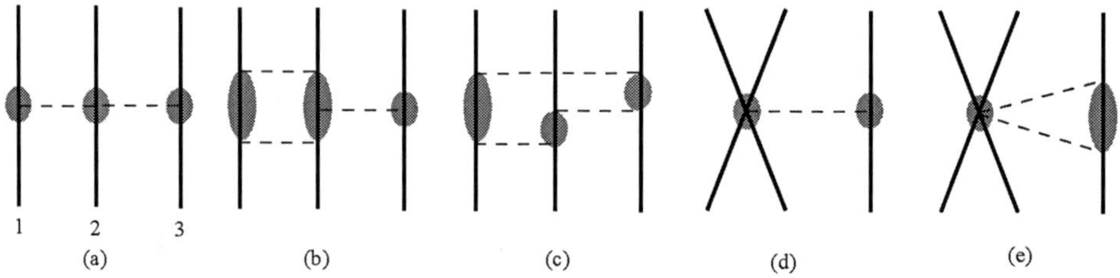

FIGURE 1. Various topologies that appear in the 3NF at N^3LO. Solid and dashed lines represent nucleons and pions, respectively. Shaded blobs are the corresponding amplitudes. The long-range part of the 3NF considered consists of (a) 2π exchange graphs, (b) 2π-1π diagrams and the so-called ring diagrams (c). The topologies (d) and (e) involve four-nucleon contact operators and are considered of shorter range.

[27, 28] H. Witała in this conference. Because of this situation we are motivated to study another type of 3NF.

The Tucson-Melbourne 3NF has other pieces which contain diagrams where additional mesons propagate and there are irreducible diagrams of double isobar excitation etc. These new 3NFs are investigated in the triton [29]. The Illinois 3NF [30] is also extended by the idea of irreducible diagrams of double isobar excitation and relativistic corrections. The 2-π exchange 3NF has recently been considered based on the so-called infrared regularization[31].

The 3NF in chiral effective field theory was developed in [32, 33]. The power ν of Q^ν, where Q is a typical external momentum in the diagram determines the order of the chiral potentials. Higher orders are required with increasing energy. Up to the next leading order (NLO) there is no 3NF in the nonrelativistic approximation. A 3NF of the Fujita-Miyazawa type first appears in next-to-next leading order (NNLO). There are in addition 2 terms of a new 3NF at NNLO which link to new low-energy-coefficients (LEC) of the chiral Lagrangian. These LECs are fixed by the nd doublet scattering length [32] and the nuclei binding energies [34, 35].

The structure of the 3NF at N^3LO[33] is visualized in Fig. 1 and can be written as

$$V_{3N} = V_{2\pi} + V_{2\pi\text{-}1\pi} + V_{\text{ring}} + V_{1\pi\text{-cont}} + V_{2\pi\text{-cont}} + V_{1/m}. \tag{1}$$

Namely, they consist of (a) two-pion-exchange, (b) two-pion-one-pion-exchange, (c) ring diagram, (d) one-pion-exchange-contact topologies, (e) two-pion-exchange-contact contribution and $1/m$ correction terms. The terms of $V_{2\pi}$, $V_{2\pi-1\pi}$ and V_{ring} are the long-range parts because these topologies consist of only pion propagating with LEC (c_i and g_A). Partial wave decomposition for $V_{2\pi}$ and $V_{1\pi-\text{cont}}$ at NNLO was already given in [36, 37, 32]. In the next section we show details of the $V_{2\pi-1\pi}$ 3NF at N^3LO.

TWO-PION-ONE-PION EXCHANGE 3NF

The 2π-1π 3NF corresponding to topology (b) in Fig. 1 can be written[33]:

$$V_{2\pi-1\pi} = \frac{\vec{\sigma}_2 \cdot \vec{Q}'}{Q'^2 + m_\pi^2} \Big[(\vec{\tau}_2 \cdot \vec{\tau}_3) \Big\{ (\vec{Q} \cdot \vec{Q}')(\vec{\sigma}_1 \cdot \vec{Q}) F_1(Q) + (\vec{\sigma}_1 \cdot \vec{Q}) F_2(Q)$$
$$+ (\vec{\sigma}_1 \cdot \vec{Q}') F_3(Q) \Big\} + (\vec{\tau}_1 \cdot \vec{\tau}_2) \Big\{ (\vec{Q} \cdot \vec{Q}')(\vec{\sigma}_3 \cdot \vec{Q}) F_4(Q) + (\vec{\sigma}_3 \cdot \vec{Q}') F_5(Q)$$
$$+ (\vec{\sigma}_1 \cdot \vec{Q}) F_6(Q) + (\vec{\sigma}_1 \cdot \vec{Q}') F_7(Q) \Big\} + (\vec{\tau}_3 \times \vec{\tau}_2 \cdot \vec{\tau}_1)(\vec{\sigma}_3 \times \vec{\sigma}_1 \cdot \vec{Q}) F_8(Q) \Big] \quad (2)$$

where we take a notation of [36] from [33]. Namely, permutation $(ijk)_{[36]} = (312), (ijk)_{[33]} = (123)$ and, $\vec{Q}[36] = -\vec{q}_1[33]$, $\vec{Q}'[36] = -\vec{q}_3[33]$ and the subscripts refer to the nucleon labels and $\vec{q}_i = \vec{k}'_i - \vec{k}_i$ being the final and initial momenta of the nucleon i. The scalar function $F_{1...8}$ are combinations of the LEC's:

$$F_1(Q) = \frac{1}{256\pi F_\pi^6} \Big[-g_A^6 \Big\{ \frac{m_\pi}{4m_\pi^2 + Q^2} + \frac{2m_\pi}{Q^2} - \frac{8m_\pi^2 + Q^2}{Q^2} A(Q) \Big\}$$
$$+ g_A^4 \Big\{ \frac{m_\pi}{Q^2} + \frac{Q^2 - 4m_\pi^2}{Q^2} A(Q) \Big\} \Big],$$
$$F_2(Q) = \frac{g_A^4}{128\pi F_\pi^6} \left[m_\pi + (Q^2 + 2m_\pi^2) A(Q) \right],$$
$$F_3(Q) = \frac{1}{256\pi F_\pi^6} \Big[-g_A^6 \left\{ 3m_\pi + (8m_\pi^2 + 3Q^2) A(Q) \right\}$$
$$+ g_A^4 \left\{ m_\pi + (Q^2 + 4m_\pi^2) A(Q) \right\} \Big],$$
$$F_4(Q) = -\frac{1}{Q^2} F_5(Q) = -\frac{g_A^6}{128\pi F_\pi^6} A(Q),$$
$$F_6(Q) = 2F_7(Q) = \frac{g_A^4}{64\pi F_\pi^6} \left[m_\pi + (Q^2 + 2m_\pi^2) A(Q) \right],$$
$$F_8(Q) = -\frac{g_A^4}{512\pi F_\pi^6} \left[m_\pi + (Q^2 + 4m_\pi^2) A(Q) \right] \quad (3)$$

with the loop function

$$A(Q) = \frac{1}{2Q} \arctan \frac{Q}{2m_\pi}. \quad (4)$$

Further in this paper the results are always given for one particular choice of nucleon labels. Using the permutation operator $P = P_{12}P_{23} + P_{13}P_{23}$ the full expression results:

$$V^{\rm full}_{2\pi-1\pi} = V_{2\pi-1\pi}(1+P) + V^{\rm mirror}_{2\pi-1\pi}(1+P) \quad (5)$$

The second term $V^{\rm mirror}_{2\pi-1\pi}$ corresponds to the reverse diagram (b') in Fig. 2.

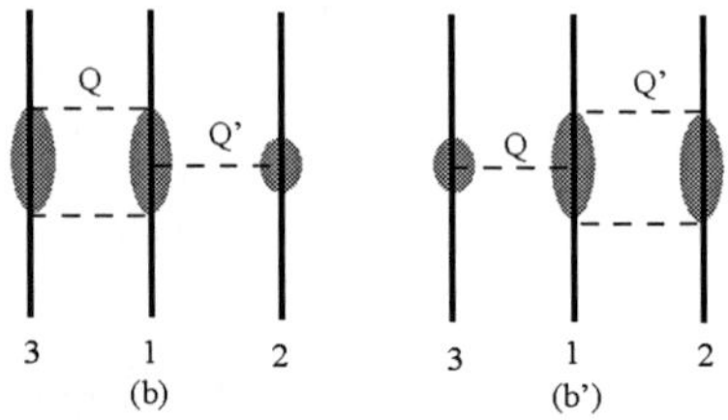

FIGURE 2. Two diagrams of two-pion-one-pion-exchange 3NF ($V_{2\pi-1\pi}$ and $V^{\rm mirror}_{2\pi-1\pi}$). The particle label is changed as [36]. Pion propagating of the diagram (b') turns over to (b).

PARTIAL WAVE DECOMPOSITION

We need the matrix elements of the 3NF in a partial wave basis $|pq\alpha\rangle_i$, where p and q are Jacobi momenta ($\vec{p}$ is the relative momenta between particle j and k, and $\vec{q}$ is the momentum of particle i relative to the pair $j-k$) and α denotes discrete quantum numbers. In detail the partial wave is

$$|pq\alpha\rangle_1 = |p(ls)j\; q(\lambda\frac{1}{2})I\; (jI)JM\rangle_1 |(t\frac{1}{2})TM_T\rangle_1 = |pq\alpha_J\rangle_1 |\alpha_T\rangle_1 \tag{6}$$

where l,s, and j are orbital angular momentum, spin, and total angular momentum in the two-body subsystem 2-3, λ, $\frac{1}{2}$, and I are the corresponding quantities for particle 1, and JM denote the conserved 3N total angular momentum and its magnetic quantum number. In isospin space, t refers to particle 2 and 3 and couples with $\frac{1}{2}$ to total isospin T and its magnetic quantum number M_T.

Following the scheme of [36] we calculate the partial wave elements of the 3NF as

$$\begin{aligned}
{}_1\langle pq\alpha|V_{2\pi-1\pi}(1+P)|p'q'\alpha'\rangle_1 &= \sum\!\!\!\!\!\!\int^{\prime\prime} {}_1\langle pq\alpha|V_{2\pi-1\pi}|p''q''\alpha''\rangle_3 \\
&\times {}_1\langle p''q''\alpha''|(1+P)|p'q'\alpha'\rangle_1 .
\end{aligned} \tag{7}$$

and

$$\begin{aligned}
{}_1\langle pq\alpha|V_{2\pi-1\pi}|p'q'\alpha'\rangle_3 &= \sum_{n=1}^{8} {}_1\langle pq\alpha|V^{(n)}_{2\pi-1\pi}|p'q'\alpha'\rangle_3 \\
&= \sum_{n=1}^{8} \sum\!\!\!\!\!\!\int^{\prime\prime} \sum\!\!\!\!\!\!\int^{\prime\prime\prime} \sum\!\!\!\!\!\!\int^{\prime\prime\prime\prime} {}_1\langle pq\alpha|p''q''\alpha''\rangle_2 \\
&\quad \times {}_2\langle p''q''\alpha''|w^{(n)}_{2\pi-1\pi}|p'''q'''\alpha'''\rangle_2\, {}_2\langle p'''q'''\alpha'''|p''''q''''\alpha''''\rangle_3 \\
&\quad \times {}_3\langle p''''q''''\alpha''''|w^{(n)}_{2\pi-1\pi}|p'q'\alpha'\rangle_3 \\
&= \sum_{n=1}^{8} \sum\!\!\!\!\!\!\int^{\prime\prime} \sum\!\!\!\!\!\!\int^{1} \sum\!\!\!\!\!\!\int^{2} {}_1\langle pq\alpha|p''q''\alpha''\rangle_2 \\
&\quad \times \Big({}_2\langle p''q''\alpha''_J|w^{(n)}_{2\pi-1\pi}|p_1q_1\alpha_{1J}\rangle_2\, {}_2\langle p_1q_1\alpha_{1J}|p_2q_2\alpha_{2J}\rangle_3 \\
&\quad \times {}_3\langle p_2q_2\alpha_{2J}|w^{(n)}_{2\pi-1\pi}|p'q'\alpha'_J\rangle_3 \Big)
\end{aligned}$$

$$\times\ {}_2\langle\alpha_T''|I^{(n)}|\alpha_T'\rangle_3. \tag{8}$$

where the label n of $w^{(n)}_{2\pi-1\pi}$ corresponds to the number of the scalar function F_n in Eq. (3) and the isospin parts ${}_2\langle\alpha_T''|I^{(n)}|\alpha_T'\rangle_3$ are given in Eqs (33) and (34) in [36].

In Eq. (8) we introduced intermediate sums $\sum^1$ and $\sum^2$, where the related quantities carry the indices 1 and 2. Note that the label numbers 1 and 2 of the variables $p_1p_2q_1q_2\alpha_1\alpha_2$ are not connected to a particle number. The particle number i is usually implicitly controlled by ${}_i\langle|$ and $|\rangle_i$ e.g. $|p\rangle_3 = p_{\text{“}3\text{”}}$. Therefore, the momenta $\vec{Q}$ and $\vec{Q}'$ in Eq. (2) have the relations:

$$\begin{aligned}\vec{Q} &= \vec{p}\,'' - \vec{p}_1,\\ \vec{Q}' &= \vec{p}' - \vec{p}_2.\end{aligned} \tag{9}$$

Now [1], our aim is to have the matrix elements ${}_2\langle pq\alpha_J|w^{(n)}_{2\pi-1\pi}|p_1q_1\alpha_{1J}\rangle_2$ and ${}_3\langle p_2q_2\alpha_{2J}|w^{(n)}_{2\pi-1\pi}|p'q'\alpha_J'\rangle_3$:

$$\begin{aligned}
&{}_2\langle pq\alpha_J|w^{(1),\mu}_{2\pi-1\pi}|p_1q_1\alpha_{1J}\rangle_2 \equiv {}_2\langle pq\alpha_J|\sqrt{\frac{4\pi}{3}}QY_1^{\mu}(\hat{Q})F_1(Q)(\vec{\sigma}_1\cdot\vec{Q})|p_1q_1\alpha_{1J}\rangle_2\\
&= \frac{\delta(q-q_1)}{q^2}\delta_{\lambda\lambda_1}\delta_{II_1}C(1\mu J_1M_1JM)\\
&\times(-)^{I+J}(-1)^{s+s_1}\sqrt{\hat{j}\hat{j}_1\hat{s}\hat{s}_1\hat{J}_1}\begin{Bmatrix}\frac{1}{2} & \frac{1}{2} & s_1\\ 1 & s & \frac{1}{2}\end{Bmatrix}\begin{Bmatrix}1 & j_1 & j\\ I & J & J_1\end{Bmatrix}\\
&\times\Bigg[\delta_{ll'}\frac{2\pi}{3}\sqrt{6}(-)^{l+1}\tilde{H}^{(1)}_l(p,p_1)\begin{Bmatrix}j & j_1 & 1\\ s_1 & s & l\end{Bmatrix}\\
&-40\pi\sqrt{6}(-)^{s+j_1}\begin{Bmatrix}2 & 1 & 1\\ l_1 & s_1 & j_1\\ l & s & j\end{Bmatrix}\\
&\times\sum_{\bar{l}}\hat{\bar{l}}\tilde{H}^{(1)}_{\bar{l}}(p,p_1)\sum_{a+b=2}\frac{p^ap_1^b}{\sqrt{(2a)!(2b)!}}\begin{Bmatrix}b & a & 2\\ l & l_1 & \bar{l}\end{Bmatrix}C(a0\bar{l}0,l0)C(b0\bar{l}0l_10)\Bigg],
\end{aligned} \tag{10}$$

$${}_3\langle p_2q_2\alpha_{2J}|w^{(1),\mu}_{2\pi-1\pi}|p'q'\alpha_J'\rangle_3 \equiv {}_3\langle p_2q_2\alpha_{2J}|\sqrt{\frac{4\pi}{3}}Q'Y_1^{-\mu}(\hat{Q}')\frac{\vec{\sigma}_2\cdot\vec{Q}'}{Q'^2+m_\pi^2}|p'q'\alpha_J'\rangle_3. \tag{11}$$

The result of (11) is the same as in Eq. (73) of [36] but $F^2(Q'^2) = 1$.

$$\begin{aligned}
&{}_2\langle pq\alpha_J|w^{(2)}_{2\pi-1\pi}|p_1q_1\alpha_{1J}\rangle_2 \equiv {}_2\langle pq\alpha_J|(\vec{\sigma}_1\cdot\vec{Q})F_2(Q)|p_1q_1\alpha_{1J}\rangle_2\\
&= \frac{\delta(q-q_1)}{q^2}\delta_{jj_1}\delta_{\lambda\lambda_1}\delta_{II_1}\delta_{JJ_1}\delta_{MM_1}\delta_{|l-l_1|,1}\\
&\times 2\pi\sqrt{6}(-)^{j+1}\sqrt{\hat{s}\hat{s}_1}\begin{Bmatrix}\frac{1}{2} & \frac{1}{2} & s_1\\ 1 & s & \frac{1}{2}\end{Bmatrix}\begin{Bmatrix}l_1 & s_1 & j\\ s & l & 1\end{Bmatrix}
\end{aligned}$$

---

[1] Caution: p'' is already replaced by p, $\vec{Q} = \vec{p} - \vec{p}_1$ for ${}_2\langle pq\alpha|w^{(n)}_{2\pi-1\pi}|p_1q_1\alpha_1\rangle_2$.

$$\times\sqrt{\max(l,l_1)}(pH_{l_1}^{(2)}(p,p_1)-p_1H_l^{(2)}(p,p_1))(-)^{\max(l,l_1)}. \tag{12}$$

$$_3\langle p_2q_2\alpha_{2J}|w_{2\pi-1\pi}^{(2)}|p'q'\alpha_J'\rangle_3 \equiv_3 \langle p_2q_2\alpha_{2J}|\frac{\vec{\sigma}_2\cdot\vec{Q}'}{Q'^2+m_\pi^2}|p'q'\alpha_J'\rangle_3. \tag{13}$$

The result of (13) is the same as in Eq. (57) of [36] but $F^2(Q'^2)=1$.

$$\begin{aligned}&_2\langle pq\alpha_J|w_{2\pi-1\pi}^{(3)}|p_1q_1\alpha_{1J}\rangle_2 \equiv_2 \langle pq\alpha_J|F_3(Q)|p_1q_1\alpha_{1J}\rangle_2\\ &=\frac{\delta(q-q_1)}{q^2}\delta_{jj_1}\delta_{ll_1}\delta_{ss_1}\delta_{\lambda\lambda_1}\delta_{II_1}\delta_{JJ_1}\delta_{MM_1}2\pi H_l^{(3)}(p,p_1).\end{aligned} \tag{14}$$

$$\begin{aligned}&_3\langle p_2q_2\alpha_{2J}|w_{2\pi-1\pi}^{(3)}|p'q'\alpha_J'\rangle_3 \equiv_3 \langle p_2q_2\alpha_{2J}|\frac{(\vec{\sigma}_1\cdot\vec{Q}')(\vec{\sigma}_2\cdot\vec{Q}')}{Q'^2+m_\pi^2}|p'q'\alpha_J'\rangle_3\\ &=\frac{\delta(q_2-q')}{q'^2}\delta_{J'J_2}\delta_{M'M_2}\delta_{j'j_2}\delta_{\lambda'\lambda_2}\delta_{I'I_2}\\ &\times[\delta_{l'l_2}\delta_{s's_2}4\pi(-)^{s'+1}\tilde{H}_{l'}^{(3)}(p'p_2)\begin{Bmatrix}\frac{1}{2} & \frac{1}{2} & s'\\ \frac{1}{2} & \frac{1}{2} & 1\end{Bmatrix}\\ &+240\pi(-)^{j_2+l_2+s_2}\sqrt{\hat{s}'\hat{s}_2}\begin{Bmatrix}l_2 & s_2 & j'\\ s' & l' & 2\end{Bmatrix}\begin{Bmatrix}1 & 1 & 2\\ \frac{1}{2} & \frac{1}{2} & s'\\ \frac{1}{2} & \frac{1}{2} & s_2\end{Bmatrix}\\ &\times\sum_{\bar{l}}\bar{l}H_{\bar{l}}^{(3)}(p'p_2)\sum_{a+b=2}\frac{p'^a(-p_2)^b}{\sqrt{\hat{a}!\hat{b}!}}\begin{Bmatrix}l' & l_2 & 2\\ b & a & \bar{l}\end{Bmatrix}C(a0\bar{l}0,l'0)C(b0\bar{l}0,l_20)]\end{aligned} \tag{15}$$

The result of (15) is the same as in Eq. (A.16) of [37] but $F^2(Q'^2)=1$.

$$_2\langle pq\alpha_J|w_{2\pi-1\pi}^{(4),\mu}|p_1q_1\alpha_{1J}\rangle_2 \equiv_2 \langle pq\alpha_J|\sqrt{\frac{4\pi}{3}}QY_1^\mu(\hat{Q})F_4(Q)|p_1q_1\alpha_{1J}\rangle_2 \tag{16}$$

The result of (16) is the same $M_b^{J,2}$ as in Eq. (72) of [36] but $F_4(Q)$ instead of $F^2(\vec{Q}^2)/(Q^2+m_\pi^2)$.

$$_3\langle p_2q_2\alpha_{2J}|w_{2\pi-1\pi}^{(4),\mu}|p'q'\alpha_J'\rangle_3 \equiv_3 \langle p_2q_2\alpha_{2J}|\sqrt{\frac{4\pi}{3}}Q'Y_1^{-\mu}(\hat{Q}')\frac{\vec{\sigma}_2\cdot\vec{Q}'}{Q^2+m_\pi^2}|p'q'\alpha_J'\rangle_3 \tag{17}$$

The result of Eq.(17) is the same $M_b^{J,3}$ as in Eq. (73) of [36] but $F^2(Q'^2)=1$.

In order to obtain $_2\langle pq\alpha_J|w_{2\pi-1\pi}^{(5),\mu}|p_1q_1\alpha_{1J}\rangle_2$ and $_3\langle p_2q_2\alpha_{2J}|w_{2\pi-1\pi}^{(5),\mu}|p'q'\alpha_J'\rangle_3$ we have the following identity:

$$\begin{aligned}&(\vec{\sigma}_2\cdot\vec{Q}')(\vec{\sigma}_3\cdot\vec{Q}')=4\pi|p_2-p'|^2\left\{\{\sigma_2,Y_1(\widehat{p_2-p'})\}^0,\{\sigma_3,Y_1(\widehat{p_2-p'})\}^0\right\}^0\\ &=\sum_\mu(-)^\mu 4\pi\left[-\frac{1}{\sqrt{3}}|p_2-p'|^2\sigma_3^\mu\{\sigma_2,\mathscr{Y}_{00}^0(\hat{p}_2,\hat{p}')\}^{1,-\mu}\right.\end{aligned}$$

$$+4\sqrt{5}\sum_{a+b=2}\frac{p_2{}^a(-p')^b}{\sqrt{\hat{a}!\hat{b}!}}\sigma_3^\mu\{\sigma_2,\mathscr{Y}^2_{ab}(\hat{p}_2,\hat{p}')\}^{1,-\mu}\Big]. \tag{18}$$

Then,

$$\begin{aligned}
&{}_2\langle pq\alpha_J|w^{(5),\mu}_{2\pi-1\pi}|p_1q_1\alpha_{1J}\rangle_2 \equiv_2 \langle pq\alpha|\sigma_3^\mu F_5(Q)|p_1q_1\alpha_{1J}\rangle_2\\
&= \frac{\delta(q-q_1)}{q^2}\delta_{ll_1}\delta_{\lambda\lambda_1}\delta_{II_1}\sqrt{6}\,2\pi H^{(5)}_l(pp_1)\\
&\times(-)^{1+l+I+J}\sqrt{\hat{j}\hat{j}_1\hat{s}\hat{s}_1\hat{J}_1}\begin{Bmatrix} s & j & l\\ j_1 & s_1 & 1\end{Bmatrix}\begin{Bmatrix} \frac{1}{2} & \frac{1}{2} & s\\ 1 & s_1 & \frac{1}{2}\end{Bmatrix}\\
&\times\begin{Bmatrix} 1 & j & j_1\\ I & J_1 & J\end{Bmatrix}C(1\mu J_1M_1,JM).
\end{aligned} \tag{19}$$

This is the same $\tilde{M}^{J,2}_{d_2}$ as in Eq. (A.38) [2] of [37] but $F_5(Q)$ instead of $F(Q^2)/(Q^2+m_\pi^2)$.

$$\begin{aligned}
&{}_3\langle p_2q_2\alpha_{2J}|w^{(5),\mu}_{2\pi-1\pi}|p'q'\alpha'_J\rangle_3 \equiv_3 \langle p_2q_2\alpha_{2J}|4\pi\Big[-\frac{1}{\sqrt{3}}|p_2-p'|^2\{\sigma_2,\mathscr{Y}^0_{00}(\hat{p}_2,\hat{p}')\}^{1,-\mu}\\
&+4\sqrt{5}\sum_{a+b=2}\frac{p_2{}^a(-p')^b}{\sqrt{\hat{a}!\hat{b}!}}\{\sigma_2,\mathscr{Y}^2_{ab}(\hat{p}_2,\hat{p}')\}^{1,-\mu}\Big]\frac{1}{Q'^2+m_\pi^2}|p'q'\alpha'_J\rangle_3
\end{aligned} \tag{20}$$

Comparing Eq.(18) and Eq.(64) of [36] we have the result:

$${}_3\langle p_2q_2\alpha_{2J}|w^{(5),\mu}_{2\pi-1\pi}|p'q'\alpha'_J\rangle_3 = -\sqrt{3}\,{}_3\langle p_2q_2\alpha_{2J}|w^{(4),\mu}_{2\pi-1\pi}|p'q'\alpha'_J\rangle_3 \tag{21}$$

$${}_2\langle pq\alpha_J|w^{(6),\mu}_{2\pi-1\pi}|p_1q_1\alpha_{1J}\rangle_2 \equiv_2 \langle pq\alpha_J|\sqrt{\frac{4\pi}{3}}QY^\mu_1(\hat{Q})F_6(Q)(\vec{\sigma}_1\cdot\vec{Q})|p_1q_1\alpha_{1J}\rangle_2 \tag{22}$$

This equals Eq. (10) but $F_6(Q)$ instead of $F_1(Q)$.

$${}_3\langle p_2q_2\alpha_{2J}|w^{(6),\mu}_{2\pi-1\pi}|p'q'\alpha'_J\rangle_3 = {}_3\langle p_2q_2\alpha_{2J}|w^{(2)}_{2\pi-1\pi}|p'q'\alpha'_J\rangle_3 \tag{23}$$

$${}_2\langle pq\alpha_J|w^{(7)}_{2\pi-1\pi}|p_1q_1\alpha_{1J}\rangle_2 \equiv_2 \langle pq\alpha_J|F_7(Q)|p_1q_1\alpha_{1J}\rangle_2 \tag{24}$$

Again, this equals Eq.(14) but $F_7(Q)$ instead of $F_3(Q)$.

$${}_3\langle p_2q_2\alpha_{2J}|w^{(7)}_{2\pi-1\pi}|p'q'\alpha'_J\rangle_3 = {}_3\langle p_2q_2\alpha_{2J}|w^{(3)}_{2\pi-1\pi}|p'q'\alpha'_J\rangle_3 \tag{25}$$

$$\begin{aligned}
&{}_2\langle pq\alpha_J|w^{(8)}_{2\pi-1\pi}|p_1q_1\alpha_{1J}\rangle_2 \equiv_2 \langle pq\alpha_J|(\vec{\sigma}_3\times\vec{\sigma}_1\cdot\vec{Q})F_8(Q)|p_1q_1\alpha_{1J}\rangle_2\\
&= \frac{\delta(q-q_1)}{q^2}\delta_{jj_1}\delta_{\lambda\lambda_1}\delta_{JJ_1}\delta_{MM_1}
\end{aligned}$$

---

[2] Errata: $\delta(p_2-p')/p'^2$ of Eq. (A.37[37]) and $\delta(p-p_1)/p^2$ of (A.38[37]) need not. In the first line of Eq. (A.6[37]) two ")" are missed and the last σ_1 should be replaced by σ_3. Eq. (A.17[37]) needs $(-)^{\bar{l}}$. Eq.(87[36]) needs $(-)^{i_1}$, then Eq.(88[36]) needs not $(-)^{i_1}$. (see former erratta of [36] in [37].) Calculations including the 3NF by the Bochum-Kraków-KIT group until now will not change because these errata were discarded during computer program coding.

$$
\begin{aligned}
&\times i\sqrt{2}\,2\pi \sum_{a+b=1} p^a p_1^b \sum_{\bar{l}} \tilde{l}\sqrt{\hat{a}\hat{b}}(-)^{l_1} \begin{Bmatrix} l_1 & l & 1 \\ a & b & \bar{l} \end{Bmatrix} \\
&\times C(\bar{l}0a0,l0)C(\bar{l}0b0,l_10)H_{\bar{l}}^{(8)}(p,p_1)\sqrt{\hat{s}}(-)^{j+s_1} \begin{Bmatrix} l_1 & s_1 & j \\ s & l & 1 \end{Bmatrix} \\
&\times 6\sqrt{3\hat{s}_1} \begin{Bmatrix} 1 & 1 & 1 \\ \frac{1}{2} & \frac{1}{2} & s_1 \\ \frac{1}{2} & \frac{1}{2} & s \end{Bmatrix}
\end{aligned} \tag{26}
$$

$$
{}_3\langle p_2 q_2 \alpha_{2J} | w^{(8)}_{2\pi-1\pi} | p' q' \alpha'_J \rangle_3 = {}_3\langle p_2 q_2 \alpha_{2J} | w^{(2)}_{2\pi-1\pi} | p' q' \alpha'_J \rangle_3 \tag{27}
$$

The functions $H_l^{(n)}(p,p_1)$ and $\tilde{H}_l^{(n)}(p,p_1)$ for the element ${}_2\langle pq\alpha_J | w^{(n)}_{2\pi-1\pi} | p_1 q_1 \alpha_{1J}\rangle_2$ are defined as:

$$
\begin{aligned}
H_l^{(n)}(p,p_1) &\equiv \int_{-1}^{1} P_l(x) F_n(Q) dx \\
\tilde{H}_l^{(n)}(p,p_1) &\equiv \int_{-1}^{1} P_l(x) F_n(Q) Q^2 dx
\end{aligned} \tag{28}
$$

with

$$
Q = \sqrt{p^2 + p_1^2 - 2pp_1 x} \tag{29}
$$

where $F_n(Q)$ is given in Eq.(3) and $P_l(x)$ is the Legendre polynomial. Similarly, functions $H_l^{(n)}(p_2,p')$ and $\tilde{H}_l^{(n)}(p_2,p')$ for the element ${}_3\langle p_2 q_2 \alpha_{2J} | w^{(n)}_{2\pi-1\pi} | p' q' \alpha'_J\rangle_3$ are defined as:

$$
\begin{aligned}
H_l^{(n)}(p_2,p') &\equiv \int_{-1}^{1} P_l(x) \frac{1}{Q'^2 + m_\pi^2} dx = \frac{1}{p' p_2} Q_l\Big(\frac{p'^2 + p_2^2 + m_\pi^2}{2p' p_2}\Big) \\
\tilde{H}_l^{(n)}(p_2,p') &\equiv \int_{-1}^{1} P_l(x) \frac{1}{Q'^2 + m_\pi^2} Q'^2 dx = 2\delta_{l,0} - \frac{m_\pi^2}{p' p_2} Q_l\Big(\frac{p'^2 + p_2^2 + m_\pi^2}{2p' p_2}\Big)
\end{aligned} \tag{30}
$$

with

$$
Q' = \sqrt{p_2^2 + p'^2 - 2p_2 p' x} \tag{31}
$$

where $Q_l(x)$ is the Legendre function (second kind).

For the matrix elements of $n = 1,4,5$ and 6 we need to sum over μ instead of Eq. (8):

$$
\begin{aligned}
{}_1\langle pq\alpha | V^{(n)}_{2\pi-1\pi} | p'q'\alpha'\rangle_3 &= \sum_\mu (-)^\mu \sum_{J''}\sum_{J_1}\sum_{J_2} {}_1\langle pq\alpha | p''q''\alpha''\rangle_2 \\
&\times \Big({}_2\langle p''q''\alpha''_J | W^{(n),\mu}_{2\pi-1\pi} | p_1 q_1 \alpha_{1J}\rangle_2 \, {}_2\langle p_1 q_1 \alpha_{1J} | p_2 q_2 \alpha_{2J}\rangle_3 \\
&\quad \times {}_3\langle p_2 q_2 \alpha_{2J} | W^{(n),\mu}_{2\pi-1\pi} | p'q'\alpha'_J\rangle_3 \Big) \\
&\times {}_2\langle \alpha''_T | I^{(n)} | \alpha'_T\rangle_3 .
\end{aligned} \tag{32}
$$

The useful identity for this summation is

$$
\sum_\mu (-)^\mu C(1\mu J_1 M_1, JM) C(1-\mu J'M', J_1 M_1) = \delta_{JJ'}\delta_{MM'}\sqrt{\frac{\hat{J}_1}{\hat{J}}}. \tag{33}
$$

SUMMARY AND OUTLOOK

We demonstrated the partial wave decomposition for the two-pion-one-pion-exchange 3NF $V_{2\pi-1\pi}$ at N^{3}LO [33] in chiral effective field theory. As shown in Eq.(10) to Eq. (27) the full matrix elements are generated by the usual scheme[36].

In Eq. (2.9) of [33] the two-pion-exchange 3NF $V_{2\pi}$ no linking with the LEC's ($\bar{c}_1$, $\bar{c}_3$ and $\bar{c}_4$) is exhibited

$$
\begin{aligned}
V_{2\pi} = & \frac{g_A^4}{256\pi F_\pi^6} \frac{\vec{\sigma}_3 \cdot \vec{Q}\, \vec{\sigma}_2 \cdot \vec{Q}'}{[Q^2+m_\pi^2]\,[Q'^2+m_\pi^2]} \Big[\vec{\tau}_2 \cdot \vec{\tau}_3 \Big(m_\pi (m_\pi^2 + 3Q^2 + 3Q'^2 + 4\vec{Q}\cdot\vec{Q}') \\
& + (2m_\pi^2 + Q^2 + Q'^2 + 2\vec{Q}\cdot\vec{Q}')(3m_\pi^2 + 3Q^2 + 3Q'^2 + 4\vec{Q}\cdot\vec{Q}')A(Q'') \Big) \\
& + \vec{\tau}_1 \times \vec{\tau}_2 \cdot \vec{\tau}_3\, \vec{Q} \times \vec{Q}' \cdot \vec{\sigma}_1 \Big(m_\pi + (4m_\pi^2 + Q^2 + Q'^2 + 2\vec{Q}\cdot\vec{Q}')A(Q'') \Big) \Big]
\end{aligned}
\tag{34}
$$

with $\vec{Q}'' = -\vec{Q} - \vec{Q}'$. The terms including the loop function $A(Q'')$ require another technique for the partial wave expansion. Furthermore, the ring 3NF shown in Fig. 1(c) may require a more complicated scheme.

As an idea for the solution one may extend Eq. (8) into the following triple-permutation expansion:

$$
\begin{aligned}
{}_1\langle pq\alpha | V_{\text{ring}} | p'q'\alpha' \rangle_3 = & \sum\!\!\!\!\!\!\int_{\check{}} \sum\!\!\!\!\!\!\int_{''} \sum\!\!\!\!\!\!\int_{'''} \sum\!\!\!\!\!\!\int_{''''} {}_1\langle pq\alpha | w_{\text{ring}}(\vec{Q}'') | \check{p}\check{q}\check{\alpha} \rangle_1\, {}_1\langle \check{p}\check{q}\check{\alpha} | p''q''\alpha'' \rangle_2 \\
& \times_2 \langle p''q''\alpha'' | w_{\text{ring}}(\vec{Q}) | p'''q'''\alpha''' \rangle_2\, {}_2\langle p'''q'''\alpha''' | p''''q''''\alpha'''' \rangle_3 \\
& \times_3 \langle p''''q''''\alpha'''' | w_{\text{ring}}(\vec{Q}') | p'q'\alpha' \rangle_3 .
\end{aligned}
\tag{35}
$$

ACKNOWLEDGMENTS

Partial financial support from the EU I3HP Hadron Physics Project (contract number RII3-CT-2004-506078), by the DFG (TR 16) and by BMBF (research grant 06BN411) is gratefully acknowledged. This work is partially supported by the Helmholtz Association through funds provided to the virtual institute "Spin and strong QCD"(VH-VI-231) and to the young investigator group "Few-Nucleon Systems in Chiral Effective Field Theory" (VH-NG-222).

REFERENCES

1. J. I. Fujita, and H. Miyazawa, *Prog. Theor. Phys.* **17**, 360 (1957).
2. S. A. Coon, M. D. Scadron, P. C. McNamee, B. R. Barrette, D. W. E. Blatt, and B. H. J. McKellar, *Nucl. Phys. B* **317**, 242 (1979).
3. S. A. Coon, and W. Glöckle, *Phys. Rev. C* **23**, 1790 (1981).
4. H. T. Coelho, T. K. Das, and M. R. Robilotta, *Phys. Rev. C* **28**, 1812 (1983).
5. M. R. Robilotta, and H. T. Coelho, *Nucl. Phys. A* **460**, 645 (1986).
6. J. Carlson, V. R. Pandharipande, and R. B. Wiringa, *Nucl. Phys. A* **401**, 59 (1983).
7. U. van Kolck, *Phys. Rev. C* **49**, 2932 (1994).
8. J. A. Eden, and M. F. Gari, *Phys. Rev. C* p. 1510 (1996).
9. H. Sakai *et al.*, *Phys. Rev. Lett.* **84**, 5288 (2000).
10. K. Sekiguchi *et al.*, *Phys. Rev. C* **65**, 034003 (2002).
11. K. Hatanaka *et al.*, *Phys. Rev. C* **66**, 044002 (2002).
12. K. Ermisch *et al.*, *Phys. Rev. Lett.* **86**, 5862 (2001).
13. Y. Maeda *et al.*, *Phys. Rev. C* **76**, 014004 (2007).
14. T. Sasakawa, and S. Ishikawa, *Few-Body Systems* **1**, 3 (1986).
15. W. Glöckle, H. Witała, D. Hüber, H. Kamada, and J. Golak, *Phys. Rep.* **274**, 107 (1966).
16. Y. Koike, and J. Haidenbauer, *Nucl. Phys. A* **463**, 365 (1987).
17. K. Sagara *et al.*, *Phys. Rev. C* **50**, 576 (1994).
18. J. Strate *et al.*, *Nucl. Phys. A* **501**, 51 (1989).
19. K. Sagara *et al.*, *Nucl. Phys. A* **790**, 348c (2007).
20. Y. Koike, and S. Ishikawa, *Nucl. Phys. A* **631**, 683c (1998).
21. H. Witała *et al.*, *Phys. Rev. Lett.* **81**, 1183 (1998).
22. S. Nemoto *et al.*, *Phys. Rev. C* **58**, 2599 (1998).
23. K. Sekiguchi *et al.*, *Phys. Rev. C* **70**, 014001 (2004).
24. K. Ermisch *et al.*, *Phys. Rev. C* **68**, 051001(R) (2003).
25. S. Kistryn *et al.*, *Phys. Rev. C* **72**, 044006 (2005).
26. E. Stephan *et al.*, *Phys. Rev. C* **76**, 057001 (2007).
27. H. Witała *et al.*, *Phys. Rev. C* **71**, 054001 (2005).
28. H. Witała *et al.*, arXiv : 0801.0367v1[nucl – th] (2007).
29. J. Adam *et al.*, *Phys. Rev. C* **69**, 034008 (2004).
30. S. C. Pieper *et al.*, *Phys. Rev. C* **64**, 014001 (2001).
31. S. Ishikawa, and M. R. Robilotta, *Phys. Rev. C* **76**, 014006 (2007).
32. E. Epelbaum *et al.*, *Phys. Rev. C* **66**, 064001 (2002).
33. V. Bernard, E. Epelbaum, H. Krebs, and U.-G. Meißner, arXiv : 0712.1967v1[nucl – th] (2007).
34. A. Nogga, P. Navrátil, B. R. Barrett, and J. P. Vary, *Phys. Rev. C* **73**, 064002 (2006).
35. P. Navrátil, V. G. Gueorguiev, J. P. Vary, O. W. E, and A. Nogga, *Phys. Rev. Lett.* **99**, 042501 (2007).
36. D. Hüber, H. Witała, A. Nogga, W. Glöckle, and H. Kamada, *Few-Body Systems* **22**, 107 (1997).
37. D. Hüber, J. L. Friar, A. Nogga, H. Witała, and U. van Kolck, *Few-Body Systems* **30**, 95 (2001).

Cross Sections of the Deuteron-Proton Breakup as a Probe of Three-Nucleon System Dynamics

St. Kistryn[*], E. Stephan[†], N. Kalantar-Nayestanaki[**], A. Biegun[†,**], B. Kłos[†], R. Sworst[*], K. Bodek[*], I. Ciepał[*], A. Deltuva[‡], E. Epelbaum[§], A. C. Fonseca[‡], W. Glöckle[¶], J. Golak[*], H. Kamada[‖], M. Kiš[**], A. Kozela[††], M. Mahjour-Shafiei[**,‡‡], A. Micherdzińska[†,§§], A. Nogga[§], P. U. Sauer[¶¶], R. Skibiński[*], H. Witała[*], J. Zejma[*] and W. Zipper[†]

[*] *Jagiellonian University, PL-30059 Kraków, Poland*
[†] *University of Silesia, PL-40007 Katowice, Poland*
[**] *Kernfysisch Versneller Instituut, NL-9747 Groningen, The Netherlands*
[‡] *Centro de Física Nuclear da Universidade de Lisboa, P-1649-003 Lisboa, Portugal*
[§] *Forschungszentrum Jülich, Institut für Kernphysik (Theorie), D-52425 Jülich, Germany*
[¶] *Institut für Theoretische Physik II, Ruhr Universität Bochum, D-44780 Bochum, Germany*
[‖] *Department of Physics, Kyushu Institute of Technology, Kitakyushu 804-8550, Japan*
[††] *Institute of Nuclear Physics PAN, PL-31342 Kraków, Poland*
[‡‡] *Departament of Physics, Univ. of Teheran, Iran*
[§§] *University of Winnipeg, Winnipeg MB R3B 2E9, Canada*
[¶¶] *ITP, Universität Hannover, D-30167 Hannover, Germany*

Abstract. The three-nucleon system is the simplest, non-trivial testing ground which allows one to probe quantitatively modern nucleon-nucleon interaction models. The rigorous technique of solving the Faddeev equations for the three-nucleon system allows one to test the influence of additional dynamical ingredients, referred to as three-nucleon forces, as well as the Coulomb interaction and relativistic effects. A large set of high precision, exclusive cross-section data for the ^{1}H(d,pp)n breakup reaction at 130 MeV was obtained at KVI, Groningen. This allowed us to establish for the first time clear evidence for three-nucleon force contributions to the cross sections of the breakup process and to confirm recent predictions of a sizable influence of the Coulomb force in this reaction.

Keywords: NUCLEAR REACTIONS ^{1}H(d,pp)n, E=130 MeV; measured differential cross sections $d^5\sigma/d\Omega_1\, d\Omega_2\, dS$ in function of $\theta_1, \theta_2, \varphi_{12}, S$. Studied effects of three-nucleon forces and Coulomb interaction.
PACS: 21.30.Fe, 21.45.Ff, 25.10+s, 25.45.-z

INTRODUCTION

Exploring the details of the nucleon-nucleon (*NN*) interaction is one of the most important activities of nuclear physicists. An exact knowledge of all features of the two nucleon (2*N*) system dynamics would provide a natural basis for understanding the properties and interactions of nuclei. This optimistic presumption has to be verified by applying models of the *NN* interaction to describe properties of many-nucleon systems with increasing complexity. Obviously, the least complicated non-trivial environment is the one composed of three nucleons.

Dynamics of the three-nucleon (3*N*) system can be comprehensively studied by means of the nucleon-deuteron (*Nd*) breakup reaction. Its final state, constrained by only gene-

CP1011, *New Facet of Three Nucleon Force—50 Years of Fujita Miyazawa Three Nucleon Force (FM50), The International Symposium,* edited by H. Sakai, K. Sekiguchi, and B. F. Gibson

ral conservation laws, provides a rich source of information to test the $3N$ Hamiltonian details. It is of particular importance when components of the models which account for subtle effects, like three-nucleon force (3NF) contributions to the potential energy of the $3N$ system, are under investigation. Also, other dynamical ingredients: the Coulomb interaction and relativistic effects, might influence the breakup observables in various ways, therefore calling for detailed studies. Nowadays, precise predictions for observables in the $3N$ system can be obtained via exact solutions of the $3N$ Faddeev equations for any nucleon-nucleon interaction, even with the inclusion of a 3NF [1] and other dynamical components. Physical input to those calculations is assumed in several ways: as semi-phenomenological realistic NN potentials (like CD Bonn, Nijmegen or Argonne AV18) combined with a model 3NF (represented by Tucson-Melbourne TM99), the two- and three-nucleon interactions obtained with the explicit treatment of the Δ-isobar excitation within the coupled-channel framework [2], or forces obtained via Chiral Perturbation Theory methods [3]. It should be mentioned that in the pure $2N$ sector all of these approaches lead to an almost equally perfect description of the measured $2N$ observables.

Verification of the continuously developing theoretical models of the $3N$ system dynamics requires reliable, precise and rich breakup data sets, covering large fractions of the phase space. Unfortunately, it still remains difficult to perform such measurements at the required level of accuracy. Until recently, the database remained rather poor. Moreover, the experimental studies were usually confined to just a few specific kinematic configurations. Consequently, their conclusions were to a large extent ambiguous, not supported by a systematic probing of large sets of data.

EXPERIMENT

The project, dedicated to improving the status of the breakup database, has been undertaken by the Polish-Dutch collaboration. Our new approach to the breakup research assumed a simultaneous measurement in a large part of the phase space by using a high-acceptance, position-sensitive detection system.

Measurements of the $^{1}H(\vec{d},pp)n$ reaction were carried out at KVI Groningen, The Netherlands. Polarized deuteron beam with the energy of 130 MeV was focused to a spot of 2 mm diameter on a liquid hydrogen target of 4 mm thickness. The experimental setup consisted of a three-plane multi-wire proportional chamber (MWPC) and of two layers of a segmented scintillator hodoscope: horizontal transmission ΔE and vertical stopping E detectors, each containing 24 elements. Position information from the MWPC was used for precise reconstruction of the particle emission angles, while the hodoscope allowed to identify the charged reaction products (protons vs . deuterons) and to determine their energies. The ΔE-E wall covered a substantial fraction of the phase-space: from about $10°$ to $35°$ for the polar angles θ and the full (2π) range of the azimuthal angles φ.

Breakup cross-section as well as vector and tensor analyzing power data were acquired for the whole covered phase space. Data collected simultaneously for the elastic scattering channel were used for normalization and determination of the beam polarization. High quality of the experiment allowed also to use those data to extract a new set of

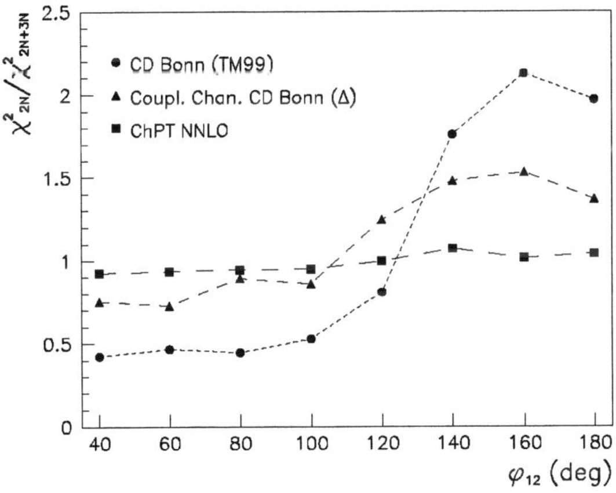

FIGURE 1. Ratios of χ^2 values for calculations without and with the 3NF contributions. The cross-section data with the experimental absolute normalization were used in computing the χ^2_{2N} and χ^2_{2N+3N} values for groups of kinematic configurations with the same φ_{12} angle. The data are compared with calculations using realistic potentials, the coupled-channels model, and the ChPT approach, shown by dots, triangles and squares, respectively. Lines connecting the points are to guide the eye only.

the elastic scattering analyzing powers [4]. More details on the experimental setup and procedures, as well as on the data analysis are given in Refs. [5, 6]

CROSS SECTION RESULTS

The worldwide first extensive set of the breakup cross-section data provides a *systematic* base for comparisons with the theoretical predictions. In a representative scan of the phase space the data are presented [5, 6, 7] on a regular grid of polar and azimuthal angles with a constant step in the arc-length variable S. Polar angles of the two protons θ_1 and θ_2 are changed between 15° and 30° with the step of 5° (with an additional set for $\theta_1 = \theta_2 = 13°$) and their relative azimuthal angle φ_{12} is taken in the range from 40° to 180°, with the step of 20°. For each combination of the central values θ_1, θ_2 and φ_{12} the experimental data were integrated within the limits of $\pm$ 1° for the polar angles and of $\pm$ 5° for the relative azimuthal angle. The bin size along the kinematic curve S was 4 MeV. Such defined set comprises 80 kinematical configurations, with a total of nearly 1800 experimental points.

These results allowed to conclude on importance of the 3NF effects for the breakup reaction – only inclusion of this additional dynamics in the calculations leads generally to a better description of the cross sections. The detailed discussion on this subject is given in Refs. [5, 7]. A summary is shown in Fig. 1 in form of a global comparison, as the ratio of χ^2_{2N} to χ^2_{2N+3N} in order to magnify the influence of the 3NF effects. The χ^2 values are calculated for all data points with respect to various theoretical predictions, without and with inclusion of 3NF effects. One finds that the consistency between the predictions of the CD Bonn potential and the data is improved by adding the TM99 3NF

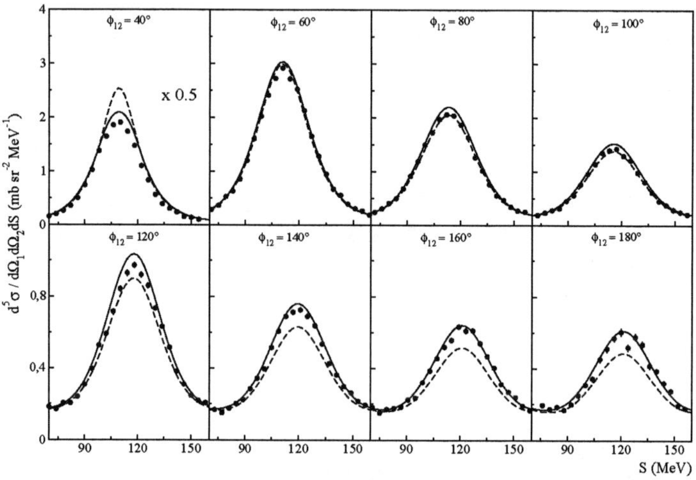

FIGURE 2. Differential cross sections for the deuteron-proton breakup at 130 MeV deuteron energy, plotted as a function of the arc-length S along the kinematic curve. The data are shown for 8 kinematic configurations characterized by the proton polar angles $\theta_1 = \theta_2 = 13°$ and various relative azimuthal angles ϕ_{12}, as indicated in the individual panels. Experimental data are compared to the results of calculations with the coupled-channels CD Bonn + Δ potential, without (dashed lines) and with (solid lines) inclusion of the Coulomb interaction. The data and calculations in the left top panel are normalized to the common vertical axis by the indicated factor.

in configurations with relatively large φ_{12} angles (ratio above 1). On the other hand, for $\varphi_{12} < 120°$ including the 3NF into the calculations drives the results away from the data. This behavior is qualitatively confirmed by the coupled-channels calculations, however the amplitude of the changes induced by including the Δ-isobar excitation contributions is smaller. For the ChPT calculations essentially no effect is present, the ratio stays close to 1 for all values of φ_{12}, what is due to large theoretical uncertainties of the ChPT predictions at NNLO. We were able to conclude that inclusion of 3NF (in the realistic potential approach) leads to a global better description of the data, quantified by a decrease of the χ^2 value by about 40%.

In the kinematical region of small proton polar angles significant discrepancies between the measured and predicted cross sections were observed [5, 7]. This has been attributed to the fact that in these configurations the Coulomb interaction plays an important role, while the calculations are valid for the neutron-deuteron system. Only recently the Coulomb force was successfully included in the coupled-channel calculations for the breakup reaction [2]. It has been shown [8] that this improvement strongly reduces the discrepancies for the cross-section data. Both, suppressing the cross-section values at small angles φ_{12} and its enhancement at large φ_{12} values are generally correctly reproduced - see Fig. 2. Even the most spectacular cases of the cross-section distributions distortion at still smaller values of φ_{12} are well discribed with the inclusion of the Coulomb effects [8].

The role of additional (i.e. more than just $2N$) dynamics in the breakup cross sections

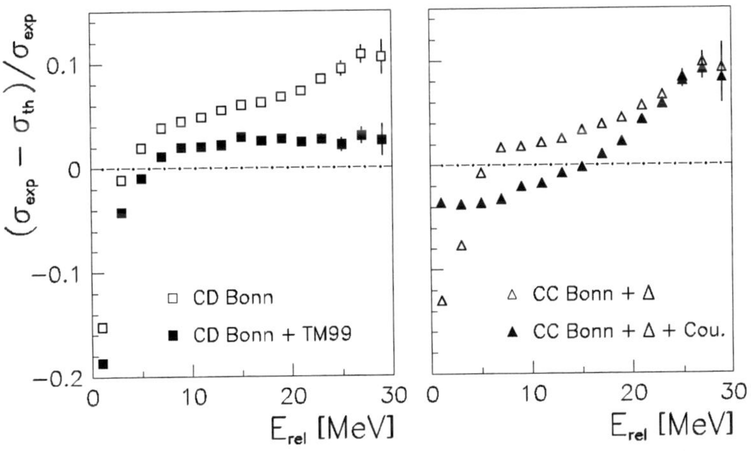

FIGURE 3. Relative discrepancies between the experimental data and the theoretical predictions of the breakup cross sections as a function of the relative energy of the two breakup protons. *Left panel:* Empty squares show the results obtained for the calculations with the CD Bonn potential alone, while the solid squares represent the comparison with the calculations including also the TM99 3NF. *Right panel:* Empty triangles show the results obtained for the CD Bonn + Δ potential calculations without the Coulomb force. The solid triangles represent the results for the same calculations but with the inclusion of the long-range Coulomb interaction.

is recapitulated in a global approach in Fig. 3. The relative differences of the experimental and theoretical cross sections, $(\sigma_{exp} - \sigma_{th})/\sigma_{exp}$, were determined and plotted as a function of E_{rel}, the kinetic energy of the relative motion of the two protons. The σ's stand here for the full, five-fold differential cross section $d^5\sigma/d\Omega_1 d\Omega_2 dS$.

Only with a rich set of data it is possible to observe the qualitative difference in action of 3NF and Coulomb force. Inclusion of 3NF significantly improves the data description at larger values of E_{rel}, while slightly driving the predictions away from the data at low relative energies. The Coulomb force influence is practically restricted to the region of small values of E_{rel}, where its inclusion leads to a significantly improved reproduction of the data. Arising from such comparison is a suggestion that combination of the two effects might allow for correct data description in the whole range of kinematical variables. Final conclusion in that point, however, will be possible only when calculations treating simultaneously all the effects become available.

CONCLUSIONS

The described project is the first extensive and systematic study, covering a large fraction of the breakup phase space. Global comparisons of the whole cross section data set with various theoretical predictions show that the present day models of the $3N$ system dynamics reproduce the majority of the data with satisfactory precision. In many cases in which the predicted effects due to a genuine 3NF are non-negligible, their inclusion tends to improve the agreement with the data. However, thanks to the applied experimental technique of covering a significant fraction of the breakup phase space with a highly symmetric detection system, it has been shown that there are also systematic regularities in discrepancies of the measured cross sections and the predictions of all the theoretical approaches. These established disagreements are to a large extent explained by including Coulomb effects into the calculations. This is an important step towards a precise and complete description of the breakup observables, which should eventually include all aspects of the medium-energy reaction mechanism.

The theoretical predictions show that the effects of the Coulomb force and of the $3N$ interaction affect the breakup observables in different ways and with varying strength when inspecting the full reaction phase space. Also first attempts to include relativistic effects into the theoretical treatment of the breakup reaction [9] reveal still different sensitivity of the observables. Such selectivity makes possible tracing the details of certain effects in regions where the others are proved to have relatively small influences. This is e.g. true for studies of the $3N$ forces - even if the Coulomb effects in the breakup cross sections are large, there are regions in which their influence is much smaller than the expected effects of the additional nuclear dynamics. With even larger experimental coverage of the breakup phase space with respect to several observables (the analyzing power results from the described project are just beginning to appear [10, 11]) and for various beam energies (see [12] for an announcement of a first preliminary set), the eventually established pattern of discrepancies between the data and the calculations might help to improve our understanding of the full dynamics of the $3N$ system.

ACKNOWLEDGMENTS

Financial support for the project has been provided in parts by several Polish, Dutch and European Institutions. We would like to express our gratitude to the AGOR accelerator group for developing an excellent beam. The hospitality of KVI is warmly acknowledged.

REFERENCES

1. W. Glöckle et al., *Phys. Rep.* **274**, 107 (1996) and references therein.
2. A. Deltuva, A. C. Fonseca and P. U. Sauer, *Phys. Rev. C* **73**, 057001 (2006).
3. E. Epelbaum, *Rep. Prog. Nucl. Phys.* **57**, 654 (2006).
4. E. Stephan et al., *Phys. Rev. C* **76**, 057001 (2007).
5. St. Kistryn, *Three-Nucleon Force Effects in the Deuteron-Proton Breakup Reaction,* Habilitation Thesis, DjaF Kraków, 2005, ISBN 83-86774-42-8.
6. St. Kistryn et al., *Phys. Rev. C* **68**, 054004 (2003).
7. St. Kistryn et al., *Phys. Rev. C* **72**, 044006 (2005).
8. St. Kistryn et al., *Phys. Lett. B* **641**, 23 (2006).
9. H. Witała, J. Golak and R. Skibiński, *Phys. Lett. B* **634**, 374 (2006).
10. E. Stephan et al., contribution to these Proceedings.
11. R. Sworst et al., to appear in *Acta Phys. Pol. B*.
12. H. Mardanpour et al., *Nucl. Phys.* **A790**, 426c (2007).

A large, precise set of polarization observables for deuteron-proton breakup at 130 MeV

E. Stephan*, St. Kistryn†, N. Kalantar-Nayestanaki**, A. Biegun*, K. Bodek†, I. Ciepał†, A. Deltuva‡, E. Epelbaum§, A. C. Fonseca‡, W. Glöckle¶, J. Golak†, H. Kamada||, M. Kiš**, B. Kłos*, A. Kozela††, M. Mahjour-Shafiei**, A. Micherdzińska*, A. Nogga§, P. U. Sauer‡‡, R. Skibiński†, R. Sworst†, H. Witała†, J. Zejma† and W. Zipper*

*University of Silesia, PL-40007 Katowice, Poland
†Jagiellonian University, PL-30059 Kraków, Poland
**Kernfysisch Versneller Instituut, NL-9747 Groningen, The Netherlands
‡Centro de Física Nuclear da Universidade de Lisboa, P-1649-003 Lisboa, Portugal
§Forschungszentrum Jülich, Institut für Kernphysik (Theorie), D-52425 Jülich, Germany
¶Institut für Theoretische Physik II, Ruhr Universität Bochum, D-44780 Bochum, Germany
||Department of Physics, Kyushu Institute of Technology, Kitakyushu 804-8550, Japan
††Institute of Nuclear Physics PAN, PL-31342 Kraków, Poland
‡‡ITP, Universität Hannover, D-30167 Hannover, Germany

Abstract. High precision vector A_x, A_y and tensor A_{xx}, A_{xy}, A_{yy} analyzing powers for the $^1H(\vec{d},pp)n$ breakup reaction were measured at 130 MeV beam energy with the detection system covering a large part of the phase space. Results are compared with rigorous theoretical calculations based on realistic nucleon-nucleon potentials, also with a so-called three-nucleon force included, as well as on chiral perturbation theory. Theoretical predictions generally describe the data quite well, but in some regions discrepancies have been observed, which indicate incompleteness of the present-day treatment of three nucleon dynamics.

Keywords: NUCLEAR REACTIONS $^1H(\vec{d},pp)n$, E=130 MeV; measured vector A_x, A_y and tensor A_{xx}, A_{xy}, A_{yy} analyzing powers in function of $\theta_1, \theta_2, \varphi_{12}, S$. Studied effects of three nucleon forces.
PACS: 21.30.-x, 21.45.-v, 21.45.Ff, 24.70.+s

Properties of systems composed of three nucleons (3N) are, at not-too-high energies, determined mainly by pairwise nucleon-nucleon (NN) interactions. However, at a certain level of precision, subtle effects of dynamics extending beyond the dominant NN contribution can be studied. Theoretical approaches to describe those additional dynamical ingredients incorporate models of three nucleon force (3NF) to supplement the realistic NN potentials, or use a coupled-channel formalism with explicit treatment of a Δ isobar. Within the chiral perturbation theory 3NF contributions occur naturally at a certain order of the calculations.

Since the predicted effects are usually small, the data set used to verify the calculations has to be very precise and also possibly large, to exclude cases of accidental improvements achieved when including additional dynamical effects. Deuteron-proton breakup experiments that explore a significant part of phase space provide such a large data base. Moreover, studies with polarized deuterons provide access to a number of spin observables: vector A_x, A_y and tensor A_{xx}, A_{xy}, A_{yy} analyzing powers, some of which va-

CP1011, *New Facet of Three Nucleon Force—50 Years of Fujita Miyazawa Three Nucleon Force (FM50), The International Symposium,* edited by H. Sakai, K. Sekiguchi, and B. F. Gibson

nish in the case of the elastic scattering process. Due to development of highly polarized beams and accessibility of fast data acquisition systems, it has recently became possible to achieve statistical precision of the many-parameter analyzing power data sufficient to motivate studies of subtle dynamic effects.

Analyzing powers for the ${}^1H(\vec{d},pp)n$ reaction were measured at KVI, with the use of a polarized deuteron beam with an energy of 130 MeV [1]. The experimental setup SALAD covered azimuthal angles between 12° and 40° and the full range of polar angles. In the experiment seven different beam polarization states were used, with various combinations of vector and tensor polarizations $P \equiv (P_Z, P_{ZZ})$: (0, 0), (2/3, 0), (-2/3, 0), (0, 1), (0,-2), (1/3, 1), (1/3, -1). The actual beam polarizations were determined on the basis of the elastic scattering data. To that purpose asymmetries of rates measured for the polarized, $N_P^\theta(\phi)$, and unpolarized, $N_0^\theta(\phi)$, beam states were constructed for every scattering angle θ as functions of the angle ϕ, which for the transversally polarized beam corresponds to the angle between the polarization direction and the scattering plane vector:

$$f_P^\theta(\phi) = \frac{N_P^\theta(\phi) - N_0^\theta(\phi)}{N_0^\theta(\phi)} \tag{1}$$

Knowing the analyzing power values for elastic scattering, measured in a dedicated experiment at RIKEN [2], the P_Z and P_{ZZ} values were obtained from the fit of the function

$$f_P^\theta(\phi) = iT_{11}(\theta)\sqrt{3}P_Z\cos\phi - T_{22}(\theta)\frac{\sqrt{3}}{2}P_{ZZ}\cos 2\phi - T_{20}(\theta)\frac{\sqrt{2}}{4}P_{ZZ}.$$

to the experimental $f_P^\theta(\phi)$ distribution. It was found that the polarization values were between 50% and 80% of the maximal (nominal) values listed above. It was also possible, by studying the elastic scattering events registered at various θ angles, to extend significantly the angular range of experimental data for vector and tensor analyzing powers of the deuteron-proton elastic scattering at 130 MeV. Details of the procedure and the results obtained are presented in [3].

In order to study polarization observables for the reaction of interest, the breakup events collected for each of the polarization states P were analyzed. For each configuration $\theta_1, \theta_2, \varphi_{12}$ of the two outgoing protons and for each 8 MeV wide bin in the S variable (arc-length along the kinematical curve) events were sorted with respect to the angle ϕ with the binning of 10°. The ϕ angle is defined by the polarization direction and the vector normal to the plane spanned by the momentum vectors of the beam and of the first outgoing proton. The obtained numbers of events have been normalized to the beam current collected in the Faraday Cup and corrected for dead time losses. After normalization, the numbers of events $N_P^\xi(\phi)$ for the selected polarization state P and the kinematical point $\xi = (\theta_1, \theta_2, \varphi_{12}, S)$ were obtained as a function of the angle ϕ. On the basis of these numbers a ratio

$$f_P^\xi(\phi) = \frac{N_P^\xi(\phi) - N_0^\xi(\phi)}{N_0^\xi(\phi)} \tag{2}$$

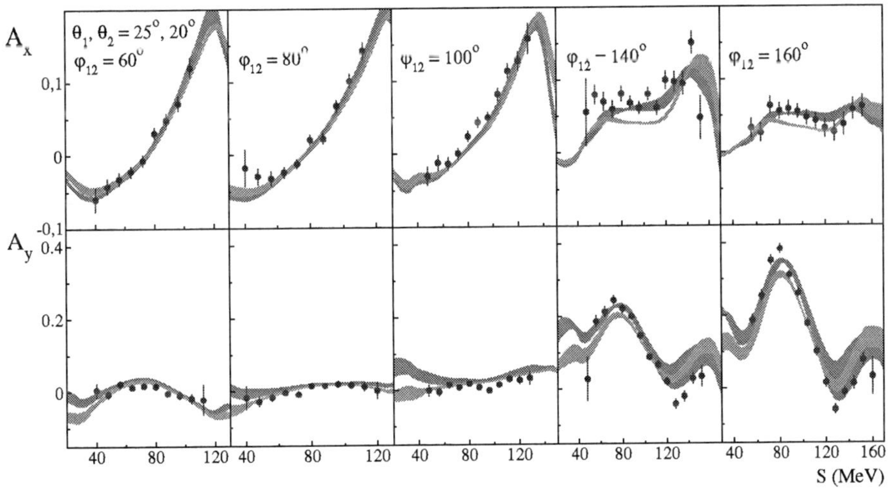

FIGURE 1. Examples of vector analyzing power results obtained for configurations characterized by the same combination of the polar angles and various relative azimuthal angles of the two protons (listed in panels). Only predictions of ChPT are shown: the dark-gray and light-gray bands represent calculations performed at NNLO and N3LO, respectively. Currently available N3LO calculations are incomplete, missing 3NF graphs at the highest included order. All other theoretical approaches lead to results consistent with the ones obtained within ChPT at N3LO.

was constructed. $N_0^{\xi}(\phi)$ denotes the analogously obtained number of events for the unpolarized beam. Applying the cross section formula for a three-particle final state of the reaction with a polarized deuteron projectile [4], $f_P^{\xi}(\phi)$ can be written as:

$$f_P^{\xi}(\phi) = -\frac{3}{2}\sin\phi P_Z A_x(\xi) + \frac{3}{2}\cos\phi P_Z A_y(\xi) - \cos\phi\sin\phi P_{ZZ} A_{xy}(\xi) + \frac{1}{2}\sin^2\phi P_{ZZ} A_{xx}(\xi) + \frac{1}{2}\cos^2\phi P_{ZZ} A_{yy}(\xi). \quad (3)$$

Vector and tensor analyzing powers were extracted from the fit of the function (3) to the experimental distributions of $f_P^{\xi}(\phi)$. Final results were obtained as weighted averages of the analyzing power values obtained for various polarization states. More details about the analysis procedure can be also found in [5].

The analyzing power values are evaluated for a set of kinematical configurations of the two outgoing protons, spanned on a grid of polar angles θ_1, θ_2 between 15° and 30° with the step of 5° and relative azimuthal angle φ_{12} between 40° and 180° with the step of 20°. Taking into account the step of 8 MeV in S, such a choice leads to the total of above 800 data points for each observable. Angular ranges of data integration around the central values of angles have been chosen as $\Delta\theta = \pm 2°$ for the polar and $\Delta\varphi_{12} = \pm 10°$ for the relative azimuthal angles. In this way small statistical uncertainties of the analyzing power results are assured, but, alas, the effects of averaging of the observables are not

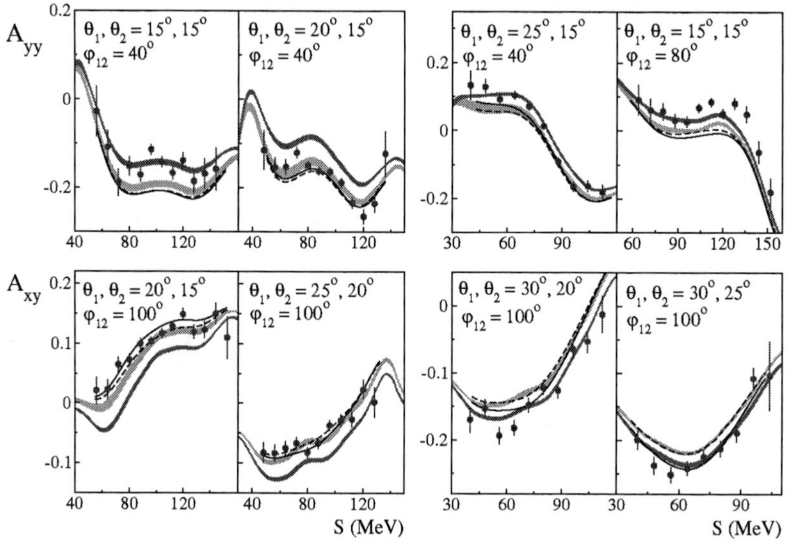

FIGURE 2. Examples of tensor analyzing power results obtained for configurations in which significant effects of the TM99 3NF are expected. The light-gray and dark-gray bands represent 2N and 2N+TM99 calculations, respectively. The results of the coupled-channel calculations with the explicit Δ-excitation, with and without Coulomb interaction are shown as solid and dashed lines, respectively.

negligible. Therefore, before comparisons, the theoretical predictions had to be averaged over the same angular ranges.

The data are compared to predictions based on various approaches in modeling the three nucleon system dynamics. Rigorous Faddeev calculations are performed with the use of the modern realistic nucleon-nucleon potentials alone (referred to in the following by 2N) as well as including the Tucsone-Melbourne three nucleon force model (2N+TM99) [6]. The experimental results are also confronted with predictions for the potential generated within the chiral perturbation theory (ChPT NNLO and ChPT N3LO [7]) as well as by the coupled-channel approach with the explicit treatment of a single Δ isobar (CDB+Δ) and with the Coulomb interaction included (CDB+Δ+C) [8].

For a majority of the studied configurations results of all theoretical predictions are consistent with each other and describe the experimental results quite well. This means, in particular, that pure NN potentials are sufficient to reproduce the analyzing power data in a certain part of the phase space. Several sample results are shown in Fig. 1, for other examples see [5]. There are, however, also regions, where discrepancies between the data and the calculations are observed. For the vector analyzing powers at low relative azimuthal angles predictions given by all the theoretical approaches are very similar, but the situation changes with the growing values of φ_{12}: all but ChPT NNLO predictions are still consistent with each other and drift away from the data, while ChPT NNLO calculations reproduce the experimental results quite well (see example in Fig. 1). It is important to consider, that the ChPT N3LO calculations are still incomplete: they miss 3NF contributions at the N3LO order. The question of how well the full N3LO calculations will reproduce the data remains open.

In other regions where some discrepancies between the data and the theoretical calculations occur, especially for the case of the tensor analyzing powers, there are configurations in which inclusion of the TM99 3NF improves the description (see examples in Fig. 2). Alas, in several other configurations the calculations including the 3NF contributions lead to a worse agreement with the experimental data (cf. Fig. 2). This suggests an existence of some still missing ingredients in the spin part of the 3NF model. Effects of the Δ excitation in the coupled-channel calculations are practically negligible, so this piece of the 3N system dynamics also does not explain the problem.

Predicted Coulomb effects are rather small in all analyzing powers, but in some regions their magnitude is non-negligible and they can be comparable to the effects of the TM99 3NF (e.g. in A_{xy} for the configuration θ_1=30°, θ_2=25°, φ_{12}=100°, shown in Fig. 2). Therefore calculations incorporating both, the TM99 and the Coulomb effects would be also of great interest and advantage.

Precise data for analyzing powers of the dp breakup reaction at 130 MeV confirm that the current methods of calculating these observables lead to generally correct results within the phase space region studied, and that the quality of the description provided by various approaches is very similar. However, there are also observables and configurations where the agreement between the data and theory is not so satisfactory. The discrepancy is not always removed by adding the TM99 3NF, which is a hint of problems in the spin part of this model of the 3NF. Systematic, quantitative comparisons over the whole data set, which will provide a better overview of the pattern of (dis-)agreements between the data and the presently available calculations, are under way. On the other hand, possibly complete calculations, including all ingredients of the 3N system dynamics (3NF, Coulomb interaction, relativistic effects), as well as developments in ChPT are very important for better understanding of the breakup data. To enable conclusive comparisons further rich data sets obtained at other energies, are being worked upon.

ACKNOWLEDGMENTS

We would like to express our gratitude to AGOR accelerator group for developing excellent beam and to Rob Kremers for his work on the POLIS source. A part of the numerical calculations has been performed on the IBM Regatta p690+ of the NIC in Jülich, Germany.

REFERENCES

1. St. Kistryn et al., *Phys. Rev. C* **72**, 044006 (2005).
2. H. Mardanpour *et al.*, Eur. Phys. J. **31**, 383 (2007).
3. E. Stephan et al., *Phys. Rev. C* **76**, 057001 (2007).
4. G.G.Ohlsen et al., *Nucl. Instr. and Meth.* **179**, 283 (1981).
5. R. Sworst et al., to appear in *Acta. Phys. Pol. B.*
6. W. Glöckle et al., *Phys. Rep.* **274**, 107 (1996) and references therein.
7. E. Epelbaum, *Prog. Part. Nucl. Phys.* **57**, 654 (2006).
8. A. Deltuva, A. C. Fonseca, and P. U. Sauer, *Phys. Rev. C* **73**, 057001 (2006).

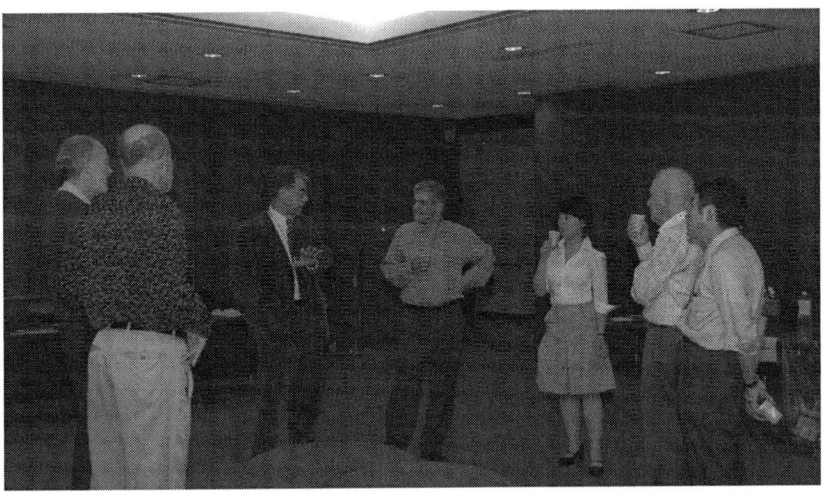

The Three-Nucleon Analyzing Power Puzzle — The Past 20 Years

W. Tornow

Department of Physics, Duke University, Durham, North Carolina 27708-0305, USA and Triangle Universities Nuclear Laboratory, Durham, North Carolina 27708-0308, USA

Abstract. The history of the three-nucleon analyzing power puzzle is described from the point of view of an experimentalist who has been collaborating closely with few-body theoreticians in trying to unravel the physics behind this long-standing low-energy scattering phenomenon. Although a number of very interesting and surprising effects have been found along the way, the puzzle is still unexplained. Currently, the hope is that long-range three-nucleon force terms predicted by Chiral Effective Field Theory in N^3LO will eventually solve the puzzle.

Keywords: Three-nucleon reactions, Polarization, Faddeev calculations
PACS: 21.45.-v, 21.45.Ff, 21.60.De, 24.70.+s, 25.40.Cm, 25.40.Dn

In 1982/83 the Tübingen-TUNL group published their first data for the analyzing power $A_y(\theta)$ in neutron-deuteron (n-d) elastic scattering [1, 2]. These high-precision data obtained at TUNL at E_n = 10.0 and 14.1 MeV were compared to state-of-the-art three-nucleon (3N) calculations of Doleschall. As can be seen from Fig. 1 these calculations were in good agreement with the experimental data [1, 2].

In 1986 Koike and Haidenbauer [3] presented a paper with the title "The Neutron-Deuteron Elastic Scattering with the Paris Potential" at the 11th International Conference on Few-Body Problems in Tokyo/Sendai. Although no comparison was given between data and calculations, the authors point out that their present result for $A_y(\theta)$ is much too small compared to experimental data, and that some work is still needed to improve the convergence of their calculations.

In collaboration with Koike, the TUNL group published in 1987 a paper [4] in which our data at E_n = 10.0 and 14.1 MeV and new TUNL data at 12.0 MeV were compared to Koike's calculations using finite-rank approximations of the Paris nucleon-nucleon (NN) potential model (see Fig. 2). At that time Koike was convinced that the small difference between data and calculations observed in the angular region of the A_y maximum was not an artifact of the finite rank approximation of the ^3P_j NN interactions. Koike also showed [4] that the calculations given in Fig. 2 are the result of a complicated interplay between the ^3P_j NN interactions, while $^1P_1$ and higher partial-wave NN interactions hardly play any role in the energy range investigated.

The discrepancy noticed in Fig. 2 between data and calculations in the region of the A_y maximum can be considered as the first evidence ($\sim$2.5 σ effect) for a phenomenon which was later coined the three-nucleon analyzing power puzzle [5]. Perhaps, even as early as 1976 first hints may have existed when Benayoun *et al.* [6] published the first Faddeev calculation for n-d elastic scattering performed with a realistic local nucleon-nucleon (NN) potential. However, the data and the NN potential model available at the

CP1011, *New Facet of Three Nucleon Force—50 Years of Fujita Miyazawa Three Nucleon Force (FM50), The International Symposium,* edited by H. Sakai, K. Sekiguchi, and B. F. Gibson

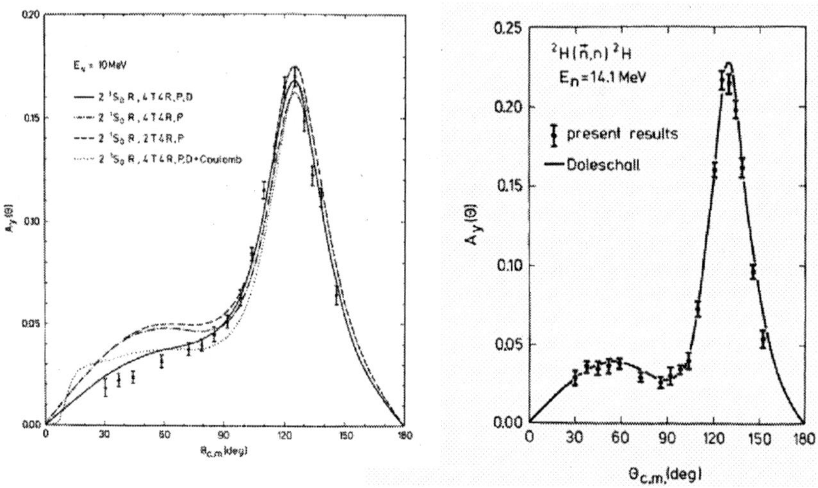

FIGURE 1. Neutron-deuteron analyzing power $A_y(\theta)$ calculations by Doleschall in comparison to data at E_n = 10.0 MeV (left hand side) [1] and 14.1 MeV (right hand side) [2].

time make it difficult to draw a definitive conclusion whether or not the comparison between data and calculations shown in Ref. [6] indicated a real physics effect.

The discovery (> 5σ effect) of the A_y puzzle took place also in 1987, when Witała, Glöckle and Cornelius presented their paper with the title "Three-Nucleon Continuum Calculations with Realistic NN Potentials" [7] at the 11th European Conference on Few-Body Physics in Fontevraud, France. Figure 3 shows the main result of Ref. [7], the failure of rigorous 3N calculations to reproduce the magnitude of the n-d $A_y(\theta)$, a finding that still persists even with today's high-precision NN potential models and 3N forces. The calculations had to be compared to n-d data, rather than to the more accurate proton-deuteron (p-d) data, because at the time it was not possible to accurately incorporate the Coulomb interaction in rigorous 3N calculations.

In 1991 we showed that the relative difference between calculations and data in the angular region of the A_y maximum remained constant at the 25% level in the energy range between 5 and 14 MeV [8]. A confirmation of the observed energy independence of the A_y puzzle came in 1993 when Witała, Hüber and Glöckle compared their calculations [5] to the high-precision n-d $A_y(\theta)$ data obtained by the Wisconsin group at E_n = 3.0 MeV [9]. Except for the data at 8.5 MeV published by the Tübingen group in 1988 [10] and the 1989 work of the Erlangen group [11] at E_n = 13.0 MeV, all previous n-d $A_y(\theta)$ data were obtained at TUNL. Witała et al. used the new Bonn B, Nijmegen, and the Paris potentials and basically obtained the same result for the three NN potential models. They also showed that the inclusion of the Tuscon-Melbourne (TM) 3N force does not solve the A_y puzzle. This observation sparked renewed interest in earlier studies aimed at finding out whether the A_y puzzle can be solved by modifying the ^3P_j NN interactions used in rigorous 3N calculations [12]. Although it has been shown [13] that the low-energy ^3P_j NN phase-shifts can be modified considerably without sacrificing

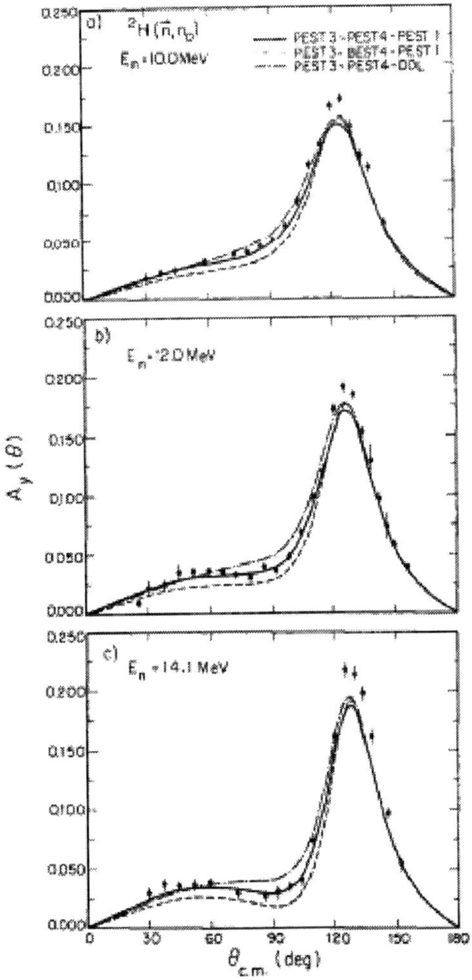

FIGURE 2. Neutron-deuteron analyzing power $A_y(\theta)$ data at E_n = 10.0, 12.0 and 14.1 MeV in comparison to calculations by Koike [4].

their description of the NN database, it is technically difficult to implement this information in 3N calculations unless a NN potential can be constructed that reflects these specific modifications.

In 1995 the Pisa group managed to solve the Coulomb problem in rigorous p-d calculations [14] below the deuteron breakup threshold, and in 1999 they extended their calculations to energies above the deuteron breakup threshold [15]. This important breakthrough opened the opportunity to use the large amount of high-precision p-d and d-p data and compare them to accurate 3N calculations. Not only was the A_y puzzle confirmed for p-d scattering, but it was also found that a similar problem exists for the

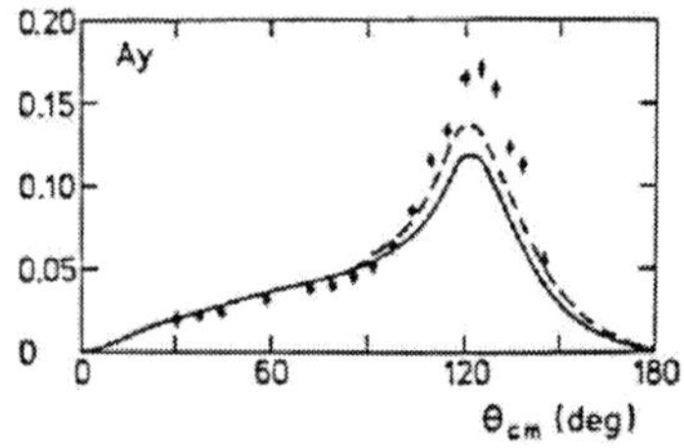

FIGURE 3. Neutron-deuteron analyzing power $A_y(\theta)$ data at $E_n = 10.0$ MeV [1] in comparison to rigorous calculations by Witała, Glöckle, and Cornelius [7] using the then new Bonn NN potential (solid curve) and the Paris NN potential (dashed curve).

vector analyzing power $iT_{11}(\theta)$ in d-p scattering, as generally expected, while the tensor analyzing power data were described fairly well by theory. Figure 4 shows a comparison between data and calculations for $A_y(\theta)$ with the Argonne v18 NN potential (Av18), with and without the Urbana-IX 3N force [16]. For completeness, calculations are also shown for the n-d $A_y(\theta)$ obtained with Av18 only. The work of the Pisa group made it possible to more accurately study the energy dependence of the A_y puzzle than was possible with n-d data alone. It also allowed p-d phase-shift analyses to be performed over a fairly large energy range with theoretical phases used as starting values in the search procedure. It was found that the ${}^4P_{1/2}$, ${}^4P_{3/2}$, and ${}^4P_{5/2}$ 3N phase shifts were the ones that deviated most [19] from the Av18 based starting values (except for the ${}^2S_{1/2}$ phase shift). Because the ${}^4P_j$ 3N interactions are closely related to the NN ${}^3P_J$ interactions, this information is crucial in evaluating the origin of the A_y puzzle, i.e., whether it is caused by presently unknown, long-range 3N forces, or by deficiencies in the NN ${}^3P_J$ interactions used in NN potential models.

Figure 5 shows the relative difference between calculations and data for A_y in nucleon-deuteron (N-d) scattering in the energy range between 650 keV and 16 MeV. The top panel shows the relative difference for n-d and p-d scattering using again the Av18 NN potential model. Comparing these two data sets a difference is noticed for nucleon energies below about 5 MeV, with p-d scattering providing evidence for a larger relative difference than observed for n-d scattering. In Ref. [20] we speculated that this new twist to the 3N analyzing power puzzle might be caused by the so-called Mott-Schwinger or magnetic-moment (MM) interaction, *i.e.*, the interaction of the nucleon magnetic moment in the Coulomb field of the deuteron. Subsequently, this interaction was included into the 3N calculations first by Witała *et al.* [24] and shortly afterwards also by Kievsky et al. [25]. As can be seen from Fig. 6, this interaction decreases (increases) the relative difference for p-d (n-d) scattering at very low incident nucleon energies. Finally, the bottom panel of Fig. 5 shows the relative differences for p-d and n-d scattering, now corrected for the MM interaction, providing a basically energy independent relative difference of 25% for both p-d and n-d scattering in the entire energy range investigated.

There is another subtle effect that has been ignored so far. Only very recently, Witała *et al.* [26] pointed out that relativity, *i.e.*, the Wigner spin rotation, has a significant effect on the magnitude of $A_y(\theta)$ at very low energies, while at energies above $E_n = 30$

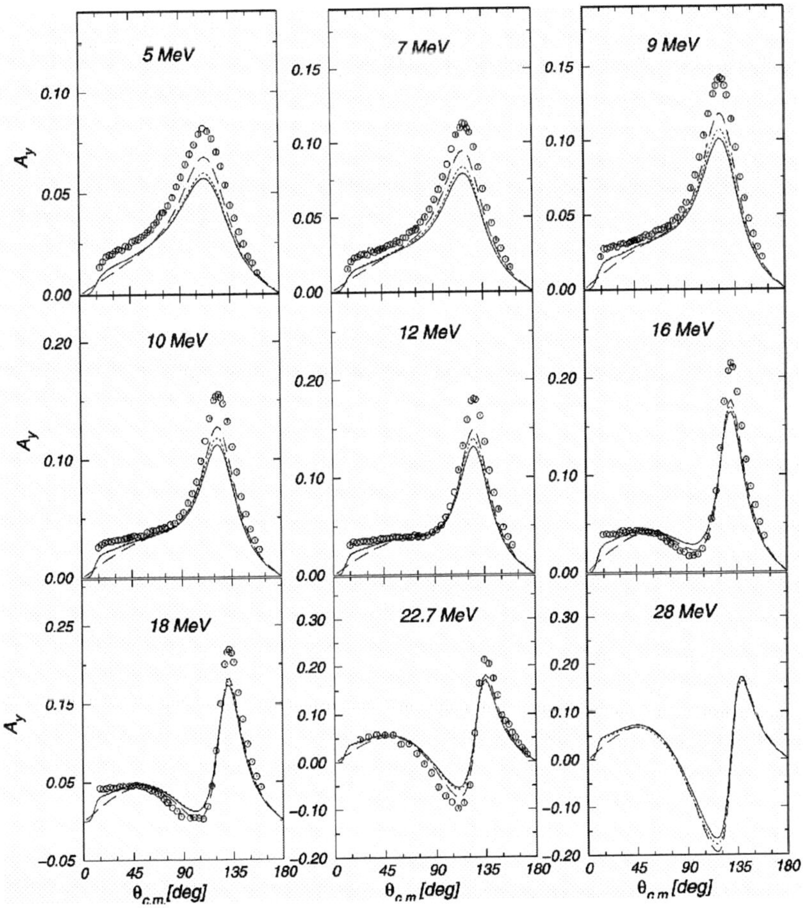

FIGURE 4. Proton-deuteron analyzing power $A_y(\theta)$ calculated by Kievsky *et al.* [16] between 5 and 28 MeV using the Argonne v18 (Av18) NN potential model (solid curve) and Av18 + Urbana IX three-nucleon force (dotted curve). The dashed curve is for n-d scattering calculated with Av18. The p-d data are from Sagara et al. (E_p = 5 MeV through 18 MeV [17]) and Grüebler et al. (E_p = 22.7 MeV [18]).

MeV relativity plays no role anymore. This counterintuitive result reduces the calculated magnitude of $A_y(\theta)$ in the region of the A_y maximum even further (by almost 10% at E_n = 5 MeV). As a result, the relative difference between calculations and data has increased to a total of about 35%, making the A_y puzzle an even more astonishing phenomenon.

Although the bottom panel of Fig. 5 shows a constant energy dependence of the A_y puzzle, it is known from p-d and n-d $A_y(\theta)$ data and calculations that the A_y puzzle disappears at energies above about 35 MeV. At these relatively high energies the shape of the $A_y(\theta)$ angular distribution is dominated by a pronounced negative minimum (see Fig. 7) in contrast to low energies, where $A_y(\theta)$ is governed by a positive maximum. As

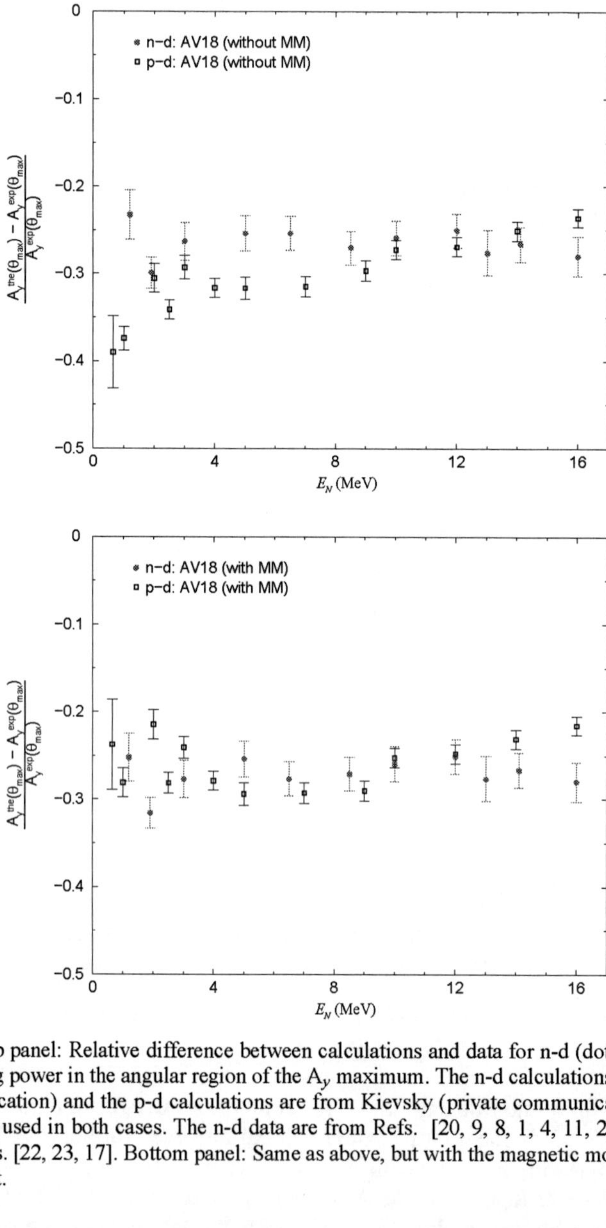

FIGURE 5. Top panel: Relative difference between calculations and data for n-d (dots) and p-d (open squares) analyzing power in the angular region of the A_y maximum. The n-d calculations are from Witała (private communication) and the p-d calculations are from Kievsky (private communication). The Av18 NN potential was used in both cases. The n-d data are from Refs. [20, 9, 8, 1, 4, 11, 2, 21] and the p-d data are from Refs. [22, 23, 17]. Bottom panel: Same as above, but with the magnetic moment interaction taken into account.

can be concluded from Fig. 4, the negative minimum seen at high energies is starting to develop already at energies of slightly above 15 MeV. Therefore, in determining the relative difference for energies above 15 MeV, we take into account the difference

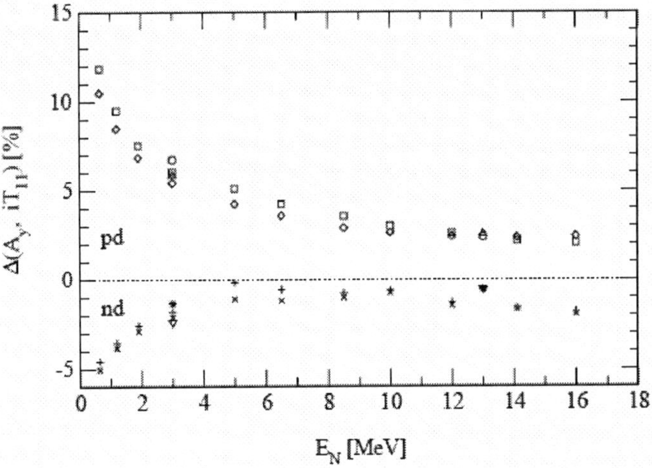

FIGURE 6. Calculated relative difference due to the magnetic moment interaction on the magnitude of the analyzing power in the angular region of the A_y maximum for p-d scattering (top panel, diamonds) and n-d scattering (bottom panel, pluses). The squares and crosses are for iT_{11}. The CD-Bonn potential was used in these calculations of Witała *et al.* [24].

between calculations and data in the minimum and in the maximum of the $A_y(\theta)$ angular distribution. Figure 8 summarizes the present status in the energy range up to 35 MeV. Clearly, there is not a smooth transition of the relative difference between 18 and 30 MeV. When we first noticed this unexpected behavior it was solely based on the p-d data at 20.15 and 22.7 MeV. We then decided to measure n-d $A_y(\theta)$ data in the same energy range. Although the error bars of the new and preliminary TUNL data of Weisel *et al.* at 19, 21, and 22.5 MeV [29] are larger than the uncertainties of the p-d data at 22.7 MeV, they confirm the trend observed for p-d scattering. This second twist to the A_y puzzle is completely unexpected and presently unexplained. Clearly, accurate p-d data are needed to fill the gap between 24 and 30 MeV.

Attempts have been made to explain the 3N A_y puzzle using novel 3N force approaches. In 1999 Kievsky [36] proposed a phenomenological spin-orbit type 3N force. Using a two-parameter fit he was able to describe the data using Av18 and a long-range spin-orbit 3N force acting in the NN channels with spin $S_{ij} = 1$ and isospin $T_{ij} = 1$. Two years later Canton and Schadow [37] successfully described the $A_y(\theta)$ data using a one-pion exchange 3N force. Here, one of the two nucleons involved in the exchange rescatters from a third nucleon while the pion is in flight. Currently, the hope is that the 3N force terms predicted by Chiral Effective Field Theory in N^3LO will eventually provide the correct explanation of the A_y puzzle [38]. However, one has to keep in mind that the A_y puzzle is a low-energy phenomenon. Therefore, the range required for these new 3N force terms has to be of the order of 3 fm, otherwise they will not be able to

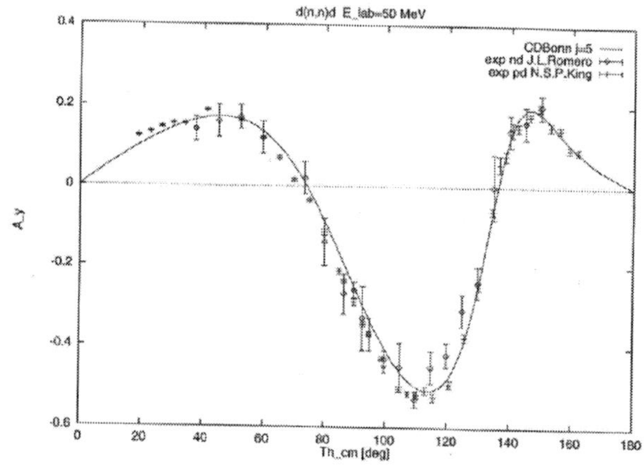

FIGURE 7. Nucleon-deuteron analyzing power at $E_N = 50$ MeV. The p-d data are from King *et al.* [27] and the n-d data are from Romereo *et al.* [28]. The solid curve is a n-d calculation of Witała (private communication) using the CD-Bonn NN potential model.

explain data for $A_y(\theta)$ and $iT_{11}(\theta)$ at $E_{c.m.} = 432$ keV [22].

Looking into the future, we always felt that the 4N system will provide the ultimate test of our understanding of the 3N analyzing power puzzle. Clearly, there exists an analyzing power puzzle for p-^{3}He scattering [39, 40]. However, the recent calculations of the Lisbon group [41] and new TUNL data for $A_y(\theta)$ in n-^{3}He scattering [42] clearly show that such a problem does not exist for this 4N isospin channel at low energies. In addition, the other mixed isospin channel, p-^{3}H scattering, does not show an analyzing power puzzle either. Unfortunately, there are no $A_y(\theta)$ data available for the n-^{3}H channel.

Apparently, much work is still ahead of us in order to explain the 3N analyzing power puzzle. Koike and Haidenbauer [3] were certainly right when they wrote in 1986 in reference to their comparison of n-d calculations and experimental data that "Some interesting physics seems to be there..."

ACKNOWLEDGMENTS

This work was supported in part by the US Department of Energy, Office of Nuclear Physics under grant DE-FG02-97ER41033. The author acknowledges valuable contributions from his TUNL collaborators J.H. Esterline, G.J. Weisel, and A.S. Crowell. This work would not have been possible without the theoretical contributions from H. Witała and A. Kievsky.

REFERENCES

1. W. Tornow *et al.*, *Phys. Rev. Lett.* **49** (1982) 312.
2. W. Tornow *et al.*, *Phys. Rev.* **C27** (1983) 2439.

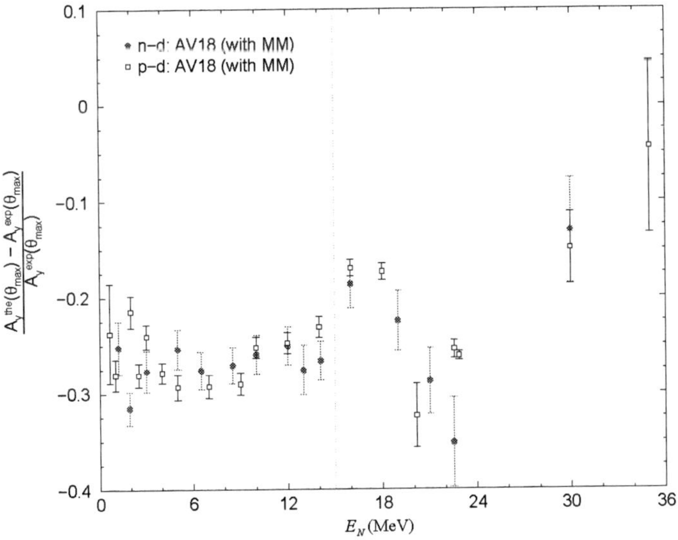

FIGURE 8. Same as the bottom panel of Fig. 5 but with the nucleon energy range extended to 35 MeV. The n-d data above 16 MeV are from Refs. [29, 30] while the p-d data are from Refs. [17, 31, 18, 32, 33, 34]. The n-d calculations above 16 MeV are from Witała (private communication), while the p-d calculations above 16 MeV are from Kievsky (private communication), and Deltuva and Fonseca [35]. At E_p = 22.7 MeV we averaged the theoretical results of Kievsky and Deltuva.

3. Y. Koike and J. Haidenbauer, *Nucl. Phys.* **A463** (1987) 3653.
4. C.R. Howell *et al.*, *Few-Body Systems* **2** (1987) 19.
5. H. Witała, D. Hüber, W. Glöckle, *Phys. Rev.* **C49** (1994) R14.
6. Benayoun *et al.*, *Phy. Rev. Lett.* **36** (1976) 1438.
7. H. Witała, W. Glöckle, T. Cornelius, *Few-Body Systems* Suppl. **2** (1987) 555.
8. W. Tornow *et al.*, *Phys. Lett.* **B257** (1991) 273.
9. E.J. McAninch *et al.*, *Phys. Lett.* **B307** (1993) 13.
10. W. Tornow *et al.*, *Phys. Lett.* **B203** (1988) 341.
11. J. Cub *et al.*, *Few-Body Systems* **6** (1989) 151.
12. W. Tornow, H. Witała, A. Kievsky, *Phys. Rev.* **C57** (1998) 555 and references therein.
13. T.Tornow and W. Tornow, *Few-Body Systems* **26** (1999) 1.
14. A. Kievsky, M. Viviani, S. Rosati, *Phys. Rev.* **C52** (1995) R15.
15. A. Kievsky, M. Viviani, S. Rosati, *Phys. Rev. Lett.* **82** (1999) 3759.
16. A. Kievsky, M. Viviani, S. Rosati, *Phys. Rev.* **C64** (2001) 024002.
17. K. Sagara *et al.*, *Phys. Rev.* **C50** (1994) 576.
18. W. Grüebler *et al.*, *Nucl. Phys.* **A398** (1983) 445; F. Sperisen *et al.*, *ibid.*. **A422** (1984) 81.
19. Z.M. Chen, W. Tornow, A. Kievsky, *Few-Body Systems* **35** (2004) 15 and references therein.
20. E.M. Neidel *et al.*, *Phys. Lett.* **B552** (2003) 29.
21. K. Sagara, private communication.
22. C.R. Brune *et al.*, *Phys. Lett.* **B428** (1998) 13.
23. H. Shimizu *et al.*, *Phys. Rev.* **C52** (1995) 1193.
24. H. Witała *et al.*, *Phys. Rev.* **C67** (2003) 064002.
25. A. Kievsky, M. Viviani, L.E. Marcucci, *Phys. Rev.* **C69** (2004) 014002.
26. H. Witała *et al.*, *Phys. Rev.* **C71** (2005) 054001.
27. N.S.P. King *et al.*, *Phys. Lett.* **B69** (1977) 151.
28. J.L. Romero *et al.*, *Phys. Rev.* **C25** (2005) 2214; J.W. Watson *et al.*, *ibid.* 2219.

29. G. Weisel *et al.*, private communication
30. H. Dobiasch *et al. Phys. Lett.* **B76** (1978) 195.
31. J.C. Faivre *et al. Nucl. Phys.* **A127** (1969) 169.
32. S.J. Hall, A.R. Johnston, R.J. Griffiths, *Phys. Lett.* **14** (1965) 212.
33. A.R. Johnson *et al. Phys. Lett.* **19** (1965) 289.
34. S.N. Bunker *et al. Nucl. Phys.* **A113** (1968) 461.
35. A. Deltuva and A.C. Fonseca, private communication.
36. A. Kievsky, *Phys. Rev.* **C60** (1999) 034001.
37. L. Canton and W. Schadow, *Phys. Rev.* **C64** (2001) 031001(R).
38. E. Epelbaum, *arXiv:nuc-th/0509032*, v1 (14 Sept. 2005).
39. M. Viviani *et al.*, *Phy. Rev. Lett.* **86** (2001) 3739.
40. A. Deltuva and A.C. Fonseca, *Phys. Rev. Lett.* **98** (2007) 162502.
41. A. Deltuva and A.C. Fonseca, *Phys. Rev.* **C76** (2007) 021001(R).
42. J.H. Esterline *et al.*, private communication.

Remaining problems in the low-energy *pd* breakup reaction

K. Sagara, T. Sueta, T. Sugimoto, H. Shimoda, S. Kuroita and Y. Maeda

Department of Physics, Kyushu University, Hakozaki, Fukuoka 812-8581, Japan

Abstract. Space star anomalies in *nd* and *pd* breakup reactions are peculiar phenomena, different from other discrepancies in the 3N systems. Recently progress on the *pd* star anomaly has been made both in experiment and in theoretical calculations. We have started systematic experiments on the *pd* star, and some preliminary data are presented.

Keywords: Few-body systems
PACS: 21.30.-x,21.45.-v,21.45.Ff

INTRODUCTION

In 3-nucleon (3*N*) breakup reactions, the kinematic condition in which the 3N go out forming vertexes of an equilateral triangle is called a star configuration. When the equilateral triangle is perpendicular to a beam axis, the condition is called the "space star configuration" (SS). In the star configuration, the relative energies between two any outgoing nucleons have the same value. A final-state interaction (FSI) between two outgoing nucleons is strongest when their relative energy is zero, and the FSI becomes weaker as the relative energy increases. Therefore, the star lies farthest from the FSI peaks, and cross sections in the star configuration are small and sensitive to anomalous phenomena.

First, the *nd* breakup cross section in the SS at E_n = 13 MeV was found to be about 25% higher than predicted by Faddeev calculations[1], and the disagreement was called the SS anomaly. Another experiment for *nd* breakup at E_n = 13 MeV confirmed the SS anomaly[2], although the data disagree a little with previous measurements. Cross sections for *nd* breakup in the SS have been measured at several energies. The SS anomaly in *nd* breakup is most prominent at 13 MeV and decreases with energy till it vanishes around 30 MeV.

Cross sections for *pd* breakup in the SS also have been measured at several energies. Contrary to the *nd* case, the measured *pd* breakup cross section in the SS at 13 MeV is about 15% smaller than the calculated *nd* breakup cross section[3], and the disagreement exists in a wide angular region around the SS[4]. Recently Deltuva et al. have succeeded in performing *pd* breakup calculations with a screened Coulomb force[5], and have found that Coulomb effects decrease the breakup cross section in the SS by only a few percent and effects of the Δ-isobar are very small. Thus, the SS anomaly does exist also in *pd* breakup.

The SS anomaly is a curious phenomenon. First, it is remarkably charge asymmetric; the *nd* cross section and the *pd* cross section in the SS are larger and smaller, respec-

CP1011, *New Facet of Three Nucleon Force—50 Years of Fujita Miyazawa Three Nucleon Force (FM50), The International Symposium,* edited by H. Sakai, K. Sekiguchi, and B. F. Gibson

tively, than corresponding theoretical calculations. Second, at present we know of no possible origins for the SS anomaly. Below 30 MeV, effects of the 2π3NF are very small and effects of any other heavy-meson 3NF may be even smaller; relativistic effects may be negligible; and a sufficient number of angular-momentum partial waves are included in the 3N calculations below 30 MeV to insure accuracy. In the same energy region, there is also the long-standing A_y puzzle. The A_y puzzle has not been solved yet, but at least it is known that the A_y puzzle has some relation with P-wave interactions.

Recently the Köln group performed a *pd* breakup experiment at E_d = 19 MeV (equivalent to E_p = 9.5 MeV) in star configurations in which the equilateral triangle was inclined to the beam axis, and they found that the *pd* breakup cross section is about 25% smaller than the *pd* calculations[6]. We made an inclusive measurement of the $D(p,p)pn$ cross section at 13 MeV to see the global features of the breakup cross section, and found that the data agree well with *pd* calculations[7]. This indicates that star anomaly is a localized and minor discrepancy.

We then started a systematic experiment for the *pd* breakup cross section in inclined star configurations. A few preliminary experiments have already been made.

DEFINITION OF THE INCLINATION ANGLE OF THE STAR, α

Consider the $D(p,p_1p_2)n$ breakup reaction in the star configuration in the c.m. system, as shown in Fig. 1. We define the Cartesian coordinates by taking the z-axis along the p-beam, and the x-axis in the star plane on which p_1, p_2 and n form an equilateral triangle. We call the sum of the momenta of the two outgoing protons $\vec{p}_{1+2}$. In our star experiment $\vec{p}_{1+2}$ is set in the yz-plane, and the inclination angle of the star, α, is defined as the angle between $\vec{p}_{1+2}$ and the z-axis. As far as we know, all the star experiments have been performed so far with $\vec{p}_{1+2}$ in the yz-plane.

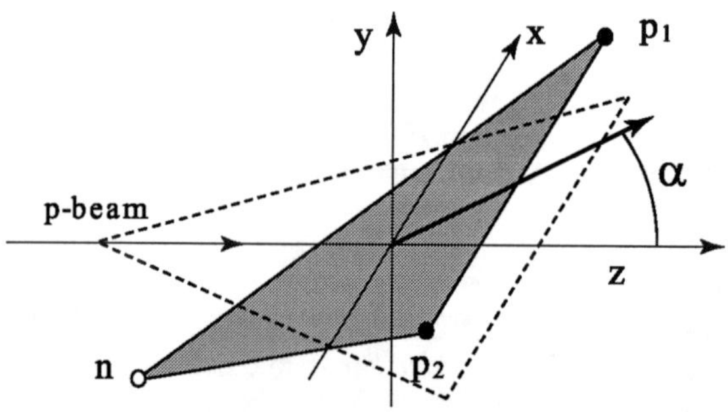

FIGURE 1. A star configuration inclined by an angle α for the $D(p,p_1p_2)n$ reaction in the c.m. system.

An $H(d,p_1p_2)n$ experiment using a d-beam and a $D(p,p_1p_2)n$ experiment using a p-beam are the same in the c.m. system, and we always take the z-axis along the p-beam in the c.m. system, even if a d-beam is in practice used in the experiment. In our definition, the $H(d,pp)n$ experiment in Köln[6] was made in a range of α = 124°-180°.

If the two restrictions on $\vec{p}_{1+2}$ lying in yz-plane and the x-axis being in the star plane are discarded, the star triangle can rotate, for example, around the normal to the star plane, around $\vec{p}_{1+2}$, or around $\vec{p}_1$. When we perform experiments in such rotated configurations, we need to define additionally an "azimuthal" angle.

PRELIMINARY EXPERIMENT FOR AN INCLINED STAR

A proton beam from the Kyushu University tandem accelerator was incident on a CD_2 target of about 0.3-0.6 mg/cm^2 in thickness, and two the protons from the $D(p,p_1p_2)n$ reaction were detected in coincidence by Si detectors in an inclined star configuration. The beam intensity was 120 nA. The CD_2 target foil was moved by about 3mm every several hours before the D content of the target decreased to 70%.

The energy (E) and the arrival time (T) of particles were measured by each detector. Assuming that detected particles were protons, time-of-flight from the target to the detector (TOF) was calculated from the energy for each particle. In a two-dimensional time spectrum of (T_1-T_2) vs. (TOF_1-TOF_2), true coincidence protons formed a clear line. Using the time spectrum, accidentally coincident events were estimated and subtracted from the total (true + accidental) events to obtain true events.

A monitor detector was placed at 40° to count the number of protons scattered by the target deuterons. From the proton counts and precise knowledge of the *pd* scattering cross section[8], we obtained absolute values for the *pd* breakup cross section.

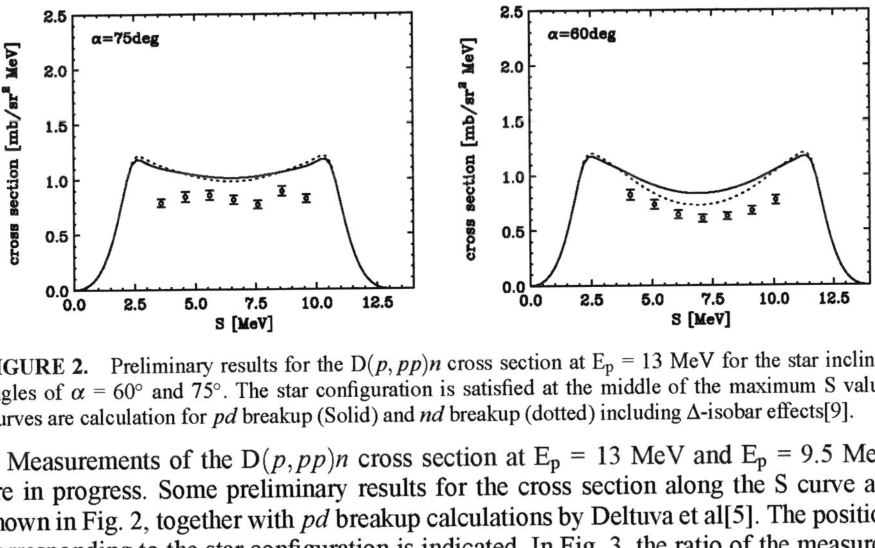

FIGURE 2. Preliminary results for the $D(p,pp)n$ cross section at E_p = 13 MeV for the star inclined angles of α = 60° and 75°. The star configuration is satisfied at the middle of the maximum S value. Curves are calculation for *pd* breakup (Solid) and *nd* breakup (dotted) including Δ-isobar effects[9].

Measurements of the $D(p,pp)n$ cross section at E_p = 13 MeV and E_p = 9.5 MeV are in progress. Some preliminary results for the cross section along the S curve are shown in Fig. 2, together with *pd* breakup calculations by Deltuva et al[5]. The position corresponding to the star configuration is indicated. In Fig. 3, the ratio of the measured cross section to the calculated one for the star at 13 MeV and 9.5 MeV is displayed, together with the previous data from KUTL and Köln. At present, we have nothing to say about the α dependence of the cross section ratio (anomaly).

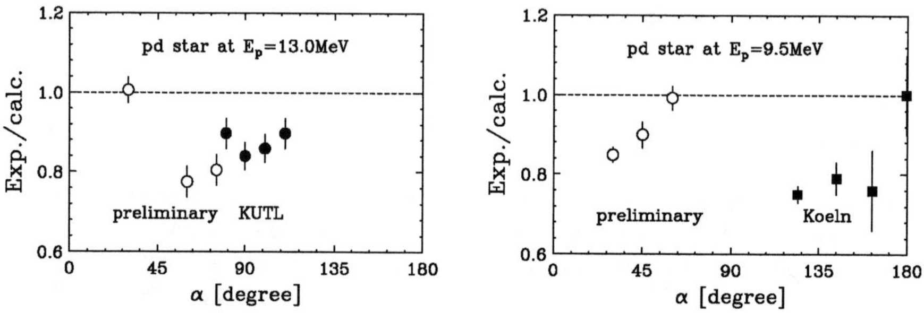

FIGURE 3. Ratio of the measured D$(p,pp)n$ cross section to the calculated one[9] at the star inclined angle α. Open circles, closed circles, and closed rectangles show present preliminary data, KUTL data[4], and Köln data[6].

PLAN FOR SYSTEMATIC EXPERIMENTS FOR THE STAR CONFIGURATION

Measurements of the D$(p,pp)n$ cross section at E_p = 13 MeV and 9.5 MeV in a wide range of inclination angles, $\alpha = 0°$-$120°$, are in progress at KUTL. Experimental methods have been established. Data taking needs significant beam time due to a large number of accidental coincidence events. The experiments will be completed in 2008.

We have another plan of measurement of the H$(d,pp)n$ cross section at E_d = 26 MeV for a wide range of inclination angles α of 105°-180°. The experimental proposal has been approved, and the experiment will be run in 2008. A rotating foil (CH_2) target is being developed.

PROSPECTS

Systematic data on the α-dependence of the star anomaly in the *pd* breakup cross section will be obtained in 2008. The database will be helpful to clarify characteristics of the star anomaly and to understand the origin of the star anomaly. Recent success of *pd* breakup calculations[5] has motivated an extensive investigation of the star anomaly using precise and reliable *pd* data.

We took 40 years to confirm Fujita-Miyazawa type of 3NF. We have now three kinds of remaining problems in the 3N system, $\pi\rho$3NF and $\rho\rho$3NF, the A_y puzzle, and the star anomaly. For investigation of $\pi\rho$3NF and $\rho\rho$3NF, extensive data of *pd* scattering, *pd* breakup, and *pd* capture are being acuumulated at intermediate energies. The data set for the A_y puzzle is almost complete. Accumulation of systematic low-energy data on the star anomaly has just started.

The authors thank Dr. A. Deltuva for his *pd* breakup calculations.

REFERENCES

1. J. Strate et al., Nucl. Phys. A **501**, 51 (1989).

2. H. R. Setze et al., Phys. Rev. C. **71**, 034006 (2005).
3. G. Rauprich et al., Nucl. Phys. **A535**, 313 (1991).
4. T. Ishida et al., Mod. Phys. Letters **A18**, 436 (2003).
5. A. Deltuva et al., Phys. Rev. C **71**, 054005 (2005).
6. J. Ley et al., Phys. Rev. C **73**, 064001 (2006).
7. S. Kuroita et al., this symposium's talk.
8. K. Sagara et al., Phys. Rev. C **50**, 576 (1994).
9. A. Deltuva, private communication.

Measurements of inclusive *pd* breakup cross sections at low and intermediate energies

S. Kuroita*, K. Sagara*, T. Kudoh*, T. Sueta*, H. Shimoda*, T. Sugimoto*, Y. Yamada*, T. Wakasa*, T. Noro*, Y. Tameshige†, A. Tamii† and K. Hatanaka†

*Department of Physics, Kyushu University, Hakozaki, Fukuoka 812-8581, Japan
†Research Center for Nuclear Physics, Osaka University, Ibaraki, Osaka 567-0047, Japan

Abstract. Inclusive experiments for D$(p,p)pn$ breakup were performed at incident proton energies of 247 MeV and 13 MeV. At 247 MeV, the cross sections and analyzing power A_y measured at angles from 7° to 20° largely disagree with calculations even if a 2π-exchange three-nucleon force (2π3NF) effects are included, and the disagreement increases at forward angles. At 13 MeV, cross section measured at angles from 10° to 70° agrees well with *pd* calculation and 2π3NF effects are small. The large disagreement in *pd* breakup at 247 MeV may have some relationship with disagreements in *pd* scattering at 250 MeV.

Keywords: Few-body systems
PACS: 21.30.-x,21.45.-v,21.45.Ff

INTRODUCTION

Experiments on *Nd* scattering have been performed extensively and the existence of a 2π-exchange three-nucleon force (2π3NF), such as that first proposed by Fijita and Miyazawa in 1957 [1], has been confirmed from discrepancies between calculations without a 3NF and measurements of the ^{3}H binding energy and the cross section minimum in *Nd* scattering[2] (the Sagara discrepancy[4]). After a 2π3NF is included in 3N calculations, there still remain discrepancies with *Nd* scattering, such as the A_y puzzle below 30 MeV, multiple disagreements in the analyzing powers above 50 MeV, and discrepancies in the cross section at 250 MeV.

To investigate the origins of these discrepancies, experiments in other 3N reaction channels, i.e., in *Nd* breakup and in *Nd* capture, may be necessary. We started experiments for *pd* breakup in a wide energy range. Exclusive D$(p,pp)n$ or H$(d,pp)n$ experiments are useful for detailed study of discrepancies, but they need much time. For the first step, we started inclusive D$(p,p)pn$ experiments to see global features of *pd* breakup. The *pd* breakup experiments were performed at proton beam energies of 247 MeV and 13 MeV. The space star anomaly in *nd* breakup[5] is most prominent at 13 MeV and a large discrepancy in *Nd* scattering has been reported at 250 MeV[6, 7].

CP1011, *New Facet of Three Nucleon Force—50 Years of Fujita Miyazawa Three Nucleon Force (FM50), The International Symposium,* edited by H. Sakai, K. Sekiguchi, and B. F. Gibson

EXPERIMENT AT 247 MEV

A polarized proton beam of 247 MeV from a ring cyclotron of the Research Center for Nuclear Study (RCNP) was incident on a liquid deuterium target of about 150 mg/cm^2 in thickness having a window of $4.4\mu m$ thick aramid foil. Continuum-energy protons from *pd* breakup as well as protons from *pd* scattering were detected by the large acceptance spectrometer (LAS). The magnetic field strength of LAS was changed three or four times to detect protons in a wide energy range from 75 MeV to 250 MeV. The thickness of the the liquid deuterium target was monitored by detecting *p* and *d* from *pd* scattering in coincidence in the vertical plane. We prepared a CD_2 foil target and measured its thickness by using precisely-known cross section for *pd* scattering at 12 MeV[3]. The absolute thickness of the liquid target was determined using a CD_2 target. The cross section for *pd* scattering at 247 MeV was simultaneously measured and the result agreed well with previous data[6]. Polarization of the *p*-beam was measured using *pp* scattering during the experiment.

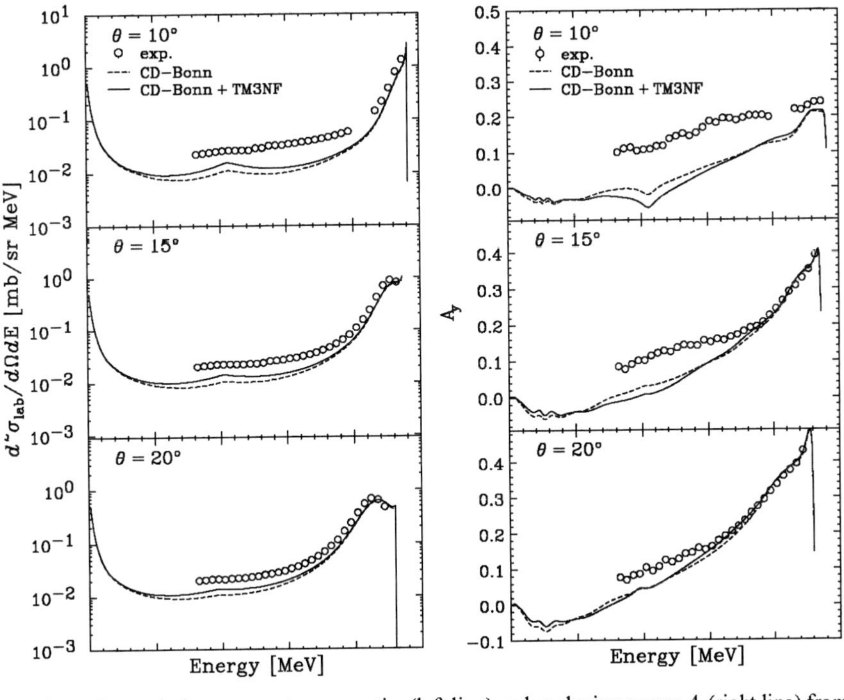

FIGURE 1. Open circles represent cross section(left line) and analyzing power A_y(right line) from the $D(\vec{p},p)pn$ reaction at 247 MeV. Solid and dashed curves stand for 3N calculations using CD-Bonn 2NF with and without TM3NF[8], respectively.

The window foil of the target produced continuum proton backgrounds, which were measured by decreasing the D_2 target density to about 1/1000 by heating the target up to 80K. The background level was sufficiently low owing to use of the liquid target.

The cross section and analyzing power A_y for the $D(p,p)pn$ reaction were measured at 7°, 10°, 15° and 20°. Experimental results at 10°-20° are shown in Fig. 1 with *nd* calculations performed with and without 2π3NF by Witala[8]. In Fig. 2 the experimental results at 10° are compared with *nd* and *pd* calculations with and without Δ-effects by Deltuva[9].

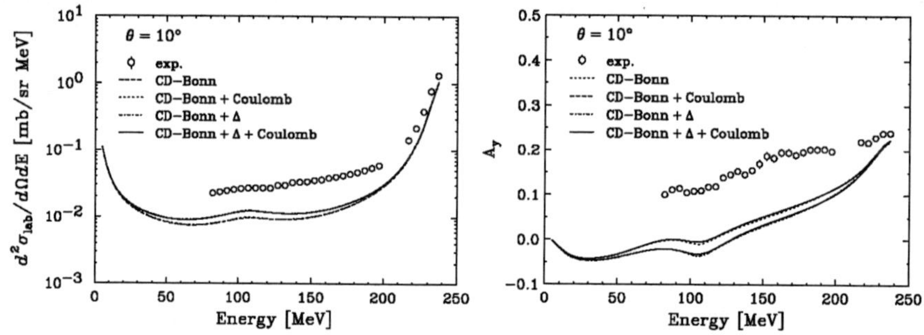

FIGURE 2. Open circles represent cross section(left) and analyzing power A_y(right) for the $D(\vec{p},p)pn$ reaction at 247 MeV. Solid and dot-dash curves show *pd* calculations using CD-Bonn 2NF with and without Δ-isobar, and dash and dot curves show *nd* calculations using CD-Bonn 2NF with and without Δ-isobar[9], respectively.

The experimental cross section is a few times larger than the calculated values. The large disagreement is seen also in A_y. Inclusion of a 2π3NF or Δ-effects reduces the disagreement in the cross section but the effects are too small to reproduce the data. Inclusion of a 2π3NF or Δ-effects increased the disagreement in A_y a little. Coulomb effects are very small at this high energy. Disagreements in cross section and A_y become larger at forward angles. A possible cause of the disagreement may be 3NF components other than the 2π3NF.

EXPERIMETS AT 13 MEV

At Kyushu University tandem accelerator laboratory (KUTL), the cross section for the $D(p,p)pn$ reaction was measured at 13 MeV in a wide angular range, 10°-70°. A 13 MeV proton beam was incident on a D_2 gas target and continuum protons were detected using ΔE-E Si-detectors and a double-slit system. The ΔE-E system identified *p*, *d* and heavier particles.

The target region was defined by the double slit and the beam path, and the target was 99.75% pure D_2 gas and contaminant gas. Proton backgrounds (BG) were produced by beam halo, multiple scattering on cell walls and slit edges, contaminants in the target, and so on. Beam halo and multiple scattering were reduced by a well designed gas cell and a beam slit system, and the BG level was decreased to below 50% even at forward angles. The remaining BG was carefully estimated from proton spectra taken with vacuum, H_2, He and air targets.

Recoiling deuterons from *pd* scattering were detected simultaneously with protons from *pd* breakup. Using precise *pd* scattering cross section measured at KUTL[3], the absolute value of *pd* breakup cross sections was obtained without using the solid angle of the detector, detection efficiency, and an integrated beam charge.

Experimental data at 10°, 20°, 40° and 60° are shown in Fig. 3 with calculations by Deltuva[9]. Experimental data are well reproduced by *pd* calculations. Coulomb effects are large at this low energy and effects of Δ-isobar (also of 2π3NF) are very small.

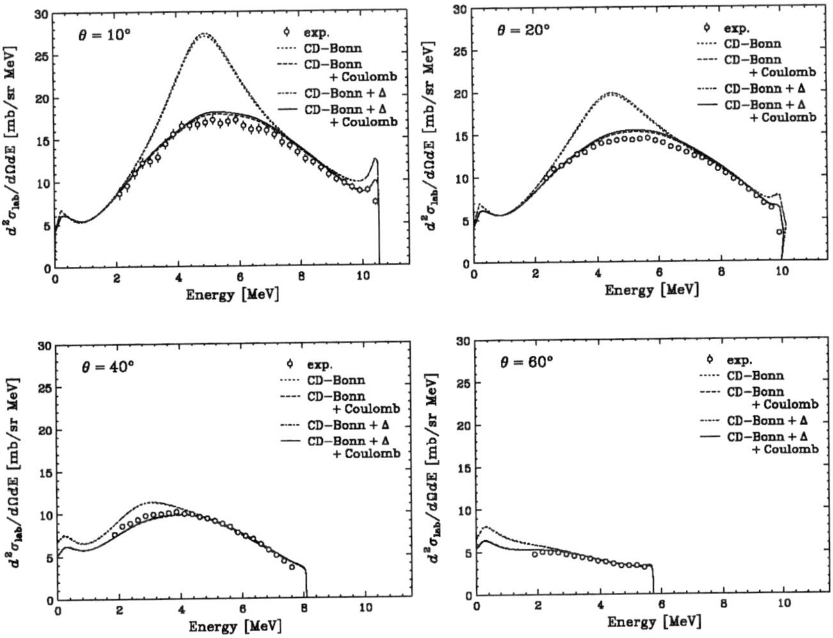

FIGURE 3. Open circles represent the cross section at 13 MeV. Solid and dot-dash curves show *pd* calculations using the CD-Bonn 2NF with and without a Δ-isobar, and dash and dot curves show *nd* calculations using the CD-Bonn 2NF with and without the Δ-isobar[9], respectively.

SUMMARY AND DISCUSSION

Global features of *pd* breakup have been studied by means of an inclusive measurement of the $D(p,p)pn$ reaction at 13 MeV and 247 MeV. At 13 MeV, the inclusive cross sections agree well with *pd* calculations. It indicates that the space star anomaly is a localized phenomenon and no global anomaly exists in *pd* breakup cross sections at low energy. This is consistent with the fact that the cross section from *pd* scattering agree with *pd* calculations including a 2π3NF.

At 247 MeV, the inclusive cross section and analyzing power A_y for *pd* breakup largely disagree with *pd* and *nd* calculations, even if a 2π3NF and the Coulomb force are included. The cross section and A_y for *pd* elastic scattering at 250 MeV also largely disagree with calculations at backward angles.

One possible origin of these phenomena in the $\vec{p}d$ system may be a short-range 3NF which is not included in the calculations. Search for the 3NF origins is an interesting subject. It is especially interesting to see whether disagreements in $\vec{p}d$ breakup and in $\vec{p}d$ elastic scattering appear at the same energy, and also to see whether the disagreement varies around the π threshold of about 210 MeV. The energy dependence of the disagreement will be studied in the near future.

ACKNOWLEDGMENTS

The authors thank Prof. H. Witala and Dr. A. Deltuva for their 3N calculations to compare with the present data.

REFERENCES

1. J. Fujita and H. Miyazawa, Prog. Theor. Phys. **17**, 260 (1957).
2. H. Witala et al., Phys. Rev. Lett. **81**, 1183 (1998).
3. K. Sagara et al., Phys. Rev. C **50**, 576 (1994).
4. Y. Koike and S. Ishikawa, Nucl. Phys. A **631**, 683c (1998).
5. J. Strate et al., Nucl. Phys. A **501**, 51 (1989).
6. K. Hatanaka et al., Phys. Rev. C **66**, 044002 (2002).
7. Y. Maeda et al., Phys. Rev. C **76**, 014004 (2007).
8. H. Witala, private communication.
9. A. Deltuva, private communication.

$N-d$ Elastic Scattering at Low Energy with Realistic Interactions

L.E. Marcucci, L. Girlanda, A. Kievsky, S. Rosati, M. Viviani

Department of Physics "E. Fermi", University of Pisa, I-56127 Pisa, Italy
INFN-Sezione di Pisa, I-56127 Pisa, Italy

Abstract. The structure of $N-d$ low-energy scattering states is described using the hyperspherical harmonics method with realistic Hamiltonian models, consisting of two- and three-nucleon interactions. Both coordinate and momentum space two-nucleon potential models are considered. Results for the zero-energy $N-d$ scattering lengths and for the $p-d$ phase shift and mixing angle parameters at center-of-mass energy of 2 MeV are presented and discussed.

Keywords: Few-nucleon systems, hyperspherical-harmonics method, low-energy elastic scattering
PACS: 21.45.-v,21.60.De,27.10.+h

INTRODUCTION

One of the main ingredients necessary to study few-body nuclear systems is a realistic description of the nuclear interaction. A number of nucleon-nucleon (NN) potentials has been determined in recent years. They all reproduce the deuteron binding energy and fit a large set of NN scattering data below the pion-production threshold with a χ^2/datum$\simeq$1. Among these potentials, we will consider in the present study only the "phenomenological" model of Ref. [1] (AV18), and a model based on chiral symmetry derived in Ref. [2] (N3LO-Idaho). Among the many features of these models, we note only that the AV18 is a local NN potential model, expressed in coordinate space, with a strong short-range repulsion and tensor component. On the other hand, the N3LO-Idaho model is non-local and explicitly given in momentum space, and has a softer short-range repulsion and a tensor component than the AV18.

As a consequence of these different features, it is interesting to test these potential models by studying light nuclear systems. In these systems, a further contribution to the realistic nuclear Hamiltonian model comes from the three-nucleon interaction (TNI). Several models of TNI's have been proposed. They are mainly based on the exchange of pions among the three nucleons, as the Urbana IX (UIX) TNI [3], which will be considered in the present study. The more recent TNI models studied within the chiral approach [4, 5] and the extension of the UIX model known as the Illinois TNI [6] will be considered in the near future.

A second crucial ingredient in the study of light nuclear systems is the technique used to solve the A-body Schrödinger equation. Several methods have been developed in the past years (see Refs. [7, 8] for a review). Among them, we consider here the technique based on the expansion of the nuclear wave function on an appropriate basis of hyperspherical-harmonic functions (HH), known as the HH method.

The paper is organized as follows: in the next section some details on the formalism

CP1011, *New Facet of Three Nucleon Force—50 Years of Fujita Miyazawa Three Nucleon Force (FM50), The International Symposium,* edited by H. Sakai, K. Sekiguchi, and B. F. Gibson

are discussed, then the results for the $n-d$ and $p-d$ scattering lengths and low-energy $p-d$ phase shift and mixing angle parameters are presented and compared with the available experimental data. Some concluding remarks follow at the end.

THE HYPERSPHERICAL HARMONICS METHOD

The nuclear wave function for an A-body system can be written as

$$|\Psi\rangle = \Sigma_\mu c_\mu |\Phi_\mu\rangle \,, \tag{1}$$

where $|\Phi_\mu\rangle$ is a suitable complete set of states, and μ is an index denoting the set of quantum numbers necessary to completely specify the basis elements. In the present work, the functions $|\Phi_\mu\rangle$ have been written in terms of HH functions, and in configuration space they are given by [9, 10]:

$$\langle \mathbf{x}_1,\ldots,\mathbf{x}_N | \Phi_\mu \rangle = f_l(\rho) \mathscr{Y}_{\{G\}}(\Omega^{(\rho)}) \,, \tag{2}$$

where $\mathbf{x}_i$ are a set of $N = A-1$ Jacobi vectors and $\mathscr{Y}_{\{G\}}(\Omega^{(\rho)})$ are properly antisymmetrized products of HH functions and spin-isospin states. In Eq. (2), ρ is the so-called hyperradius, G the grand angular momentum quantum number, and $\Omega^{(\rho)}$ is a set of angular-hyperangular coordinates. ρ and $\Omega^{(\rho)}$ are defined as:

$$\rho = \sqrt{\sum_{i=1}^{N} \mathbf{x}_i^2} \,, \qquad \Omega^{(\rho)} = [\hat{\mathbf{x}}_1, \hat{\mathbf{x}}_2, \cdots, \hat{\mathbf{x}}_N; \phi_2, \cdots, \phi_N] \,, \tag{3}$$

$$\tan\phi_i = \frac{1}{x_i}\sqrt{\sum_{j=1}^{i-1} x_j^2} \,, \qquad i = 2,\ldots,N \,. \tag{4}$$

Finally, the suffix (ρ) recalls the use of the coordinate space and the functions $f_l(\rho)$ are a suitable set of basis functions [10].

Using the properties of the HH functions [9, 10], it is easy to obtain the momentum space version of the states $|\Phi_\mu\rangle$, namely

$$\langle \mathbf{k}_1,\ldots,\mathbf{k}_N | \Phi_\mu \rangle = g_{G,l}(Q) \mathscr{Y}_{\{G\}}(\Omega^{(Q)}) \,, \tag{5}$$

where $\mathscr{Y}_{\{G\}}(\Omega^{(Q)})$ is the same as $\mathscr{Y}_{\{G\}}(\Omega^{(\rho)})$ with $\mathbf{x}_i \to \mathbf{k}_i$, $\mathbf{k}_i$ being the Jacobi momenta, and $Q = \sqrt{\Sigma_{i=1}^N \mathbf{k}_i^2}$. Finally:

$$g_{G,l}(Q) = (-i)^G \int_0^\infty d\rho \, \frac{\rho^{3N-1}}{(Q\rho)^{3N/2-1}} J_{\mathscr{L}+1/2}(Q\rho) f_l(\rho) \qquad \mathscr{L} = G + (3N-3)/2 \,. \tag{6}$$

This last one-dimensional intergal can be evaluated analytically for the chosen $f_l(\rho)$ functions [10].

The unknown coefficients c_μ of Eq. (1) can be obtained by applying the Rayleigh-Ritz (Kohn) variational principle for the bound (scattering) state problem. The matrix elements of the different operators of the Hamiltonian are calculated, working in co-ordinate or in momentum space depending on what is more convenient, i.e. the matrix elements of the local operators are calculated using the expansion of Eq. (2), while the non-local NN interaction matrix elements are calculated using Eqs. (5) and (6). Then, the problem is reduced to an eigenvalue-eigenvector problem (system of algebraic linear equations), which can be solved with standard numerical techniques [11, 12].

RESULTS

The $n-d$ and $p-d$ doublet and quartet scattering lengths obtained with the non-local N3LO-Idaho [2] NN interaction, with or without the inclusion of the UIX TNI [3], are given in Table 1, and compared with the available experimental data [13]. Also shown are the results obtained with the local AV18 [1] NN interaction and the AV18/UIX potential model for a comparison [14]. Note that in the case of the N3LO-Idaho/UIX model, the parameter in front of the spin-isospin independent part of the UIX TNI has been rescaled by a factor of 0.384 to fit the triton binding energy. In fact, the triton, ^{3}He, and ^{4}He binding energies obtained with the AV18, AV18/UIX, N3LO-Idaho, and N3LO-Idaho/UIX potential models are listed in Table 2 and can be compared with the experimental data. A few remarks are in order: (i) the N3LO-Idaho potential model is decomposed into partial waves, and the partial wave decomposition is truncated at a certain value of the two-body total angular momentum j_{max}. In the present work j_{max} = 6 has been chosen, which allows for an accuracy of better than 1 keV in the triton binding energy. (ii) The N3LO-Idaho results for the ^{3}He and ^{4}He binding energies are obtained including the pure Coulomb electromagnetic interaction, while the AV18 results are obtained including in the electromagnetic interaction terms up to order α^2, α being the fine structure constant. (iii) The $T = 3/2$ ($T = 1$ and 2) states are included in the A=3 (4) calculations. (iv) The binding energy results are accurate at the 1 (10) keV level for the A=3 (4) nuclei. Similarly, the N3LO-Idaho and N3LO-Idaho/UIX results shown in Table 1 are accurate at the 10^{-3} fm level. In fact, the convergence of the HH expansion has been tested with the procedure of Ref. [10].

From inspection of Table 1 we can conclude that: (i) both the $n-d$ and $p-d$ quartet scattering lengths are not very model-dependent. Also, they are not affected by the inclusion of the TNI. The trend shown by the AV18 and AV18/UIX results has been found also in the case of the non-local N3LO-Idaho and N3LO-Idaho/UIX potential models. (ii) The $n-d$ doublet scattering length $^2a_{nd}$ is very sensitive to the choice of the NN potential model, when no TNI is included. However, once the TNI is included, and therefore the triton binding energy is well reproduced, $^2a_{nd}$ becomes model-independent. This is a well-known feature, related to the fact that $^2a_{nd}$ and the triton binding energy are linearly correlated. Such linear correlation, known as the Phillips line [15], can be seen in Fig. 1, where the triton binding energy is plotted as a function of $^2a_{nd}$ for the potential models considered. Note that the N3LO-Idaho/UIX and AV18/UIX data are on top of each other. (iii) The $p-d$ doublet scattering length $^2a_{pd}$

TABLE 1. Doublet and quartet $n-d$ and $p-d$ scattering lengths, in fm, obtained with the AV18, AV18/UIX, N3LO-Idaho, and N3LO-Idaho/UIX potential models. The experimental $n-d$ scattering lengths are taken from Ref. [13], while the AV18 and AV18/UIX results are taken from Ref. [14]. In the case of the N3LO-Idaho/UIX potential model, the spin-isospin independent part of the UIX TNI has been rescaled by a factor of 0.384.

	AV18	AV18/UIX	N3LO-Idaho	N3LO-Idaho/UIX	Exp.
${}^2a_{nd}$ (fm)	1.27	0.63	1.100	0.623	0.65(4)
${}^4a_{nd}$ (fm)	6.33	6.33	6.342	6.343	6.35(2)
${}^2a_{pd}$ (fm)	1.17	-0.02	0.862	-0.007	
${}^4a_{pd}$ (fm)	13.6	13.7	13.646	13.647	

TABLE 2. Triton, ^{3}He and ^{4}He binding energies, in MeV, obtained with the AV18, AV18/UIX, N3LO-Idaho, and N3LO-Idaho/UIX potential models.

Potential Model	^{3}H	^{3}He	^{4}He
AV18	7.624	6.925	24.210
N3LO-Idaho	7.854	7.128	25.380
AV18/UIX	8.479	7.750	28.462
N3LO-Idaho/UIX	8.481	7.730	28.534
Experiment	8.482	7.718	28.30

is positive and quite model-dependent, if only the two-nucleon interaction is included. Once the TNI is added, ${}^2a_{pd}$ becomes very small and negative. Some model-dependence remains, but the problem of extrapolating the experimental results to zero energy makes very difficult any meaningful comparison between theory and experiment.

Some preliminary results for $p-d$ elastic scattering at a center-of-mass energy of 2 MeV are given in Table 3. In particular, the J^π=1/2$^+$, 1/2$^-$, 3/2$^+$ and 3/2$^-$ phase shift and mixing angle parameters obtained using the N3LO-Idaho and N3LO-Idaho/UIX potential models, are compared with the corresponding AV18 and AV18/UIX results of Ref. [14]. From inspection of the table we can observe that the N3LO-Idaho and N3LO-Idaho/UIX parameters follow the same trend of the corresponding AV18 and AV18/UIX ones. In particular, the inclusion of the TNI produces sizeable effects only on the J^π=1/2$^+$ parameters, and gives less important or negligible effects for the other parameters. Some differences between the N3LO-Idaho and AV18 results exist, but a meaningful comparison will be made only once all the important partial waves are included and the corresponding scattering observables are calculated. Work in this direction is currently underway [16].

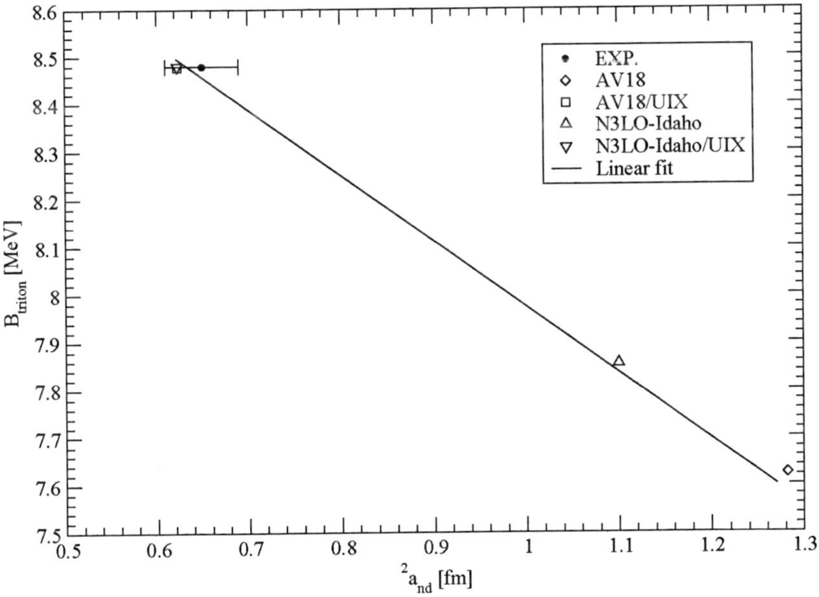

FIGURE 1. The triton binding energy (MeV) as a function of the $n-d$ doublet scattering length ${}^2a_{nd}$ (fm) of Table 1. The straight line (Phillips line) is a linear χ^2 fit.

CONCLUSIONS

The $N-d$ low-energy scattering problem has been studied with local and non-local NN interaction models using the HH method. The variational wave function is expanded over a complete set of basis functions, which can be expressed either in coordinate or in momentum space, depending on which is more convenient. The inclusion of the three-nucleon and Coulomb interaction does not represent any additional difficulty.

The results for the A=3 doublet and quartet scattering lengths obtained using the non-local N3LO-Idaho two-nucleon and N3LO-Idaho/UIX two- and three-nucleon potential models have been presented and discussed. A similar investigation for the A=4 scattering lengths has been reported in Ref. [17].

The results for the low (non-zero) energy $p-d$ elastic scattering phase shift and mixing angle parameters obtained within the HH method using the non-local N3LO-Idaho and N3LO-Idaho/UIX interaction models have also been presented and compared with the results obtained with the AV18 and AV18/UIX potentials [14]. Only the J^π=$1/2^+$, $1/2^-$, $3/2^+$ and $3/2^-$ states have been here considered, but further work to study all the

TABLE 3. $p-d$ elastic scattering phase shifts and mixing angles (in degrees) obtained from the N3LO-Idaho and N3LO-Idaho/UIX potential models at center-of-mass energy of 2 MeV. Also shown are the corresponding results obtained with the AV18 and AV18/UIX results given in Ref. [14].

J^π	(S,L)	AV18	AV18/UIX	N3LO-Idaho	N3LO-Idaho/UIX
$1/2^+$	$\delta(3/2,2)$	-3.56	-3.57	-3.569	-3.563
	$\delta(1/2,0)$	-32.2	-27.8	-31.01	-27.81
	η	1.10	1.47	1.242	1.552
$1/2^-$	$\delta(1/2,1)$	-7.36	-7.37	-7.340	-7.356
	$\delta(3/2,1)$	22.1	22.3	22.05	22.19
	ε	5.71	5.83	5.819	5.857
$3/2^+$	$\delta(3/2,0)$	-63.1	-63.1	-63.20	-63.20
	$\delta(1/2,2)$	2.15	2.15	2.161	2.161
	$\delta(3/2,2)$	-3.83	-3.83	-3.847	-3.847
	ε	0.800	0.802	0.823	0.826
	ξ	1.30	1.30	1.322	1.325
	η	-0.322	-0.316	-0.326	-0.325
$3/2^-$	$\delta(3/2,3)$	0.849	0.850	0.859	0.859
	$\delta(1/2,1)$	-7.14	-7.15	-7.101	-7.103
	$\delta(3/2,1)$	24.2	24.2	24.337	24.374
	ε	-2.20	-2.23	-2.265	-2.305
	ξ	-0.321	-0.314	-0.322	-0.318
	η	-3.11	-3.11	-3.213	-3.217

$L \leq 6$ states, L being the $p-d$ relative angular momentum, is currently underway [16].

REFERENCES

1. R.B. Wiringa, V.G.J. Stoks, and R. Schiavilla, *Phys. Rev. C* **51**, 38 (1995)
2. D.R. Entem, and R. Machleidt, *Phys. Rev. C* **68**, 041001 (2003)
3. B.S. Pudliner *et al.*, *Phys. Rev. C* **56**, 1720 (1997)
4. E. Epelbaum *et al.*, *Phys. Rev. C* **66**, 064001 (2002)
5. P. Navratil, *Few-Body Sist.* **41**, 117 (2007)
6. S.C. Pieper *et al.*, *Phys. Rev. C* **64**, 014001 (2001)
7. W. Gloeckle *et al.*, *Phys. Rep.* **274**, 107 (1996)
8. J. Carlson, and R, Schiavilla, *Rev. Mod. Phys.* **70**, 743 (1998)
9. M. Fabre de la Ripelle, *Ann. Phys.* **147**, 281 (1983)
10. M. Viviani *et al.*, *Few-Body Syst.* **39**, 159 (2006)
11. A. Kievsky *et al.*, *Nucl. Phys. A* **624**, 125 (1997)
12. A. Kievsky *et al.*, *Few-Body Syst.* **22**, 1 (1997)
13. W. Dilg, L. Koester, and W. Nistler, *Phys. Lett.* **36B**, 208 (1971)
14. A. Kievsky, M. Viviani, S. Rosati, *Phys. Rev. C* **52**, R15 (1995)
15. A.C. Phillips, *Rep. Prog. Phys.* **40**, 905 (1997)
16. L.E. Marcucci *et al.*, in preparation (2008)
17. M. Viviani *et al.*, *Nucl. Phys. A* **790**, 46c (2007)

Structure of four-baryon systems

E. Hiyama

Department of Physics, Nara Women's University, Nara 630-8506, Japan

Abstract. We discuss the structure of ^{4}He, $^4_\Lambda$H, $^4_\Lambda$He and $^{\ 4}_{\Lambda\Lambda}$H. The calculated impulse approximation transition form factor $^4\mathrm{He}(e,e')^4\mathrm{He}(0_2^+)$ agrees with the available data. It is found that a major part of the energy-weighted $E0$ sum rule value is exhausted by nonresonant, low-energy continuum states other than the second 0^+ state. Precise four-body calculations have been performed for $^4_\Lambda$He and $^4_\Lambda$H taking explicitly into account the $NNN\Lambda$ and $NNN\Sigma$ channels and for $^{\ 4}_{\Lambda\Lambda}$H with $NN\Lambda\Lambda$ and $NNN\Xi$ channels included. The role of ΛN-ΣN coupling in $^4_\Lambda$He and $^4_\Lambda$H and the role of $\Lambda\Lambda$-ΞN coupling in $^{\ 4}_{\Lambda\Lambda}$H are important.

Keywords: few-body systems, hypernuclei
PACS: 12

INTRODUCTION

One of the important and intrinsically interesting subjects in few-body physics is the study of four-body systems using realistic interactions. In the following we discuss:

(1) Why it is important to study four-body systems as such; and

(2) What new understanding do we gain by solving these systems as four-body problems.

Before discussing the physics of these four-body systems, we should review as background our calculational method. In order to solve the three- and four-body problem accurately, the Gaussian Expansion Method (GEM) was proposed by Kamimura [1, 2] some 15 years ago in order to carry out non-adiabatic three-body calculations of muonic molecules and muon-atomic collisions. The characteristics of GEM are as follows:

(i) The basis set consists of functions of Jacobian coordinates, $(\mathbf{r}_c, \mathbf{R}_c)$, for *all* three rearrangement channels ($c = 1-3$) shown in Fig. 1, each basis function being of the form $\phi_{n_c l_c}(r_c)\psi_{N_c L_c}(R_c)[Y_{l_c}(\widehat{\mathbf{r}}_c)Y_{L_c}(\widehat{\mathbf{R}}_c)]_{JM}$ with the obvious notation for angular momenta while n_c and N_c specify the radial dependence.

(ii) The radial dependence is Gaussian, $\phi_{nl}(r) = r^l e^{-\nu_n r^2}$, and similarly for $\psi_{NL}(R)$, with the range parameters forming a geometric progression, $\{\nu_n = \nu_1 a^{n-1}; n = 1 - n_{\max}\}$.

Prescription (i) for generating the basis functions in three channels makes the function space significantly larger (even if l and L are strongly restricted) than the case using the basis functions of a single channel alone, but it makes the non-orthogonality between the basis functions much less troublesome than in the latter case. These types of three-channel basis functions are particularly suitable for describing a weakly bound system along any of the R_c as well as for describing strong short-range correlations along any of the r_c ($c = 1-3$). Prescription (ii) for the Gaussian ranges has been found to be very

CP1011, *New Facet of Three Nucleon Force—50 Years of Fujita Miyazawa Three Nucleon Force (FM50), The International Symposium,* edited by H. Sakai, K. Sekiguchi, and B. F. Gibson

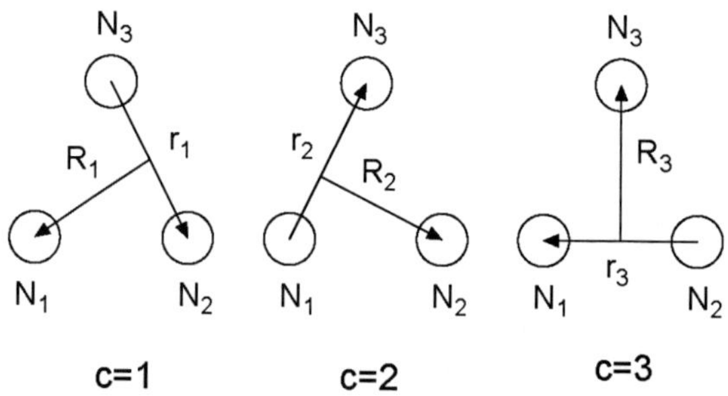

FIGURE 1. The three Jacobian coordinates of the three-body system.

advantageous in accurately describing both the short-range correlations and the long-range tail behavior in the asymptotic region of few-body wave functions. The Gaussian shape of the basis functions makes the calculation of matrix elements easy, even between basis functions of different channels. Thus, prescriptions (i) and (ii) are the essential reasons for the success of GEM in calculating various types of three-body bound and quasi-bound states such as muonic molecules [1, 5], three-nucleon bound states (^{3}H, ^{3}He) [3, 4], unstable nuclei [6, 7], and antiprotonic helium atoms [8, 9, 10, 11].

THE SECOND 0^+ STATE OF ^{4}HE

An important task in few-body nuclear physics is to solve the four-nucleon bound-state problem (^{4}He) using realistic NN interactions. A benchmark calculation by 18 co-authors including the present author was reported in Ref. [12]. Seven different few-body research groups solved for the ^{4}He ground state using a realistic interaction, the Argonne V8$'$ potential (AV8$'$) [13]. Agreement among the results of the significantly different calculational schemes was essentially perfect in terms of the binding energy, the rms radius, and the two-body correlation function.

One of the next challenging projects is to calculate properties of the second 0^+ state of ^{4}He using realistic interactions. ^{4}He is the lightest nucleus that exhibits discrete excited states. The second 0^+ state at $E_x = 20.2$ MeV has the same spin, isospin and parity as the ground state. Therefore, this excited state must have quite a different spatial structure in order to be orthogonal to the ground state. One must reproduce simultaneously the ground state and this excited state, which have the same spin, isospin, and parity, but possess quite different spatial structure.

In investigating the spatial structure of the ^{4}He 0_2^+ state, the following important questions are to be answered:

(1) Is it possible to reproduce simultaneously the energies of the first and second 0^+ states?

(2)Is it possible to explain the inelastic electron scattering data, $^4\mathrm{He}(e,e')^4\mathrm{He}(0_2^+)$?

(3) Experimentally, it was reported that the second 0^+ state exhausts only 11 % of the energy-weighted $E0$ sum rule. This means that the second 0^+ state is not a collective state. Where is the major contribution to the sum rule situated?

For this study we employed the AV8′ potential and included the Coulomb potential acting between protons. Use of realistic two-body potentials alone does not reproduce the observed binding energies of ^{3}H, ^{3}He and ^{4}He. Therefore, we employed additionally a phenomenological three-body force so as to reproduce those binding energies. Use of the AV8′, Coulomb force, and phenomenological three-body force reproduce the observed energies of ^3H, ^{3}He, and ^{4}He, and the rms charge radius of ^{4}He(0_1^+).

What results follow for the second 0^+ state? The calculated excitation energy of the state 20.25 MeV is in good agreement with the observed $E_x = 20.21$. Fig. 2 illustrates the mass densities of the ground state and the second 0^+ state. The density of the second 0^+ state in the internal region is much smaller than that of the ground state.

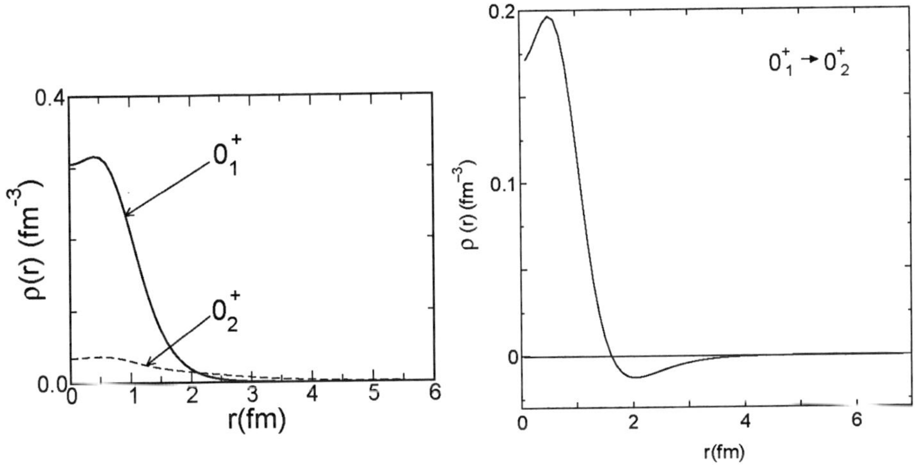

FIGURE 2. Mass densities of the 0_1^+ and 0_2^+ states of ^{4}He (left) and the transition density between them (right).

The Fourier transformation of the transition density gives the form factor for inelastic electron scattering. The observed inelastic form factor data [14, 15, 16] are compared with our calculation in Fig. 3. We see that our method is suitable for describing both compactly bound states and loosely coupled states, as well as transition between them.

Next, let's examine the energy-weighted $E0$ sum rule in ^{4}He. The $E0$ sum rule is usually written in the following manner:

$$\sum_{n\geq 2}^{\infty}(E_n - E_1)|\langle 0_n^+ | \sum_p r_p^2 | 0_1^+ \rangle|^2 = \frac{\hbar^2 Z}{m_N}\langle r_p^2 \rangle_{0_1^+}, \tag{1}$$

where E_n and $|0_n^+\rangle$ are the energy and wave function of the n-th state of the four-nucleon system, respectively, m_N being the nucleon mass, and $Z = 2$ for ^{4}He. Completeness is assumed for the set of states $\{|0_n^+\rangle; n = 1 - \infty\}$ which are normalized in a sufficiently large volume. The $E0$ strength of the second 0^+ state, 1.10 ± 0.16 fm^2, was derived from (e, e') data. However, this contribution from the second 0^+ state exhausts only 11 % of the sum rule limit. This implies that the second 0^+ state is not a collective mode. Where

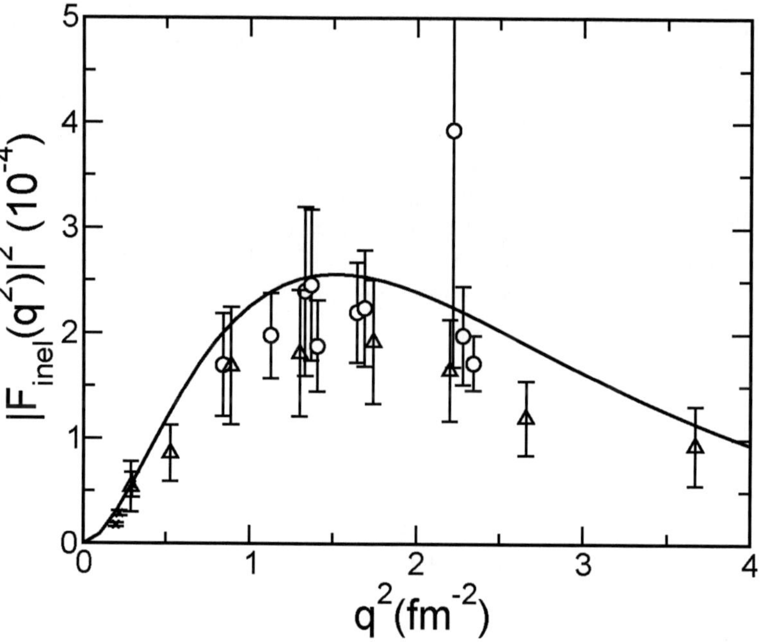

FIGURE 3. Inelastic form factor for ${}^4\mathrm{He}(e,e')^4\mathrm{He}(0_2^+)$ from the reported four-body calculation (solid line) is compared with the available data [14, 15, 16], $\triangle$, $\times$, and O, respectively.

is the major component of the $E0$ sum rule limit situated? This has been a long-standing puzzle for ^{4}He since 1970's.

We solved this puzzle as follows:

(1)Using the four-body Gaussian basis functions, we diagonalized the total Hamiltonian.

(2)Since we employed about 20,000 basis functions, we obtained the same number of four-body eigenstates. Above the second 0^+ state, we have precisely-discretized non-resonant continuum states.

(3)We calculated the energy-weighted $E0$ strength to all the discretized eigenstates.

Then, we found the following fact: The contribution from a limited number of these states, namely some 200 discretized-continuum states above the second 0^+ state but below the excitation energy 60 MeV, accounts for 70 % of the sum rule limit. Thus, we now understand that a major part of the energy-weighted $E0$ sum rule is exhausted by non-resonant, low-energy continuum states.

${}^4_\Lambda$H AND ${}^4_\Lambda$HE

Next, let's replace one nucleon in ^{4}He with a hyperon such as a Λ particle. Then, we have ${}^4_\Lambda$H and ${}^4_\Lambda$He. In non-strange nuclei, NN-$N\Delta$ coupling generates the Fujita-Miyazawa type three-body force. In $S = -1$ Λ hypernuclei, particle conversion such as ΛN-ΣN is the operative effect.

Therefore, interesting questions arise:

(1) How important is $\Lambda N - \Sigma N$ coupling in the binding of Λ hypernuclei?

(2) How large is the probability of the admixed Σ particle in Λ hypernuclei?

(3) How large is the effective three-body ΛNN force that is generated by the Σ-excitation?

In order to answer these questions, it is useful to study the structure of ${}^4_\Lambda$H and ${}^4_\Lambda$He, because both of the spin-doublet states have been observed.

For a precise study of $A = 4$ hypernuclei, it is highly desirable to perform four-body calculations taking into account both $NNN\Lambda$ and $NNN\Sigma$ channels explicitly. So far, several authors including the present author have succeeded in performing this type of difficult four-body calculation and pointed out that $\Lambda N - \Sigma N$ conversion is very important in binding the $A = 4$ hypernuclei [17, 18, 19]. To explain this, the calculational results of the present author are shown. The detail is reported in Ref.[17].

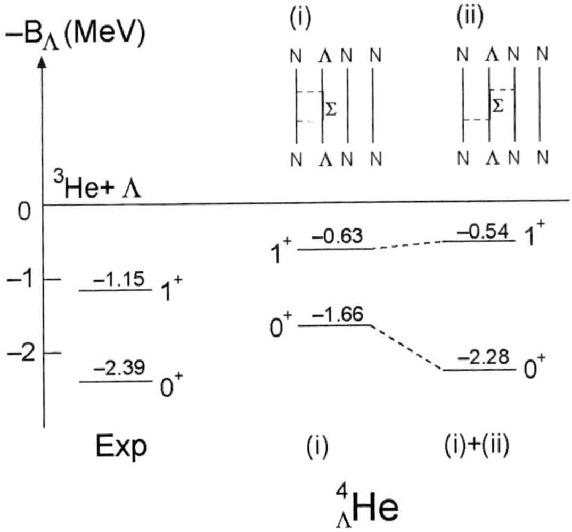

FIGURE 4. Calculated energy levels of ${}^4_\Lambda$He for the case (i) and case (i)+(ii);here, (i) denotes the two-body process and (ii) denotes the three-body process.

First, if we neglect the Σ channel and employ only the $NNN\Lambda$ channel, then both of the spin-doublet 0^+ and 1^+ states are unbound. Next, as shown in Fig. 4, when we include the $\Lambda N - \Sigma N$ coupling which can be renormalized into an effective two-body ΛN interaction, both states are bound. Finally, if we further include three-body process (ii), we obtain the energy levels shown in the right hand side of the Fig. 4. We found that the three-body process (ii) is substantial: This process results in attraction of 0.62 MeV in the 0^+ state and repulsion of 0.09 MeV in the 1^+ state of ${}^4_\Lambda$He.

${}^{\;4}_{\Lambda\Lambda}$H

In the $S = -2$ sector, it is interesting to study the structure of the bound $np\Lambda\Lambda$ (${}^{\;4}_{\Lambda\Lambda}$H) hypernucleus. In this hypernucleus, the interesting issue is $\Lambda\Lambda - \Xi N$ coupling. One of the major goals in hypernuclear physics is to study the structure of multi-strange systems. Because the threshold energy difference between ΞN and $\Lambda\Lambda$ is very small, 25 MeV, it is thought that $\Lambda\Lambda$-ΞN conversion should be strong in multi-strange systems. For the study of the $\Lambda\Lambda - \Xi N$ coupling interaction, s-shell double Λ hypernuclei such as ${}^{\;4}_{\Lambda\Lambda}$H and ${}^{\;5}_{\Lambda\Lambda}$H(${}^{\;5}_{\Lambda\Lambda}$He) are essential. So far, many authors have been studying these double Λ hypernuclei [20, 21, 22, 23]. The reason is: In the case of ${}^{\;4}_{\Lambda\Lambda}$H, one neutron and one proton plus two Λs are situated in the s-shell. Two Λs can convert into $\Xi^- p$ by $\Lambda\Lambda - \Xi N$ mixing. This additional proton can also occupy the s-shell. Because we have no observed data for ${}^{\;4}_{\Lambda\Lambda}$H an important issue is whether the YY interaction designed to reproduce the binding energy of ${}^{\;6}_{\Lambda\Lambda}$He [24] will bind ${}^{\;4}_{\Lambda\Lambda}$H.

For this study, it is desirable to perform the $nn\Lambda\Lambda$ and $NNN\Xi$ four-body calculation, which is one of the most numerically difficult four-body problems. Recently, the present author succeeded in performing such a four-body calculation. For the YY interaction, we took the Extended Soft Core '04 potential [25]. Use of this potential reproduces the observed binding energy of ${}^{\;6}_{\Lambda\Lambda}$He. First, we calculated the binding energy of ${}^{\;4}_{\Lambda\Lambda}$H using only the $NN\Lambda\Lambda$ channel, neglecting the $NNN\Xi$ channel; we found that we have no bound state. Next, we calculated it including both the $NN\Lambda\Lambda$ and $NNN\Xi$ channels. Then we obtain a weakly bound state. Thus, if the bound state of ${}^{\;4}_{\Lambda\Lambda}$H is observed in the future, we can obtain useful information about the $\Lambda\Lambda - \Xi N$ coupling mechanism.

CONCLUSIONS

We discussed four-baryon systems such as ^{4}He, ${}^4_\Lambda$H, ${}^4_\Lambda$He, and ${}^{\;4}_{\Lambda\Lambda}$H and what is interesting and important to study in these systems. The data for ^{4}He, ${}^4_\Lambda$H, and ${}^4_\Lambda$He have motivated the development of our few-body calculational technique. In the future, we expect that ${}^{\;4}_{\Lambda\Lambda}$H will be observed at J-PARC. This observation will provide the few-body community with important new information to fuel research.

ACKNOWLEDGMENTS

The computational work was supported by the Super-computer Project of the High Energy Accelerator Research Organization (KEK).

REFERENCES

1. M. Kamimura, Phys. Rev. A **38** (1988) 621.

2. M. Kamimura, Muon Catalyzed Fusion **3** (1988) 335.
3. H. Kameyama, M. Kamimura and Y. Fukushima, Phys. Rev. C **40** (1989) 974.
4. M. Kamimura and H. Kameyama, Nucl. Phys. A **508** (1990) 17c.
5. K. Nagamine and M. Kamimura, Adv. in Nucl. Phys. **24** (1998) 151.
6. S. Funada, H. Kameyama, and Y. Sakuragi, Nucl. Phys. A **575** (1994) 93.
7. E. Hiyama and M. Kamimura, Nucl. Physics A **588** (1995) 35c.
8. Y. Kino, M. Kamimura and H. Kudo, Hyp. Interact. **119** (1999) 201.
9. Y. Kino, N.Yamanaka, M. Kamimura, P. Froelich and H. Kudo, Hyp. Interact. **139** (2001) 179.
10. Y. Kino, M. Kamimura and H. Kudo, Nucl. Phys. A **31** (1998) 649.
11. Y. Kino, M. Kamimura and H. Kudo, Few-Body Syst. Suppl. **12** (2000) 40.
12. H. Kamada, A. Nogga, W. Glockle, E. Hiyama, M. Kamimura, K. Varga, Y. Suzuki, M. Viviani, A. Kievsky, S. Rosati, J. Carlson, S. C. Pieper, R. B. Wiringa, P. Navratil, B. R. Barrett, N. Barnea, W. Leidemann and G. Orlandini, Phys. Rev. C **64** (2001) 044001.
13. B.S. Pudliner, V.R. Pandaripande, J. Carlson, S. C. Pieper, and R. B. Wiringa, and Phys. Rev. C **56**, 1720 (1997).
14. R. F. Frosch, R. E. Rand, H. Crannell, J. S. McCarthy, L. R. Suelzle, and M. R. Yearian, Nucl. Phys. A **110** 637 (1968).
15. Th. Walcher, Phys. Lett. B, **31**, 442 (1970); Z. Physik **237**, 368 (1970).
16. G. Koebschall, C. Ottermann, K. Maurer, K. Roehrich, Ch. Schmitt and V. H. Walther, Nucl. Phys. A **405**, 648 (1983).
17. E. Hiyama, M. Kamimura, T. Motoba, T. Yamada and Y. Yamamoto, Phys. Rev C**65**, 011301 (R), (2001).
18. A. Nogga, H. Kamada and W. Glöckle, Phys. Rev. Lett. **88** (2002) 172501.
19. H. Nemura, Y. Akaishi, and Y. Suzuki, Phys. Rev. Lett. **89**, 142504 (2002).
20. I.N.Filikhin and A.Gal, Phys. Rev. Lett. **89**, 172501 (2002).
21. Khin Swe Myint, S. Shinmura and Y. Akaishi, Eur. Phys. J. A **16**, 21 (2003).
22. D. E. Lanscoy and Y. Yamamoto, Phys. Rev. C**69** , 014303 (2004).
23. H. Nemura, S. Shinmura, Y. Akaisshi and Khin Swe Myint, Phys. Rev. Lett. **94**, 202502 (2005).
24. H. Takahashi *et al.*, Phys. Rev. Lett. **87**, (2001) 212502.
25. Th. A. Rijken and Y. Yamamoto, Phys. Rev. C**73**, 044008 (2006).

Experimental Studies of Hypernuclear Structure and the ΛN and ΛNN Interactions

H. Tamura

Department of Physics, Tohoku University
Sendai 980-8578, Japan

Abstract. The ΛNN three-body force, through ΛN-ΣN coupling, plays an important role in the structure of Λ hypernuclei. We have been studying the ΛN interaction via hypernuclear γ-ray spectroscopy experiments and obtained the effective interaction strengths of the spin-dependent ΛN interactions. In addition, a possible effect of the ΛNN three-body force is being investigated from precise level schemes of light Λ hypernuclei. The ΛNN three-body force can also be studied from the structure of neutron-rich Λ hypernuclei by using the (π^-,K^+) reaction.

Keywords: Λ hypernuclei, γ-ray spectroscopy, ΛN interaction, neutron-rich hypernuclei, ΛN-ΣN coupling
PACS: 21.80.+a, 13.75.Ev, 23.20.Lv

INTRODUCTION

One of the important goals in modern nuclear physics is to extend our knowledge of the nuclear force to cover all the baryon-baryon interactions and to understand them in a unified way, in which quark-gluon dynamics, in addition to the conventional meson exchange picture, is expected to describe the short-range part of the interaction. In the quark-based models the short-range parts of the hyperon-nucleon (YN) and hyperon-hyperon (YY) interactions are predicted to have quite different features from that of the NN interaction. Experimental investigations of the YN and YY interactions from low-energy YN and YY scattering experiments are, however, extremely difficult due to the short lifetimes of hyperons. On the other hand, experimental data on hypernuclei provide rich information that depends crucially on the YN and YY interactions. The interaction models, which are theoretically constructed in the framework of the meson-exchange picture and the quark picture under guidance of flavor SU(3) symmetry and based upon the abundant NN scattering data as well as the sparse YN scattering data, are tested and modified so as to reproduce hypernuclear data. In particular, the structure of Λ hypernuclei provides information on the ΛN interaction including its spin-dependence and ΛN-ΣN coupling.

The three-body force in Λ hypernuclei

Because of the small ($\sim$80 MeV) mass difference between the Λ and Σ hyperons compared to the $\sim$300 MeV mass difference between a Δ and a nucleon, the probability of a Σ component in a Λ hypernucleus due to ΣN-ΛN mixing shown in Fig. 1 is much

CP1011, *New Facet of Three Nucleon Force—50 Years of Fujita Miyazawa Three Nucleon Force (FM50), The International Symposium,* edited by H. Sakai, K. Sekiguchi, and B. F. Gibson

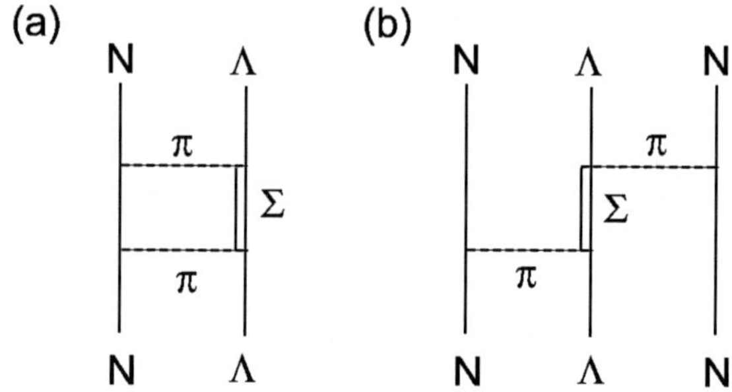

FIGURE 1. Diagrams for the (a) two-body (ΛN) and (b) three-body (ΛNN) interactions with ΛN-ΣN coupling.

larger than that of a Δ component in nonstrange nuclei. In particular, the ΛNN three-body force arising through Σ mixing in an intermediate state [Fig. 1(b)] plays an important role in the structure of Λ hypernuclei. For example, the effect of the ΛNN three-body force is essential to understand the binding energies of s-shell hypernuclei. The ΛN attractive force which reproduces the binding energies of ${}^{3}_{\Lambda}$H and ${}^{4}_{\Lambda}$H/${}^{4}_{\Lambda}$He overestimates the ${}^{5}_{\Lambda}$He binding energy. This puzzle called "overbinding problem of ${}^{5}_{\Lambda}$He" is solved [1] by taking into account the "coherent ΣN-ΛN coupling" which gives rise to the ΛNN force. This force is attractive and significantly reduces only the 0^+ ground state energy of ${}^{4}_{\Lambda}$H/${}^{4}_{\Lambda}$He, while the other levels in the A =3, 4, 5 hypernuclei are not affected. In a simple picture, the (1^+,0^+) doublet spacing is determined by the ΛN spin-spin interaction. However, the ΛNN force makes the spacing about twice as large [2, 1] which means that an accurate estimate of the effect by this force is necessary to extract the spin-spin interaction strength from this spacing. As discussed later, this force is also expected to affect the ground-state doublet spacings of p-shell hypernuclei significantly. Another interesting aspect of this force is a large attractive effect in neutron-rich hypernuclei. It is expected that this force enhances the binding of neutron-rich hypernuclei, as discussed later.

HYPERNUCLEAR γ SPECTROSCOPY

The energy levels of a Λ hyperon (or the Λ single-particle energies) in Λ hypernuclei have been measured for a wide range of mass number [3, 4], and the nuclear potential depth for a Λ in nuclear matter has been determined to be about 30 MeV, being about two thirds of that of the nucleon potential depth. This corresponds to a spin-averaged central interaction strength between a Λ and a nucleon. On the other hand, the spin dependence of the ΛN interaction has not been well determined.

Structure of a hypernucleus can be understood in terms of a weak-coupling scheme between the core nucleus and the Λ. As shown in Fig. 2, each level of the core nucleus with a spin J splits into a doublet with spins $J+1/2$ and $J-1/2$. The spacing of such a doublet is governed by the spin-dependent components of the ΛN interaction, which

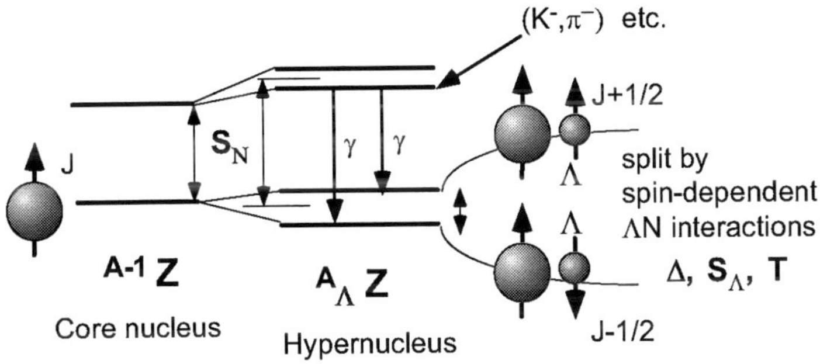

FIGURE 2. Hypernuclear fine structure and γ spectroscopy. High resolution of Ge detector is essential to resolve two levels ("hypernuclear fine structure") split by the ΛN spin-dependent interactions. See text for details.

can not easily be studied by ΛN scattering experiments. Due to small strengths of the spin-dependent ΛN interaction, the spacing is usually of the order of 10–100 keV, being generally smaller than level spacings of low-lying states of the core nucleus, and thus such a doublet spacing is called "hypernuclear fine structure". In order to separate such fine structure, precision γ spectroscopy using germanium (Ge) detectors with a few keV resolution is indispensable. It is the main reason we need to introduce precision γ spectroscopy into the investigation of hypernuclei.

The potential for the two-body ΛN effective interaction may be written as:

$$V_{\Lambda N}(r) = V_0(r) + V_\sigma(r)\, s_\Lambda s_N + V_\Lambda(r)\, \mathbf{l}_{\Lambda N} s_\Lambda + V_N(r)\, \mathbf{l}_{\Lambda N} s_N \\ + V_T(r)\, [3(\sigma_\Lambda \hat{\mathbf{r}})(\sigma_N \hat{\mathbf{r}}) - \sigma_\Lambda \sigma_N]. \quad (1)$$

Some hypernuclear data suggest strengths of the spin-spin interaction (V_σ) and the Λ-spin-dependent spin-orbit interaction (V_Λ), but due to limited experimental information they have not been well established. The nucleon-spin-dependent spin-orbit interaction (V_N) and the tensor interaction (V_T) have been undetermined.

In a shell-model description, low-lying level energies of p-shell hypernuclei with a Λ in the $0s$ orbit can be described with five radial integrals for $s_\Lambda p_N$ wave function on the five terms in Eq. 1. The four integrals of the spin-dependent terms are denoted as Δ, S_Λ, S_N, and T, respectively, for the spin-spin term, the Λ-spin-dependent spin-orbit term, the nucleon-spin-dependent spin-orbit term, and the tensor term [5, 6]. By comparing experimental level energies with shell-model calculations, these integrals (parameters) can be experimentally determined; the values of Δ, S_Λ and T are derived from spacings of the doublets, while the S_N value is given by the change in excitation energies of the core levels. The derived parameter values are then compared with those predicted from theoretical interaction models based on a meson-exchange and/or a quark picture.

For few-body hypernuclei ($A \leq 5$) and some light hypernuclei described by cluster models, their structure can be directly calculated from the ΛN interaction given by theoretical interaction models.

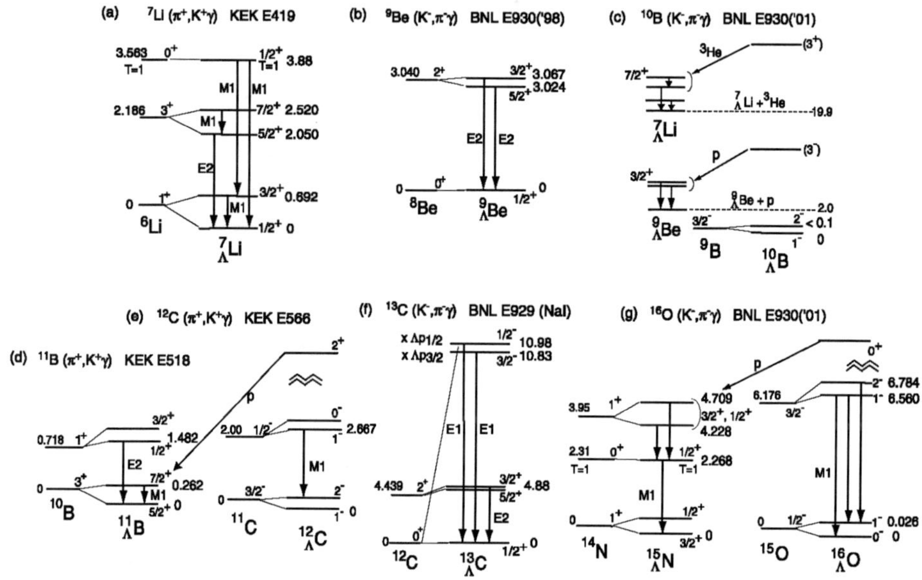

FIGURE 3. γ transitions and level schemes of p-shell Λ hypernuclei which have been investigated by γ spectroscopy since 1998. ${}^{13}_{\Lambda}$C was studied with an NaI detector array [11], while all the others were investigated with Hyperball and Hyperball2.

Hyperball and the present status of hypernuclear γ spectroscopy

In order to precisely investigate the structure of hypernuclei, we developed a technique for measuring γ-ray spectroscopy with a germanium detector array called Hyperball. Before Hyperball was constructed, hypernuclear structure was mostly studied by reaction spectroscopy of the (K^-,π^-) and (π^+,K^+) reactions using magnetic spectrometers with a few MeV(FWHM) resolution. Hyperball has drastically improved the energy resolution for Λ hypernuclear bound states to a few keV (FWHM). Figure 3 shows all the hypernuclear γ transitions observed in recent experiments since 1998 [7, 9, 10, 11, 12, 13, 14, 15]. γ rays in ${}^{13}_{\Lambda}$C were observed with NaI counter arrays [11], while all the other γ rays were observed with Hyperball. The present status of the γ-ray spectroscopy experiments is summarized in Ref. [16].

The Hyperball experiments were performed at the KEK PS using the (π^+,K^+) reaction at $p_{\pi^+} = 1.05$ GeV/c (E419, E518) and at BNL AGS with the (K^-,π^-) reaction at $p_{K^-} = 0.93$ GeV/c (E930). In the former case, we employed the K6 beam line and the Superconducting Kaon Spectrometer (SKS), which has a good hypernuclear mass resolution of ~1.5 MeV(FWHM) and a large acceptance of 100 msr [17]. In the latter case, the most important feature is the high-intensity and high-purity K^- beam only available at the D6 beam line of the BNL AGS [18].

As shown in Fig. 4(a), the Ge detector array Hyperball consists of fourteen single-type Ge detectors (with a relative efficiency of 60%), each of which is surrounded by

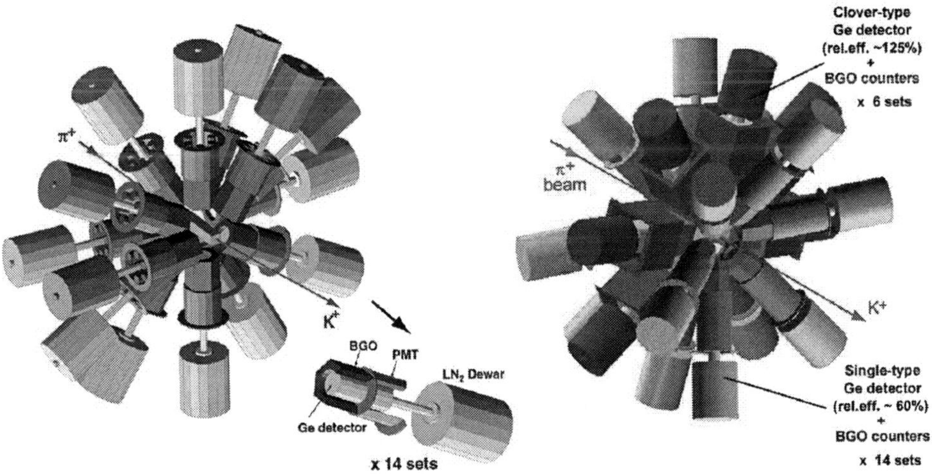

FIGURE 4. Schematic views of Ge detector arrays for hypernuclear γ-ray spectroscopy, Hyperball (Left) and Hyperball2 (Right).

a set of BGO scintillation counters for background suppression. The photoelectric peak efficiency of Hyperball is about 2.5% at 1 MeV for a point source. Recently, Hyperball was upgraded to Hyperball2 (see Fig. 4(b)), by adding six Clover-type Ge detectors (with 120% relative efficiency for each). The photoelectric peak efficiency is nearly twice as large as that of the previous Hyperball. Both Hyperball and Hyperball2 are equipped with fast readout electronics which enables detection of nuclear γ rays under severe background from high energy secondary beams. Hyperball2 is also being used for in-beam γ-ray spectroscopy experiments for ordinary nuclei at the cyclotron facility (CYRIC) of Tohoku University.

Spectroscopy of ${}^{11}_{\Lambda}$B and ${}^{12}_{\Lambda}$C at KEK

Recently, we have performed a γ-ray spectroscopy experiment for ${}^{11}_{\Lambda}$B and ${}^{12}_{\Lambda}$C via the ^{12}C(π^+,K^+) reaction at the KEK-PS K6 beam line, employing the SKS spectrometer and Hyperball2 [15]. One of the motivations of this experiment is to measure the ground-state doublet spacings of ${}^{11}_{\Lambda}$B$(7/2^+,5/2^+)$ and ${}^{12}_{\Lambda}$C$(2^-,1^-)$ to investigate the ΛN spin-spin interaction strength and the effect of the ΛNN three-body force, as described later.

When gating the p_Λ-state region of ${}^{12}_{\Lambda}$C in the missing mass spectrum of the (π^+,K^+) reaction, we observed a γ-ray peak at 261.6 ± 0.24 keV (preliminary). As shown in Fig. 3(d), it is assigned to the spin-flip $M1$ transition between the ${}^{11}_{\Lambda}$B ground-state doublet, ${}^{11}_{\Lambda}$B$(7/2^+ \rightarrow 5/2^+)$, because only the ${}^{12}_{\Lambda}$C $p_\Lambda(2^+) \rightarrow {}^{11}_{\Lambda}B(7/2^+,5/2^+) + p$ decay is energetically allowed.

When selecting the bound-state region of ${}^{12}_{\Lambda}$C, we observed a γ-ray peak at 2667.3 ± 2.8 keV (preliminary) after the Doppler shift correction. This is assigned to the ${}^{12}_{\Lambda}\mathrm{C}(1^-_2 \rightarrow 2^-)$ transition as shown in Fig. 3(e). This assignment is based on the previously measured cross sections of the 1^-_2 and other ${}^{12}_{\Lambda}$C states [4] and the expected branching ratio between the $1^-_2 \rightarrow 2^-$ and $1^-_2 \rightarrow 1^-$ transitions.

The other transition, $1^-_2 \rightarrow 1^-$, has not been observed yet, although the data analysis is still in progress. If it is observed, the energy spacing of the ground state doublet (1^-,2^-) can be determined, which will provide crucial information on the ΛN-ΣN coupling force as described in the following section.

Spin-dependent ΛN interaction

Millener made shell model calculations to describe level energies of various p-shell hypernuclei in terms of the four spin-dependent interaction parameters, by including the effects of the hypernucleus size and ΛN-ΣN coupling (see the next section) [19]. By comparing the calculated results with the experimental level energies, it is found that all the level spacings of the doublets (hypernuclear fine structure) observed so far, ${}^{7}_{\Lambda}$Li($3/2^+$,$1/2^+$), ${}^{7}_{\Lambda}$Li($7/2^+$,$5/2^+$), ${}^{9}_{\Lambda}$Be($3/2^+$,$5/2^+$), ${}^{11}_{\Lambda}$B($7/2^+$,$5/2^+$), ${}^{16}_{\Lambda}$O(1^-,0^-) and ${}^{16}_{\Lambda}$O(2^-,1^-), can be reproduced well with a common set of the effective interaction parameters (Δ, S_Λ, T), albeit with a slight modification in Δ according to the mass number. In addition, most of the core-excited level energies in the p-shell hypernuclei are also explained from a common value of the S_N parameter. The parameter values thus determined are:

$$\Delta = 0.33 \text{ or } 0.43 \text{ MeV}, \quad S_\Lambda = -0.01 \text{ MeV}, \quad S_N = -0.4 \text{ MeV}, \quad T = 0.02 \text{ MeV}. \qquad (2)$$

Here, the spin-spin strength, which was determined to be $\Delta = 0.43$ MeV for $A = 7$, needs to be changed slightly to $\Delta = 0.33$ MeV for $A = 11$ and 16 [20].

This parameter set cannot reproduce some experimental level energies in the $A = 10, 11, 12$ hypernuclei; the ${}^{10}_{\Lambda}$B data of $E(2^-) - E(1^-) < 100$ keV corresponds to $\Delta < 0.3$ MeV, and the ${}^{11}_{\Lambda}$B data of $E(5/2^-) - E(1/2^-) = 1.48$ MeV to $S_N = -0.9$ MeV, and the ${}^{12}_{\Lambda}$C data of $E(1^-_2) - E(2^-) = 2.67$ MeV to $S_N = -0.7$ MeV. To solve these inconsistencies, particularly the "${}^{10}_{\Lambda}$B puzzle" for the Δ value, the effect of a three-body force due to ΣN-ΛN coupling should be investigated for each hypernucleus, as discussed in the next section. In addition, the structure of the core-nuclear wave function might be modified by the presence of a Λ. Further theoretical studies are awaited.

The experimentally determined values (Eq. 2) are compared with theoretical predictions for various versions (ND, NF, NSC89, NSC97f, etc.) of the Nijmegen meson-exchange interaction models through G-matrix calculations [21]. We found that the predicted values for S_Λ and S_N disagree with the experimental values, indicating that the ΛN spin-orbit interaction cannot be reproduced by those meson-exchange models. On the other hand, the meson-exchange picture explains the ΛN tensor force strength well.

Another approach to extract information about the ΛN interaction from hypernuclear level data is to calculate the structure of light hypernuclei directly from theoretical baryon-baryon interaction models with variational methods, sometimes using the α

cluster model, compare the calculated results with experimental level energies, and then modify or improve the interaction models. The cluster model calculations by Hiyama *et al.*, which predicted much larger values of the ${}^{9}_{\Lambda}$Be and ${}^{13}_{\Lambda}$C spacing than the experimental data from various Nijmegen interactions [22], also indicated that the existing meson exchange models should be modified in terms of the spin-orbit interaction. In addition, a recent calculation [23] showed that the NSC97f interaction can reproduce the doublet spacings in ${}^{7}_{\Lambda}$Li ($3/2^+$,$1/2^+$) and ($7/2^+$,$5/2^+$) perfectly well, after adjusting the spin-singlet and triplet force strengths and the antisymmetric spin-orbit force (ALS) strength in NSC97f so as to reproduce the ${}^{4}_{\Lambda}$H (1^+,0^+) and ${}^{9}_{\Lambda}$Be ($3/2^+$,$5/2^+$) doublet spacings.

$\Lambda\Sigma$ coupling and the ΛNN three-body force

When the ΛN effective interaction is studied by comparing with experimental data, the contribution of the two-body diagram with ΛN-ΣN coupling [Fig. 1(a)] is incorporated in the ΛN interaction. However, the three-body diagram [Fig. 1(b)] cannot be incorporated in the two-body interaction and its effect should be estimated for each hypernuclear state. The effect of ΛN-ΣN coupling is calculated for $A = 3,4,5$ hypernuclei by incorporating ΛN-ΣN coupling explicitly into variational calculations [24, 25]. Such calculations have not been attempted for p-shell hypernuclei.

When the spin-dependent interaction strengths (Δ, S_Λ, S_N, T) are derived from low-lying hypernuclear level data measured by γ-ray spectroscopy, the effect of ΛN-ΣN coupling between a Λ in the s orbit and nucleons in the p orbit was theoretically estimated based on a shell-model calculation by Millener [19, 20]. Here, the matrix element of $v = \langle p_N^{A-5} s_\Sigma(J)|V|p_N^{A-5} s_\Lambda(J)\rangle$ was calculated using the interaction SC97f(S) used by Akaishi in Ref. [1] and the energy shift $\Lambda\Sigma = v^2/(m_\Sigma - m_\Lambda)$ was also calculated. On the other hand, the $s_N^4 s_\Sigma - s_N^4 s_\Lambda$ coupling has no spin-dependence and does not affect the spacings of low-lying hypernuclear levels.

Figure 5 shows four ground-state doublets of p-shell hypernuclei, ${}^{7}_{\Lambda}$Li, ${}^{10}_{\Lambda}$B, ${}^{11}_{\Lambda}$B and ${}^{12}_{\Lambda}$C, with results of shell model calculations by Millener [20] for the level spacing in terms of the four spin-dependent ΛN interaction parameters and for the energy shifts due to $\Lambda\Sigma$ coupling effect shown as "$\Lambda\Sigma$". Figure 5 also shows the spacing energies measured by γ-spectroscopy experiments [7, 10, 15] and the derived Δ values, $\Delta = 0.43$ MeV from ${}^{7}_{\Lambda}$Li, $\Delta < 0.3$ MeV from ${}^{10}_{\Lambda}$B, and $\Delta = 0.33$ MeV from ${}^{11}_{\Lambda}$B.

In the case of ${}^{7}_{\Lambda}$Li, for example, the energy shifts due to $\Lambda\Sigma$ coupling is large for the lower member of the doublet ($1/2^+$) and almost zero for the upper member of the doublet ($3/2^+$). It is because the two valence nucleon system in the ^{6}Li core with $T = 0$, $S = 1$ should be transformed into $T = 1$, $S = 0$ when a Λ is converted to a Σ, which is forbidden for the upper member of the doublet having $S = 3/2$. The increase in the doublet spacing due to $\Lambda\Sigma$ coupling affects the value of the spin-spin interaction strength Δ derived from these spacings, as in the case of the ${}^{4}_{\Lambda}$H/${}^{4}_{\Lambda}$He(1^+,0^+) doublet. In ${}^{7}_{\Lambda}$Li, the value is obtained as $\Delta = 0.43$ MeV [19] with the $\Lambda\Sigma$ coupling taken into account, while $\Delta = 0.50$ MeV [7] without the $\Lambda\Sigma$ coupling.

Usually, the parameter values may differ slightly among the various p-shell hypernuclei because of the different size of the Λ wave function which leads to a different overlap

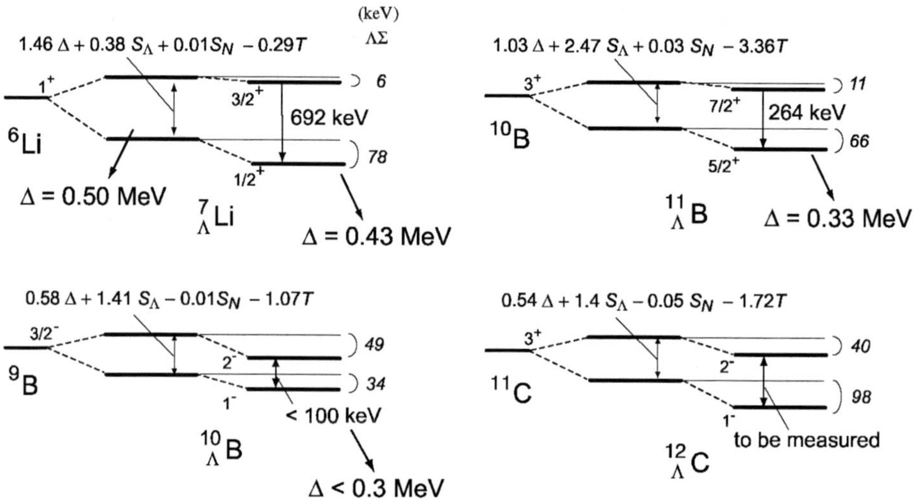

FIGURE 5. The ground state doublets for ${}^{7}_{\Lambda}$Li, ${}^{10}_{\Lambda}$B, ${}^{11}_{\Lambda}$B and ${}^{12}_{\Lambda}$C with the spacing energies in terms of the four spin-dependent ΛN interaction parameters and the energy shifts due to the $\Lambda\Sigma$ coupling effect calculated by Millener [19, 20]. The spacing energies measured by γ-spectroscopy experiments [7, 10, 15] are also shown.

with the valence nucleon wave functions. However, it is likely that the difference of the Δ values among the three hypernuclei shown in Fig. 5 is caused by inaccuracy of the theoretical calculation for the $\Lambda\Sigma$ coupling effect.

The ${}^{10}_{\Lambda}$B and ${}^{12}_{\Lambda}$C hypernuclei have similar structure, as seen from the similar coefficients in the shell model results. However, the calculation predicts that the $\Lambda\Sigma$ coupling effect decreases the ${}^{10}_{\Lambda}$B spacing, while it increases the ${}^{12}_{\Lambda}$C spacing. Therefore, the ${}^{12}_{\Lambda}$C spacing will provide a clue to understand the $\Lambda\Sigma$ coupling force and to solve the ${}^{10}_{\Lambda}$B puzzle after a complete analysis of the γ-ray spectroscopy data recently taken at KEK. In addition, we plan to further investigate this subject at the J-PARC facility by accumulating more γ-ray data for p-shell hypernuclei.

NEUTRON-RICH HYPERNUCLEI

The ΛNN force produces a particularly large effect in the binding energies of neutron-rich Λ hypernuclei, because the attraction from this interaction is coherently added to all valence neutrons [26]. This attractive force plays an essential role in making very neutron-rich hypernuclei bound. For example, ${}^{6}_{\Lambda}$H and ${}^{7}_{\Lambda}$H (called hyperheavy hydrogens) are expected to be bound according to Akaishi's calculation [26]. The ${}^{6}_{\Lambda}$H ground state is calculated to gain 1.4 MeV binding energy from the ΛNN force. This effect is also important to understand high-density neutron-rich nuclear matter with strangeness which is expected to exist in the core of neutron stars.

Experimentally, neuron-rich hypernuclei can be produced by the (K^-,π^+) or (π^-,K^+)

reactions which convert two protons into a Λ and a neutron, by two-step charge exchange reactions ($\pi^- p \rightarrow \Lambda K^0$, $K^0 p \rightarrow nK^+$, or $\pi^- p \rightarrow n\pi^0$, $\pi^0 p \rightarrow \Lambda K^+$, etc.) as well as through a Σ admixture in Λ hypernuclei ($\pi^- p \rightarrow \Sigma^- K^+$, $\Sigma^- p \rightarrow \Lambda n$).

Experimental efforts to produce neutron-rich hypernuclei have been made using the (stopped K^-, π^+) reaction [27, 28], but they have not been observed yet because of extremely small production rates. Recently, a neutron-rich hypernucleus ${}^{10}_{\Lambda}$Li was observed via the ^{10}B(π^-,K^+) reaction by employing the SKS at KEK-PS [29]. We observed about 20 events for ${}^{10}_{\Lambda}$Li bound states in the energy region of $-13 < -B_\Lambda < 0$ MeV. The observed cross section is of the order of 10 nb/sr for a 1.2 GeV/c π^- beam, being three orders of magnitude smaller than that of ${}^{10}_{\Lambda}$B via the ^{10}B(π^+,K^+) reaction. The ${}^{10}_{\Lambda}$Li cross section is much larger at $p_\pi = 1.2$ GeV/c than at 1.05 GeV/c, suggesting a significant contribution of the production through the Σ admixure. The production mechanism also provides valuable information on the ΛN-ΣN mixing in hypernuclei. An experiment to search for ${}^{6}_{\Lambda}$H and other neutron-rich hypernuclei is planned at J-PARC (E10) using the (π^-,K^+) reaction.

SUMMARY

The ΛNN three-body force through ΛN-ΣN coupling plays a particularly important role in the structure of Λ hypernuclei. From precise structure data for p-shell Λ hypernuclei obtained by γ-ray spectroscopy experiments using Hyperball and Hyperball2, the effective interaction strengths of the spin-dependent ΛN interactions have been derived. An inconsistency found in the spin-spin strength (Δ value) between the values from ${}^{7}_{\Lambda}$Li, ${}^{10}_{\Lambda}$B and ${}^{10}_{\Lambda}$B may be attributed to a possible effect of the ΛNN three-body force. A recent KEK experiment for ${}^{12}_{\Lambda}$C is expected to provide crucial information on the three-body force. The ΛNN three-body force can be also studied from the structure of neutron-rich Λ hypernuclei by using the (π^-,K^+) reaction.

ACKNOWLEDGMENTS

The author thanks to all the members of the Hyperball collaboration (KEK E419, E518, E566 and BNL E930) and of the KEK E521 collaboration. The present work was supported by Grants-in-Aid for Scientific Research from the Ministry of Education of Japan, Nos. 11440070, 15204014 and 17070001.

REFERENCES

1. Y. Akaishi *et al.*, Phys. Rev. Lett. **84** (2000) 3539.
2. B. F. Gibson and D. R. Lehman, Phys. Rev. C **37** (1988) 679.
3. T. Hasegawa *et al.*, Phys. Rev. **53** (1996) 1210.
4. H. Hotchi *et al.*, Phys. Rev. **C64** (2001) 044302.
5. R. H. Dalitz and A. Gal, Ann. Phys. **116** (1978) 167.
6. D. J. Millener, A. Gal, C. B. Dover and R. H. Dalitz, Phys. Rev. C **31** (1985) 499.
7. H. Tamura *et al.*, Phys. Rev. Lett. **84** (2000) 5963.
8. K. Tanida *et al.*, Phys. Rev. Lett. **86** (2001) 1982.

9. H. Akikawa *et al*, Phys. Rev. Lett. **88** (2002) 082501.
10. H. Tamura, Nucl. Phys. **A754** (2005) 58c.
11. S. Ajimura *et al.*, Phys. Rev. Lett. **86** (2001) 4255; H. Kohri *et al.*, Phys. Rev. C **65** (2002) 034607.
12. M. Ukai *et al.*, Phys. Rev. Lett. **93** (2004) 232501.
13. M. Ukai *et al.*, Phys. Rev. C **73** (2006) 012501(R).
14. Y. Miura *et al.*, Nucl. Phys. **A754** (2005) 75c.
15. Y. Ma *et al.*, Eur. Phys. J. **A 33** (2007) 243.
16. O. Hashimoto and H. Tamura, Prog. Part. Nucl. Phys. **57** (2006) 564.
17. T. Fukuda *et al.*, Nucl. Inst. Meth. A **361** (1995) 485.
18. P. H. Pile *et al.*, Nucl. Inst. Meth. A **321** (1992) 48.
19. D. J. Millener, Nucl. Phys. **A754** (2005) 48c.
20. D. J. Millener, Lecture Notes in Physics **724** (2007) 31.
21. D. J. Millener, *Proc. Jlab Sponsored Workshop on Hypernuclear Physics with Electromagnetic Probes*, Hampton, December 2-4, 1999, Ed. L. Tang and O. Hashimoto, p.79.
22. E. Hiyama *et al.,* Phys. Rev. Lett. **85** (2000) 270.
23. E. Hiyama, Y. Yamamoto, Th. A. Rijken, T. Motoba, Phys. Rev. C **74** (2006) 054312.
24. E. Hiyama *et al.*, Phys. Rev. **C65** (2002) 011301.
25. H. Nemura *et al.*, Phys. Rev. Lett. **89** (2002) 142504.
26. Y. Akaishi *et al.*, Frascati Phsics Series Vol.XVI pp. 59 (1999); Y. Akaishi, in these proceedings.
27. K. Kubota *et al.*, Nucl. Phys. bf A602 (1996) 327.
28. M. Agnello *et al.*, Phys. Lett. **B640** (2006) 145.
29. P. K. Saha *et al.*, Phys. Rev. Lett. **94** (2005) 052502.

A new three-baryon-force in $\Lambda\Lambda$ hypernuclei

Hidekatsu Nemura

Advanced Meson Science Laboratory, Nishina Center for Accelerator-Based Science, RIKEN, Wako, 351-0198, Japan

Abstract. We describe a few-body calculation of ${}^{5}_{\Lambda\Lambda}$H as well as ${}^{4}_{\Lambda\Lambda}$H and ${}^{6}_{\Lambda\Lambda}$He taking account fully coupled-channel two-baryon potentials acting among the octet of baryons. The wave function includes not only $pnn\Lambda\Lambda$ and $ppnn\Xi^-$ components but also $pnn\Lambda\Sigma^0$, $ppn\Lambda\Sigma^-$, $pnn\Sigma^0\Sigma^0$ and $ppn\Sigma^0\Sigma^-$. An effective YY potential based on Nijmegen model D is used. We find that the $pnn\Lambda\Sigma^0$ and $ppn\Lambda\Sigma^-$ components play an important role in producing the ${}^{5}_{\Lambda\Lambda}$H bound state. The present result requires the introduction of a new coupled-channel three-body-force, $N\Lambda\Lambda - NN\Xi$, if the intermediate $pnn\Lambda\Sigma^0$ and $ppn\Lambda\Sigma^-$ states are eliminated from the model space.

Keywords: three-body-force, few-body problem, strangeness, hypernuclei
PACS: 21.80.+a, 21.45.+v, 21.10.Dr, 13.75.Ev

INTRODUCTION

Doubly-strange, light hypernuclear systems are important to study in order to understand the baryon-baryon interactions, including the hyperon-hyperon (YY) potential as well as the hyperon-nucleon (YN) potential. In the study of strangeness $S = -1$ s-shell hypernuclei, it has been demonstrated that the $\Lambda N - \Sigma N$ coupled-channel potential plays a significant role in explaining the anomalously small binding of ${}^{5}_{\Lambda}$He[1]. The "coherent $\Lambda - \Sigma$ coupling"[1] may also be important in doubly strange hypernuclei.

A $\Lambda\Lambda$ two-body system couples to the $N\Xi$ and $\Sigma\Sigma$ states in free space. For the doubly strange ($S = -2$) hypernucleus, ${}^{5}_{\Lambda\Lambda}$H, the *core nucleus* + $\Lambda\Lambda$ system couples to the *core nucleus* + $\Lambda\Sigma$ component. This is an extension of the coherent $\Lambda - \Sigma$ coupling to the doubly strange hypernucleus, and it implies that $\Lambda N - \Sigma N$ coupling should also play an important role in the binding mechanism of the doubly strange hypernuclei. Therefore, precise calculations of the light $\Lambda\Lambda$ hypernuclei should be made in a fully coupled channel formulation taking account of explicit Σ degrees of freedom. The purpose of this study is to describe a few-body calculation of the s-shell hypernuclei with $S = -2$ by taking account of the explicit Σ degrees of freedom in a framework of a fully coupled-channel formulation.

INTERACTIONS AND METHOD

We assume that the wave function of a system with $S = -2$, comprising $A(= 4-6)$ octet baryons, has four isospin-basis components. For example, the ${}^{5}_{\Lambda\Lambda}$H has four isospin components as $pnn\Lambda\Lambda$, $NNNNN\Xi$, $NNNN\Lambda\Sigma$ and $NNNN\Sigma\Sigma$. We abbreviate these components as $\Lambda\Lambda$, $N\Xi$, $\Lambda\Sigma$ and $\Sigma\Sigma$, referring to the last two baryons. The Hamiltonian

CP1011, *New Facet of Three Nucleon Force—50 Years of Fujita Miyazawa Three Nucleon Force (FM50), The International Symposium,* edited by H. Sakai, K. Sekiguchi, and B. F. Gibson

of the system is hence given by 4×4 components as

$$H = \begin{pmatrix} H_{\Lambda\Lambda} & V_{N\Xi-\Lambda\Lambda} & V_{\Lambda\Sigma-\Lambda\Lambda} & V_{\Sigma\Sigma-\Lambda\Lambda} \\ V_{\Lambda\Lambda-N\Xi} & H_{N\Xi} & V_{\Lambda\Sigma-N\Xi} & V_{\Sigma\Sigma-N\Xi} \\ V_{\Lambda\Lambda-\Lambda\Sigma} & V_{N\Xi-\Lambda\Sigma} & H_{\Lambda\Sigma} & V_{\Sigma\Sigma-\Lambda\Sigma} \\ V_{\Lambda\Lambda-\Sigma\Sigma} & V_{N\Xi-\Sigma\Sigma} & V_{\Lambda\Sigma-\Sigma\Sigma} & H_{\Sigma\Sigma} \end{pmatrix}, \tag{1}$$

where $H_{B_1B_2}$ operates on the B_1B_2 component,

$$\begin{aligned} H_{B_1B_2} &= \sum_{i=1}^{A}\left(\frac{\mathbf{p}_i^2}{2m_i}+m_ic^2\right)-T_{c.m.}+\sum_{i<j}^{A-2}v_{NN}(\mathbf{r}_i-\mathbf{r}_j)+\sum_{j=1}^{2}\sum_{i=1}^{A-2}v_{NB_j}(\mathbf{r}_i-\mathbf{r}_{B_j}) \\ &\quad +v_{B_1B_2}(\mathbf{r}_{B_1}-\mathbf{r}_{B_2})+\sum_{i<j}^{A}\frac{q_iq_j}{|\mathbf{r}_i-\mathbf{r}_j|}, \end{aligned} \tag{2}$$

and $V_{B_1B_2-B_1'B_2'}$ is the sum of all possible two-body transition potentials connecting the B_1B_2 and $B_1'B_2'$ components:

$$V_{\Lambda\Lambda-N\Xi}=v_{\Lambda\Lambda-N\Xi},\quad V_{\Lambda\Lambda-\Lambda\Sigma}=\sum_{i=1}^{N}v_{N_i\Lambda-N_i\Sigma},\quad V_{N\Xi-\Lambda\Sigma}=v_{N\Xi-\Lambda\Sigma}, \tag{3}$$

$$V_{\Lambda\Lambda-\Sigma\Sigma}=v_{\Lambda\Lambda-\Sigma\Sigma},\quad V_{N\Xi-\Sigma\Sigma}=v_{N\Xi-\Sigma\Sigma},\quad V_{\Lambda\Sigma-\Sigma\Sigma}=\sum_{i=1}^{N}v_{N_i\Lambda-N_i\Sigma}+v_{\Lambda\Sigma-\Sigma\Sigma}. \tag{4}$$

In the present calculations, we use the Minnesota potential[2] for the NN interaction and the D2′ potential for the YN interaction. The Minnesota potential reproduces reasonably well both the binding energies and sizes of few-nucleon systems. The D2′ potential is a modified potential based upon the original D2 potential[1]. The strength of the long-range part (V_b in Table I of Ref. [1]) of the D2′ potential in the ΛN-ΛN $^3S_1$ channel is reduced by multiplying by a factor (0.954) in order to reproduce the experimental $B_\Lambda(^5_\Lambda\text{He})$ value.

For the YY interaction, we use a full-coupled channel potential among the octet baryons in both the spin triplet and the spin singlet channels. We assume that the YY potential consists of only the central component, and the effect due to the non-central force (e.g., tensor force) should be effectively included in the central part. The parameters are selected to reproduce the low-energy S matrix of the Nijmegen hard-core model D (ND)[3]. We take the hard-core radius to be $r_c = 0.56(0.45)$ fm in the spin singlet (triplet) channel. Each number is the same as the hard-core radius of the YN sector in each channel. We denote as ND_S the potential simulating the ND model.

The binding energy is calculated in a complete A-body treatment. Since all of the exotic components ($\Lambda\Lambda$, $N\Xi$, $\Lambda\Sigma$ and $\Sigma\Sigma$) are taken into account, the variational trial function must be flexible enough to accurately search the eigenenergies. The trial function is given by a combination of basis functions:

$$\Psi = \sum_{k=1}^{K} c_k\varphi_k, \qquad \text{with} \qquad \varphi_k = \mathscr{A}\{G(\mathbf{x};A_k)\chi_{kJM}\eta_{kIM_I}\}. \tag{5}$$

Here $\mathscr{A}$ is an antisymmetrizer acting on the identical particles. For the spin χ_k and the isospin η_k functions, all possible configurations are taken into account. The abbreviation $\mathbf{x} = (\mathbf{x}_1, \cdots, \mathbf{x}_{A-1})$ is a set of relative coordinates. For the spatial part, the basis function is constructed by the correlated Gaussian (CG), $G(\mathbf{x}; A_k)$, which is given by

$$G(\mathbf{x};A_k) = \exp\left\{-\frac{1}{2}\sum_{i<j}^{A} \alpha_{kij}(\mathbf{r}_i - \mathbf{r}_j)^2\right\} = \exp\left\{-\frac{1}{2}\sum_{i,j=1}^{A-1} A_{kij}\mathbf{x}_i \cdot \mathbf{x}_j\right\}. \qquad (6)$$

The stochastic variational method[4, 5] with the above CG basis produces accurate solutions. The reader is referred to Refs.[5, 6, 7] for details.

RESULTS

Table 1 lists the $B_{\Lambda\Lambda}$ values for the $S=-2$ hypernuclei. We have obtained the bound-state solutions for ${}^{4}_{\Lambda\Lambda}$H, ${}^{5}_{\Lambda\Lambda}$H and ${}^{6}_{\Lambda\Lambda}$He. For ${}^{6}_{\Lambda\Lambda}$He, the $\Lambda\Lambda$ interaction energy ($\Delta B_{\Lambda\Lambda}$) is given by

$$\Delta B^{(\mathrm{calc})}_{\Lambda\Lambda}({}^{6}_{\Lambda\Lambda}\mathrm{He}) = B^{(\mathrm{calc})}_{\Lambda\Lambda}({}^{6}_{\Lambda\Lambda}\mathrm{He}) - 2B^{(\mathrm{calc})}_{\Lambda}({}^{5}_{\Lambda}\mathrm{He}) = 1.59\mathrm{MeV}, \qquad (7)$$

which is slightly larger than the experimental value,[8]

$$\Delta B^{(\mathrm{exp})}_{\Lambda\Lambda}({}^{6}_{\Lambda\Lambda}\mathrm{He}) = 1.01 \pm 0.20^{+0.18}_{-0.11}\mathrm{MeV}.$$

We have also calculated these systems using a modified ND$_S$ (mND$_S$) YY potential; The strength of the $\Lambda\Lambda$ diagonal part of the mND$_S$ potential is reduced by multiplying the original ND$_S$ by a factor of 0.8 in order to reproduce the experimental $\Delta B_{\Lambda\Lambda}({}^{6}_{\Lambda\Lambda}\mathrm{He})$. Hereafter we use the mND$_S$ potential rather than the ND$_S$. Table 2 shows the probabilities for the $N\Xi$, $\Lambda\Sigma$ and $\Sigma\Sigma$ components of the $S=-2$ hypernuclei. We should emphasize that the probability of the $N\Xi$ component, $P_{N\Xi}({}^{5}_{\Lambda\Lambda}\mathrm{H})$, has a surprisingly large value (4.56%) in spite of the fact that the $\Lambda\Lambda - N\Xi$ coupling of the ND is rather weak as will be seen later. This is in remarkable contrast with other calculations based on $(t+\Lambda+\Lambda)$ and $(\alpha+\Xi^-)$ two-channel model[9, 10]. We obtained a sizable value for the $P_{\Lambda\Sigma}({}^{5}_{\Lambda\Lambda}\mathrm{H})$ probability. The $\Lambda\Sigma$ component couples to the $\Lambda\Lambda$ component via the $v_{N\Lambda-N\Sigma}$ potential, as can be seen in Eq. (3), whereas the $v_{\Lambda\Lambda-N\Xi-\Sigma\Sigma}$ potential in the ${}^1S_0$, $I=0$ channel does not directly connect the $\Lambda\Sigma$ component and the $\Lambda\Lambda$ component. $P_{\Sigma\Sigma}$'s are

TABLE 1. $\Lambda\Lambda$ separation energies, given in units of MeV, for the $A=4-6$, $S=-2$ s-shell hypernuclei.

YY	$B_{\Lambda\Lambda}({}^{4}_{\Lambda\Lambda}\mathrm{H})$	$B_{\Lambda\Lambda}({}^{5}_{\Lambda\Lambda}\mathrm{H})$	$B_{\Lambda\Lambda}({}^{6}_{\Lambda\Lambda}\mathrm{He})$
ND$_S$	0.107	4.05	7.94
mND$_S$	0.058	3.75	7.54
Exp			$7.25 \pm 0.19^{+0.18}_{-0.11}$

TABLE 2. Probabilities, given in percent, for the $N\Xi$, $\Lambda\Sigma$ and $\Sigma\Sigma$ components of the $A = 4-6$, $S = -2$ s-shell hypernuclei.

	${}^{4}_{\Lambda\Lambda}$H			${}^{5}_{\Lambda\Lambda}$H			${}^{6}_{\Lambda\Lambda}$He		
	$P_{N\Xi}$	$P_{\Lambda\Sigma}$	$P_{\Sigma\Sigma}$	$P_{N\Xi}$	$P_{\Lambda\Sigma}$	$P_{\Sigma\Sigma}$	$P_{N\Xi}$	$P_{\Lambda\Sigma}$	$P_{\Sigma\Sigma}$
mND$_S$	0.06	0.25	0.00	4.56	2.49	0.06	0.28	1.17	0.05

very small for both systems, due to a large mass difference between the $\Lambda\Lambda$ and $\Sigma\Sigma$ channels ($m_{\Sigma\Sigma} - m_{\Lambda\Lambda} \cong 155$ MeV).

Although the present calculation assumes no simplified structures, such as $(t+\Lambda+\Lambda)$ and $(\alpha+\Xi^-)$, this kind of model is useful to provide a clear explanation of the complicated full coupling dynamics of the $A = 5, S = -2$ hypernucleus. Let us consider a set of simple *core nucleus* $+Y(+Y)$ model wave functions for the ${}^{5}_{\Lambda\Lambda}$H:

$$|{}^{5}_{\Lambda\Lambda}\mathrm{H}\rangle = \psi_t \times \psi_{\Lambda\Lambda} \times \psi_{\Lambda\Lambda-t}, \qquad |{}^{5}_{\Xi}\mathrm{H}\rangle = \psi_\alpha \times \psi_{\Xi^-} \times \psi_{\Xi^- - \alpha}, \tag{8}$$

$$|{}^{5}_{\Lambda\Sigma}\mathrm{H}\rangle_{S_{\Lambda\Sigma}} = \sqrt{\frac{1}{3}}\left[\psi_t \times [\psi_{\Lambda\Sigma^0}]_{S_{\Lambda\Sigma}}\right] \times \psi_{\Lambda\Sigma^0 - t} - \sqrt{\frac{2}{3}}\left[\psi_h \times [\psi_{\Lambda\Sigma^-}]_{S_{\Lambda\Sigma}}\right] \times \psi_{\Lambda\Sigma^- - h}$$
$$(\text{for } S_{\Lambda\Sigma} = 0 \text{ or } 1), \tag{9}$$

where ψ_c $(c = t, h, \alpha)$ is the wave function (WF) of the core nucleus, ψ_{YY} $(YY = \Lambda\Lambda, \Xi^-, \Lambda\Sigma)$ is the WF of the hyperon(s), and ψ_{YY-c} is the WF that describes the relative motion between YY and c. We assume that all of the baryons occupy the same $(0s)$ orbit. For the ${}^{5}_{\Lambda\Sigma}$H state, we have two independent states for the WF $\psi_{\Lambda\Sigma}$, that the spin of two hyperons $(S_{\Lambda\Sigma})$ is either a singlet or a triplet. Since the $\Sigma\Sigma$ component plays a minor role, we omit the ${}^{5}_{\Sigma\Sigma}$H state. Using these WFs, we can obtain the algebraic factors for each averaged coupling potential of the allowed spin state, $\bar{v}^3$ or $\bar{v}^1$:

$$\langle V_{\Lambda\Lambda-N\Xi}\rangle = \sqrt{\frac{1}{2}}\bar{v}^{1}_{\Lambda\Lambda-N\Xi}, \tag{10}$$

$$\langle V_{\Lambda\Lambda-\Lambda\Sigma}\rangle = \begin{cases} \sqrt{\frac{9}{8}}\bar{v}^{1}_{N\Lambda-N\Sigma} + \sqrt{\frac{1}{8}}\bar{v}^{3}_{N\Lambda-N\Sigma} & (\text{for } S_{\Lambda\Sigma} = 0), \\ \sqrt{\frac{3}{8}}\bar{v}^{1}_{N\Lambda-N\Sigma} - \sqrt{\frac{3}{8}}\bar{v}^{3}_{N\Lambda-N\Sigma} & (\text{for } S_{\Lambda\Sigma} = 1), \end{cases} \tag{11}$$

$$\langle V_{N\Xi-\Lambda\Sigma}\rangle = \begin{cases} -\sqrt{\frac{3}{4}}\bar{v}^{3}_{N\Xi-\Lambda\Sigma} & (\text{for } S_{\Lambda\Sigma} = 0), \\ \frac{3}{2}\bar{v}^{1}_{N\Xi-\Lambda\Sigma} & (\text{for } S_{\Lambda\Sigma} = 1). \end{cases} \tag{12}$$

The $v_{\Lambda\Lambda-N\Xi}$ potential is suppressed by a factor of $\sqrt{1/2}$ for the $A = 5$ hypernucleus. The $v_{N\Lambda-N\Sigma}$ and $v_{N\Xi-\Lambda\Sigma}$ potentials, particularly in the spin triplet channel, play significant roles instead. Namely, these equations imply that the $\Lambda\Sigma$ component strongly couples both to the $\Lambda\Lambda$ and to the $N\Xi$ components, and the $\Lambda\Sigma$ component plays a crucial role in the hypernucleus.

The normalized energy expectation values of the Hamiltonian (1) for ${}^{5}_{\Lambda\Lambda}\mathrm{H}$ are (given in units of MeV),

$$h = \begin{pmatrix} \frac{\langle H_{\Lambda\Lambda}\rangle}{P_{\Lambda\Lambda}} & \frac{\langle V_{N\Xi-\Lambda\Lambda}\rangle}{\sqrt{P_{\Lambda\Lambda}P_{N\Xi}}} & \frac{\langle V_{\Lambda\Sigma-\Lambda\Lambda}\rangle}{\sqrt{P_{\Lambda\Lambda}P_{\Lambda\Sigma}}} \\ \frac{\langle V_{\Lambda\Lambda-N\Xi}\rangle}{\sqrt{P_{\Lambda\Lambda}P_{N\Xi}}} & \frac{\langle H_{N\Xi}\rangle}{P_{N\Xi}} & \frac{\langle V_{\Lambda\Sigma-N\Xi}\rangle}{\sqrt{P_{N\Xi}P_{\Lambda\Sigma}}} \\ \frac{\langle V_{\Lambda\Lambda-\Lambda\Sigma}\rangle}{\sqrt{P_{\Lambda\Lambda}P_{\Lambda\Sigma}}} & \frac{\langle V_{N\Xi-\Lambda\Sigma}\rangle}{\sqrt{P_{N\Xi}P_{\Lambda\Sigma}}} & \frac{\langle H_{\Lambda\Sigma}\rangle}{P_{\Lambda\Sigma}} \end{pmatrix} = \begin{pmatrix} -9.12 & -1.82 & -14.52 \\ -1.82 & 5.01 & -10.39 \\ -14.52 & -10.39 & 92.41 \end{pmatrix}. \quad (13)$$

Here, we display only the 3×3 components of the Hamiltonian (1), comprising $\Lambda\Lambda$, $N\Xi$, and $\Lambda\Sigma$, since the contributions from the $\Sigma\Sigma$ component are not large. These numbers reflect the nature of the YY potential model (ND). The ND model has a weak $\Lambda\Lambda$-$N\Xi$ coupling and a weakly attractive $N\Xi$-$N\Xi$ potential.

If we solve the eigenvalue problem, $\det(h-\lambda I)=0$, we obtain the ground state energy, $E=-11.82$ MeV, and the probability, $P_{N\Xi}=3.99\%$. On the other hand, if we ignore the third row and the third column, the eigenenergy of only the first 2×2 subspace becomes $E=-9.35$ MeV, and the probability, $P_{N\Xi}=1.57\%$. This clearly means that the couplings between the ($\Lambda\Lambda$, $N\Xi$) and $\Lambda\Sigma$ components play crucial roles to make ${}^{5}_{\Lambda\Lambda}\mathrm{H}$ bound. The large coupling potentials, $\langle V_{\Lambda\Lambda-\Lambda\Sigma}\rangle$ and $\langle V_{N\Xi-\Lambda\Sigma}\rangle$, also enhance the $P_{N\Xi}$ probability.

COUPLED-CHANNEL THREE-BARYON-FORCE

A coupled-channel Schrödinger equation for three component is written as

$$\begin{pmatrix} h_A & u_{AB} & u_{AC} \\ u_{AB} & h_B & u_{BC} \\ u_{AC} & u_{BC} & h_C \end{pmatrix} \begin{pmatrix} \psi_A \\ \psi_B \\ \psi_C \end{pmatrix} = \varepsilon \begin{pmatrix} \psi_A \\ \psi_B \\ \psi_C \end{pmatrix}. \quad (14)$$

Here the subscript (A,B,C) reads ($\Lambda\Lambda$,$N\Xi$,$\Lambda\Sigma$). The third component yields

$$\psi_C = (\varepsilon - h_C)^{-1} u_{AC}\psi_A + (\varepsilon - h_C)^{-1} u_{BC}\psi_B. \quad (15)$$

Substituting the (15) into (14), we obtain the effective 2×2 Schrödinger equation

$$\begin{pmatrix} \widetilde{h}_A & \widetilde{u}_{AB} \\ \widetilde{u}_{AB} & \widetilde{h}_B \end{pmatrix} \begin{pmatrix} \psi_A \\ \psi_B \end{pmatrix} = \varepsilon \begin{pmatrix} \psi_A \\ \psi_B \end{pmatrix}, \quad (16)$$

where

$$\widetilde{h}_X = h_X + u_{XC}(\varepsilon - h_C)^{-1} u_{XC}, \qquad (X = A,B) \quad (17)$$

$$\widetilde{u}_{AB} = u_{AB} + u_{AC}(\varepsilon - h_C)^{-1} u_{BC}. \quad (18)$$

Therefore, the effective three-baryon-force is given by

$$w_{AB} = u_{AC}(\varepsilon - h_C)^{-1} u_{BC}. \quad (19)$$

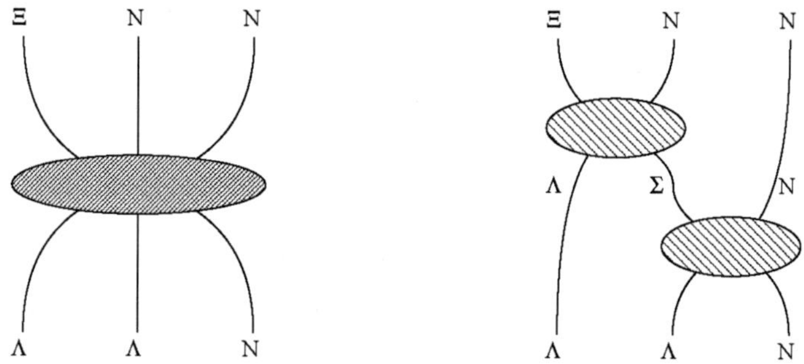

FIGURE 1. Schematic description of coupled-channel three-body-force $\Lambda\Lambda N - \Xi NN$ (left), derived from the intermediate $\Lambda\Sigma$ component (right).

Figure 1 shows a diagram of the $\Lambda\Lambda N - \Xi NN$ three-body-force due to the intermediate $\Lambda\Sigma N$ excitation. In the present case, we can deduce the energy expectation value of the coupled-channel three-baryon-force $\Lambda\Lambda N - \Xi NN$ by eliminating the $\Lambda\Sigma$ component,

$$\langle w_{\Lambda\Lambda - N\Xi} \rangle \cong -1.4 \text{ MeV}. \tag{20}$$

This has almost the same magnitude as the energy expectation value of the original $v_{\Lambda\Lambda - N\Xi}$ potential ($\cong -1.8$ MeV). The net attractive energy due to the $\Lambda\Lambda - N\Xi$ coupling thus becomes approximately three times larger than the original two-body $\Lambda\Lambda - N\Xi$ potential.

The present study is a first attempt to explore the new three-body-force arising in $\Lambda\Lambda$ hypernuclei. This kind of coupled-channel three-baryon-force should be taken into account for the study of Ξ hypernuclei since it affects the net $\Lambda\Lambda - N\Xi$ coupling and the decay width of the hypernucleus.

ACKNOWLEDGMENTS

The calculations were made using the supercomputers at RCNP and KEK.

REFERENCES

1. Y. Akaishi, et al., Phys. Rev. Lett. **84**, 3539 (2000).
2. D. R. Thompson, M. Lemere and Y. C. Tang, Nucl. Phys. A**286**, 53 (1977).
3. M. M. Nagels, T. A. Rijken and J. J. de Swart, Phys. Rev. D**15**, 2547 (1977).
4. V. I. Kukulin and V. M. Krasnopol'sky, J. Phys. G:Nucl. Part. Phys. **3**, 795 (1977).
5. Y. Suzuki and K. Varga, *Stochastic Variational Approach to Quantum-Mechanical Few-Body Problems*, Lecture Notes in Physics, Vol. m54 (Springer-Verlag, Berlin Heidelberg, 1998).
6. H. Nemura, Y. Akaishi and Y. Suzuki, Phys. Rev. Lett. **89**, 142504 (2002).
7. H. Nemura, et al.,Phys. Rev. Lett. **94**, 202502 (2005).
8. H. Takahashi, et al., Phys. Rev. Lett. **87**, 212502 (2001).
9. Khin Swe Myint, S. Shinmura and Y. Akaishi, Eur. Phys. J. A **16**, 21 (2003).
10. D. E. Lanskoy and Y. Yamamoto, Phys. Rev. C **69**, 014303 (2004).

Non-mesonic weak decays of ${}^{5}_{\Lambda}$He and ${}^{12}_{\Lambda}$C hypernuclei using the (π^+, K^+) reactions

S. Okada*, S. Ajimura†, K. Aoki**, A. Banu‡, H. C. Bhang§, T. Fukuda¶, O. Hashimoto‖, J. I. Hwang§, S. Kameoka‖, B. H. Kang§, E. H. Kim§, J. H. Kim††, M. J. Kim§, T. Maruta‖, Y. Miura‖, Y. Miyake†, T. Nagae‡‡, M. Nakamura§§, S. N. Nakamura‖, H. Noumi**, Y. Okayasu‖, H. Outa*, H. Park††, P. K. Saha¶¶, Y. Sato**, M. Sekimoto**, S. Shin***, T. Takahashi**, H. Tamura‖, K. Tanida‡‡, A. Toyoda**, K. Tsukada‖, T. Watanabe††† and H. J. Yim§

*RIKEN Nishina Center, RIKEN, Saitama 351-0198, Japan
†Department of Physics, Osaka University, Toyonaka 560-0043, Japan
**High Energy Accelerator Research Organization (KEK), Tsukuba 305-0801, Japan
‡Gesellschaft für Schwerionenforschung mbH (GSI), Darmstadt 64291, Germany
§Department of Physics, Seoul National University, Seoul 151-742, Korea
¶Laboratory of Physics, Osaka Electro-Communication University, Neyagawa 572-8530, Japan
‖Department of Physics, Tohoku University, Sendai 980-8578, Japan
††Korea Research Institute of Standards and Science (KRISS), Daejeon 305-600, Korea
‡‡Department of Physics, Kyoto University, Kyoto 606-8502, Japan
§§Department of Physics, University of Tokyo, Hongo 113-0033, Japan
¶¶Japan Atomic Energy Agency, Ibaraki 319-1195, Japan
***Department of Physics, Ewha Womans University, Seoul 120-750, Korea
†††Gifu University, Gifu 501-1193, Japan

Abstract. We have measured for the first time yields of both neutron-proton and neutron-neutron pairs emitted from the non-mesonic weak decay of the ${}^{5}_{\Lambda}$He and ${}^{12}_{\Lambda}$C hypernuclei, produced via the (π^+, K^+) reaction. From the yields of both the back-to-back *np*-pair events and *nn*-pair events, we have directly determined the ratio of the neutron- and proton-induced decay widths, $\Gamma_n(\Lambda n \to nn)/\Gamma_p(\Lambda p \to np)$. Meanwhile, we have also determined the quenching of the observed numbers of single nucleons and coincident nucleon pairs compared with theoretical calculations. Our results suggest that contributions from the two-nucleon-induced three-body process $\Lambda NN \to NNN$ are significant.

Keywords: Λ hypernuclei, non-mesonic weak decay, nucleon energy spectrum
PACS: 21.80.+a, 13.30.Eg, 13.75.Ev

1. INTRODUCTION

Non-mesonic weak decay (NWMD) of Λ hypernuclei (*e.g.* $\Lambda N \to NN$) provides a unique opportunity to study the weak interaction between baryons, because this strangeness non-conserving process is attributed purely to the weak interaction.

The ratio of the neutron- and proton-induced decay channels, $\Gamma_n(\Lambda n \to nn)/\Gamma_p(\Lambda p \to np)$, is an important observable used to study the isospin structure of the NMWD mechanism. The simplest model, a One-Pion Exchange model, suggests that the Γ_n/Γ_p ratio is much smaller than unity (~ 0.1), whereas a few experimental data indicate that

CP1011, *New Facet of Three Nucleon Force—50 Years of Fujita Miyazawa Three Nucleon Force (FM50), The International Symposium,* edited by H. Sakai, K. Sekiguchi, and B. F. Gibson

the ratio is close to or larger than unity [1, 2]. This discrepancy has been one of the main open problems in the study of the NMWD of Λ hypernuclei for more than 40 years.

In the past, there has been no coincidence measurement of the two nucleons in the NMWD, but instead the single nucleon spectra have been measured [3, 4], from which it is difficult to extract the Γ_n/Γ_p ratio without theoretical assumptions regarding the re-scattering effects in the residual nucleus, the so-called Final State Interactions (FSIs). In addition, the possible existence of a two-nucleon (2*N*) induced decay mode, $\Lambda NN \rightarrow NNN$, which has been discussed theoretically since 1991 [5], makes it difficult to extract the Γ_n/Γ_p ratio, although there has been no experimental evidence so far.

In the present experiments, a coincident measurement of the two nucleons in the NMWD was performed for the first time. From the yields of both the back-to-back *np*-pair events and *nn*-pair events, we directly determined the Γ_n/Γ_p ratio. In this paper, we report the result of an analysis for the Γ_n/Γ_p ratio and a newly discovered issue concerning the quenching of the observed numbers of single nucleons and coincident nucleon pairs from NMWD compared to the theoretical calculations.

2. EXPERIMENT

The present experiments (E462/E508) were carried out at the K6 beamline of the KEK 12-GeV proton synchrotron (PS). Hypernuclei, ${}^5_\Lambda$He (E462) and ${}^{12}_\Lambda$C (E508), were produced via the (π^+,K^+) reaction at 1.05 GeV/*c* on ^{6}Li and ^{12}C targets, respectively. The formation of these hypernuclei was identified by measuring the missing energy in the (π^+,K^+) reaction. The decay particles were detected with a plastic counter system which covers a large solid angle around the target so as to facilitate the high-statistics coincidence measurements.

We performed a coincidence measurement of two nucleons emitted from the NMWD of the lightest spin/isospin-saturated *s*-shell and *p*-shell hypernuclei, ${}^5_\Lambda$He and ${}^{12}_\Lambda$C, which were selected to minimize the FSI effect. The coincidence measurement realized through the neutron detection system covering a large solid angle and the high production yield of Λ hypernuclei, $\sim 5 \times 10^4$ events, which is one order-of-magnitude higher than that of previous experiments.

3. RESULTS AND DISCUSSION

3.1. Γ_n/Γ_p ratio

Figure 1 shows the resulting angular correlation of the *np*- and *nn*-coincident events for ${}^5_\Lambda$He [6]. The Γ_n/Γ_p ratio can be accurately determined by measuring the ratio of the back-to-back *np*- and *nn*-pair yields ($\cos\theta < -0.8$) since the $\Lambda N \rightarrow NN$ is a two-body decay process. It was confirmed that the energy sums of the *np*- and *nn*-pairs were approximately equal to their Q-values ($\sim$ 150 MeV) [6]. This confirms that the $\Lambda N \rightarrow NN$ processes were clearly identified in our analysis. As a result, the Γ_n/Γ_p ratios

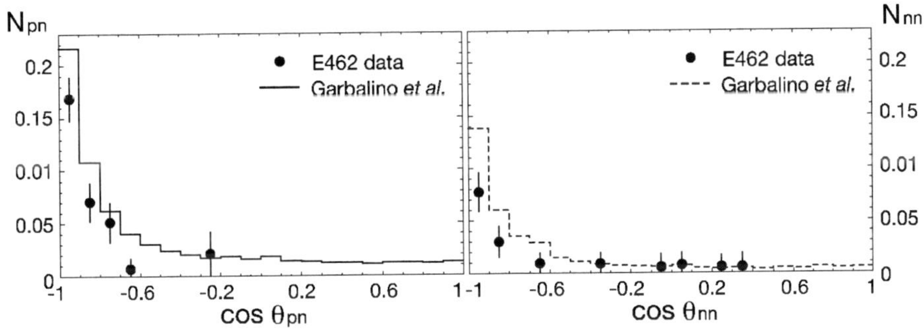

FIGURE 1. The normalized yields of the np and nn coincidence events per NMWD of ${}^{5}_{\Lambda}$He, N_{np} (left) and N_{nn} (right), plotted as a function of the opening angle between two nucleons (θ_{NN}) [6]. The histogram shows the distribution calculated by Garbarino *et al.*, in which the FSI effect and the 2N-induced process were taken into account [8].

for ${}^{5}_{\Lambda}$He [6] and ${}^{12}_{\Lambda}$C [7] were obtained as

$$\Gamma_n/\Gamma_p \sim N_{nn}/N_{np} = 0.45 \pm 0.11 \pm 0.03 \ (\text{for } {}^{5}_{\Lambda}\text{He})$$
$$\Gamma_n/\Gamma_p = 0.51 \pm 0.13 \pm 0.05 \ (\text{for } {}^{12}_{\Lambda}\text{C})$$

where N_{nn}/N_{np} denotes the yield ratio of np- and nn-pairs. The first error is statistical and the second is systematic. It is noteworthy that the reduction factors due to the FSIs being approximately canceled out when we form the N_{nn}/N_{np} ratio. The resulting N_{nn}/N_{np} ratio can be therefore regarded as the Γ_n/Γ_p ratio assuming the charge symmetry on the FSI.

It is now revealed that the ratio is significantly less than unity, which excludes the earlier claim that the ratio is close to unity. Recent theoretical calculations based on short-range interactions (*e.g.* heavy meson exchange model, direct quark model) seem to support our results, suggesting a Γ_n/Γ_p ratio toward 0.5. Our determination of the Γ_n/Γ_p ratio by means of the two-nucleon coincidence measurement has finally resolved this long-standing Γ_n/Γ_p ratio puzzle.

3.2. *NN* angular distribution and neutron energy spectrum

The normalized yields of the NN coincidence events per NMWD shown in Fig. 1 can be directly compared to those of the theoretical calculations. In the figure, we overlaid the histograms which show the angular distributions calculated by Garbarino *et al.* taking into account the FSI effect and assuming that $\Gamma_n/\Gamma_p = 0.46$ and $\Gamma_{2N}/\Gamma_{1N} = 0.20$, where Γ_{1N} and Γ_{2N} denote the decay widths of the 1N- and 2N-induced processes for ${}^{5}_{\Lambda}$He [8]. Here we found that the measured back-to-back yields are unexpectedly small compared to those of the calculation considering reductions due to the FSI effect and the 2N-induced process on the theoretical result.

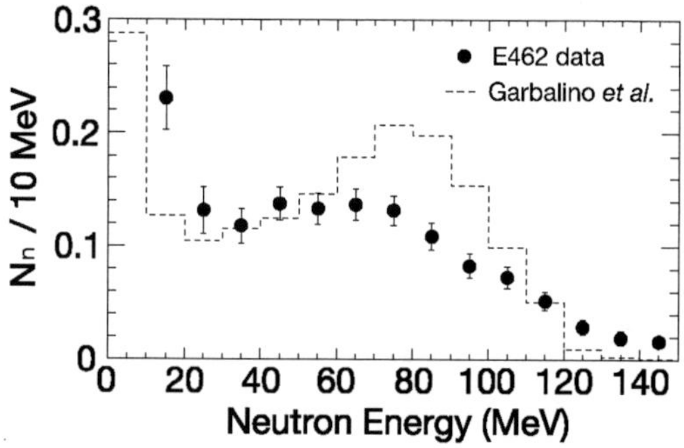

FIGURE 2. Neutron energy spectrum for NMWD of ${}^{5}_{\Lambda}$He obtained in the KEK-PS E462 experiment [10]. The dashed histogram shows the spectrum calculated by Garbarino *et al.*, in which the FSI effect and the 2N-induced process were taken into account [11].

It should be noted that there are so far two types of 2N-induced NMWD models. One is a 2N-induced three-body process in which the Q-value can be distributed over three nucleons in the final state [9]. The other is based on the correlated two-nucleon absorption of a virtual pion emitted from the weak vertex of $\Lambda N\pi$, in which the energy of the emitted nucleon is small, because the virtual pion is close to on-shell [8]. Since most of the Q-value in this 2N-induced process is distributed to two nucleons, the shape of the angular distributions should be close to that in the 1N-induced process. Thus, the model adopted in the histograms has a much smaller effect in reducing the back-to-back region than the former model. This suggests the importance of the former model, a 2N-induced three-body process, in which the Q-value is distributed over three nucleons in the final state resulting in a reduction of the back-to-back yields.

Figure 2 shows the single neutron energy spectrum from the decay of ${}^{5}_{\Lambda}$He, where the vertical scale is normalized per NMWD [10]. If the 1N-induced process ($\Lambda N \to NN$) dominates the NMWD and the FSI effect is negligible, the neutron energy spectrum should have a broadened peak, caused by Fermi motion, at about one half of the Q-value of the process ($\sim$75 MeV). We expected that this would be the case for ${}^{5}_{\Lambda}$He due to its small mass number. It is, however, clearly shown that the experimental spectra obtained are far from such a simple prediction.

Garbarino *et al.* also reported a nucleon energy spectrum for ${}^{5}_{\Lambda}$He calculated taking into account the FSI effect and with the same assumption as stated above [8]. In Fig. 2, we overlaid the calculated energy spectrum, which is significantly different from the observed spectrum. The calculated spectrum still has its maximum at about Q/2, whereas the experimental spectrum levels off in the region from 20 to 80 MeV. In their calculation, most of the Q-value in the 2N-induced process is distributed to two nucleons; thus, the shape of the energy spectra from the 2N-induced process also should be close to that in the 1N-induced process. As is the case with the angular distribution stated above, this also suggests the importance of the former model of the 2N-induced three-body process which enhances the yield in the low-energy region and reduces it in the high-energy region.

Recently, an intra-nuclear cascade (INC) calculation was performed and compared with our experimental results for ${}^{12}_{\Lambda}$C [9]. Assuming that the $2N$-induced process has a uniform phase-space distribution of the three-body final state, this calculation demonstrated that the observed yields of single nucleons and coincident nucleon pairs are reproduced with $\Gamma_{2N}/(\Gamma_{1N}+\Gamma_{2N}) = 0.4$, which suggests significant contributions from the $2N$-induced three-body process.

4. SUMMARY

We have measured the yields of the np- and nn-pairs emitted from the NMWD process, and directly determined the Γ_n/Γ_p ratio for the first time. Comparing the observed numbers of single nucleons and coincident nucleon pairs with theoretical calculations, we pointed out the importance of the $2N$-induced three-body process in which the Q-value is uniformly distributed over three nucleons in the final state. The accurate determination of Γ_{2N} is relevant for the determination of more elementary observables, such as Γ_n and Γ_p. This result calls for more detailed studies concerning the $2N$-induced NMWD processes and the FSI effect from both the theoretical and the experimental sides.

REFERENCES

1. J. J. Szymanski *et al.*, *Phys. Rev.* **C 43**, 849 (1991).
2. H. Noumi *et al.*, *Phys. Rev.* **C 52**, 2936 (1995).
3. Y. Sato *et al.*, *Phys. Rev.* **C 71**, 025203 (2004).
4. J. H. Kim *et al.*, *Phys. Rev.* **C 68**, 065201 (2003).
5. W. M. Alberico, A. De Pace, M. Ericson, and A. Molinari, *Phys. Lett.* **B 256**, 134 (1991).
6. B. H. Kang *et al.*, *Phys. Rev. Lett.* **96**, 062301 (2006).
7. M. J. Kim *et al.*, *Phys. Lett.* **B 641**, 28 (2006).
8. G. Garbarino, A. Parreño and A. Ramos, *Phys. Rev. Lett.* **91**, 112501 (2003).
9. H. Bhang *et al.*, *Eur. Phys. J.* **A 33**, 259 (2007).
10. S. Okada *et al.*, *Phys. Lett.* **B 597**, 249 (2004).
11. G. Garbarino, A. Parreño and A. Ramos, *Phys. Rev.* **C 69**, 054603 (2004).

The Illinois Extension to the Fujita-Miyazawa Three-Nucleon Force

Steven C. Pieper[1]

Physics Division, Argonne National Laboratory, Argonne, IL 60439, USA

Abstract. *Ab-initio* calculations of light nuclei require three-nucleon potentials in addition to realistic two-nucleon forces to obtain the empirical binding energies. The current status of the Illinois three-nucleon potentials used with Argonne v_{18} is presented.

Keywords: nuclear potentials
PACS: 21.60.-n,21.30.Fe,25.30.-c, 25.10.+s, 21.45.+v, 27.10.+h

INTRODUCTION

It has long been known that calculations with just realistic two-nucleon (*NN*) potentials fail to reproduce the binding energies of nuclei; three-nucleon (*NNN*) potentials are also required. These arise naturally from an underlying meson-exchange picture of the nuclear forces or from chiral effective field theories. Unfortunately, it is not yet practical to fit *NNN* potentials to *NNN* scattering data, so the parameters must be determined from calculations of light nuclei. This makes them dependent on the *NN* potential that is used. One such pair of potentials that has evolved over the years is the Argonne v_{18} (AV18) *NN* potential [1] combined with the Urbana [2, 3] or Illinois [4] *NNN* potentials.

The Urbana series of *NNN* potentials is based on the Fujita-Miyazawa potential [5, 6] with the addition of a shorter-range central repulsive core. The models evolved with improvements in the two-nucleon (*NN*) force and as few- and many-body calculational methods improved. They culminated in 1995 with the UIX model [3] which was used in conjunction with AV18 [1].

The AV18+UIX Hamiltonian reproduces the ^{3}H and ^{3,4}He binding energies well. However Ref. [3] already showed that, although this Hamiltonian produces more binding in the p-shell than AV18 alone, it still underbinds ^{6}Li and subsequent GFMC calculations showed that it underbinds all p-shell nuclei [7, 8]. This failing led to the construction of the Illinois family of *NNN* potentials in 2001 [4]. These still have the Fujita-Miyazawa potential as their principal term but also include three-pion (with one- or two-delta excitations) loops in addition to *S*-wave two-pion exchange. The three-pion loops (which are only $\sim$10% of the Fujita-Miyazawa term in $T = 0$ nuclei) introduce a markedly different isospin dependence and allow a good reproduction of all nuclear states up to A=10 when used with AV18 [9]. The current status of these potentials is presented in this paper.

---

[1] spieper@anl.gov

CP1011, *New Facet of Three Nucleon Force—50 Years of Fujita Miyazawa Three Nucleon Force (FM50), The International Symposium,* edited by H. Sakai, K. Sekiguchi, and B. F. Gibson

HAMILTONIANS

The nuclear Hamiltonian used here has the form

$$H=\sum_{i} K_i+\sum_{i<j} v_{ij}+\sum_{i<j<k} V_{ijk}\,. \tag{1}$$

Here K_i is the non-relativistic kinetic energy, including $m_n - m_p$ effects, v_{ij} is the *NN* potential and V_{ijk} is the *NNN* potential.

Argonne v_{ij}

The *NN* potential (v_{ij}) is Argonne v18 [1] which has the form

$$v_{ij}=v_{ij}^{\gamma}+v_{ij}^{\pi}+v_{ij}^{R}+v_{ij}^{CIB}. \tag{2}$$

The v_{ij}^{γ} is a very complete representation of the *pp*, *pn* and *nn* electromagnetic terms, including first- and second-order Coulomb, magnetic, vacuum polarization, etc., components with form factors.

The v_{ij}^{π} is the isoscalar one-pion exchange potential represented as a local operator:

$$v_{ij}^{\pi} = \frac{1}{3}\frac{f_{\pi NN}^2}{4\pi}\, m_{\pi}\, X_{ij}\, \tau_i\cdot\tau_j\,, \tag{3}$$

$$X_{ij} = T(m_{\pi}r_{ij})\, S_{ij}+Y(m_{\pi}r_{ij})\, \sigma_i\cdot\sigma_j\,, \tag{4}$$

$$Y(x) = \frac{e^{-x}}{x}\,\xi(r)\,, \tag{5}$$

$$T(x) = \left(\frac{3}{x^2}+\frac{3}{x}+1\right) Y(x)\,\xi(r)\,, \tag{6}$$

$$\xi(r) = (1-e^{-c_{\pi}r^2})\,. \tag{7}$$

where τ_i, σ_i and S_{ij} are isospin, spin and tensor operators, respectively. In light nuclei, $\langle v_{ij}^{\pi}\rangle$ contributes $\sim$85% of $\langle v_{ij}\rangle$.

The remaining isospin-conserving terms are

$$v_{ij}^{R} = \sum_{p=1,14} v_p(r_{ij}) O_{ij}^{p}\,, \tag{8}$$

$$O_{ij}^{p=1,14} = [1,\sigma_i\cdot\sigma_j,S_{ij},\mathbf{L}\cdot\mathbf{S},\mathbf{L}^2,\mathbf{L}^2\sigma_i\cdot\sigma_j,(\mathbf{L}\cdot\mathbf{S})^2]\otimes[1,\tau_i\cdot\tau_j]\,, \tag{9}$$

where $v_p(r)$ has short-, intermediate-, and long-range components. The long-range components are just the $Y(r)$ and $T(r)$ of the one-pion potential and are present only for those operators that have contributions from one-pion exchange. The intermediate-range components are proportional to $T^2(r)$ and the short-range component is of the Woods-Saxon form. There are also four strong charge-independence breaking terms.

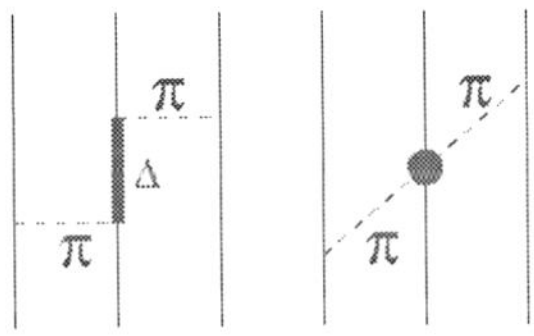

FIGURE 1. Two-pion exchange terms in the Illinois *NNN* potentials.

Urbana and Illinois V_{ijk}

The Urbana series of *NNN* potentials contains two terms:

$$V_{ijk} = V_{ijk}^{2\pi,P} + V_{ijk}^{R} . \tag{10}$$

The first term, shown in the left panel of Fig. 1, is the Fujita-Miyazawa potential [5, 6]

$$V_{ijk}^{2\pi,P} = A_{2\pi,P} \sum_{cyclic} \{X_{ij}, X_{jk}\}\{\tau_i \cdot \tau_j, \tau_j \cdot \tau_k\} + \frac{1}{4}[X_{ij}, X_{jk}][\tau_i \cdot \tau_j, \tau_j \cdot \tau_k] , \tag{11}$$

where X_{ij} is defined in Eq. (4). This is the longest-ranged nuclear *NNN* potential and is attractive in all nuclei and nuclear matter. However it is very small or even slightly repulsive in pure neutron systems.

The second term, V_{ijk}^R, represents all other diagrams including relativistic effects. It is strictly phenomenological and purely central and repulsive:

$$V_{ijk}^{R} = A_R \sum_{cyclic} T^2(m_\pi r_{ij}) T^2(m_\pi r_{jk}) . \tag{12}$$

This repulsive term is principally needed to make nuclear matter saturate at the proper density instead of a too-high density and to obtain a hard enough equation of state for neutron matter. The two coupling constants, $A_{2\pi,P}$ and A_R, were adjusted to reproduce the ^{3}H binding energy and to give reasonable nuclear-matter saturation properties [2, 3].

The Illinois *NNN* potentials [4] contain the two Urbana terms (with, of course, different coupling constants) and two additional terms:

$$V_{ijk} = V_{ijk}^{2\pi,P} + V_{ijk}^{2\pi,S} + V_{ijk}^{3\pi} + V_{ijk}^{R} . \tag{13}$$

The second panel of Fig. 1 represents the S-wave part of $V_{ijk}^{2\pi}$. This term was introduced in the Tuscon-Melbourne *NNN* potential [10] and is required by chiral perturbation theory. However, in practice it is only 3%–4% of $V_{ijk}^{2\pi,P}$ in light nuclei.

The three-pion term (Fig. 2) was introduced in the Illinois potentials. It consists of the subset of three-pion rings that contain only one Δ mass in the energy denominators. Even so it has a quite complicated form which is given in Ref. [4]. An important aspect of this structure is that there is a significant attractive term which acts only in T=3/2 triples. In $T = 0$ nuclei $\langle V_{ijk}^{3\pi} \rangle \lesssim 0.1 \langle V_{ijk}^{2\pi} \rangle$.

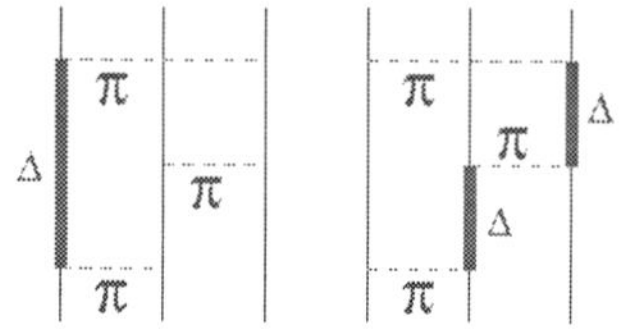

FIGURE 2. Three-pion ring terms in the Illinois *NNN* potentials.

The coupling constants $A_{2\pi,P}$, $A_{3\pi}$, and A_R were adjusted to fit 17 nuclear levels for $A \leq 8$. The $V_{ijk}^{2\pi,S}$ is too weak to be determined by fitting and its coupling was left at the value predicted by chiral perturbation theory. The constants for the final Urbana potential (UIX) and the most frequently used Illinois potential (IL2) are given in Table 1.

In light nuclei we find

$$\langle V_{ijk} \rangle \sim (0.02 \text{ to } 0.09) \langle v_{ij} \rangle \sim (0.15 \text{ to } 0.6) \langle H \rangle \tag{14}$$

where the large fraction of $\langle H \rangle$ is due to a large cancellation of K and v_{ij}. From this we expect

$$\langle V_{4N} \rangle \sim 0.06 \langle V_{ijk} \rangle \sim (0.01 \text{ to } 0.04) \langle H \rangle \sim (0.5 \text{ to } 2.) \text{ MeV} . \tag{15}$$

This is comparable to the accuracy of our calculations. Even if more accurate calculations could be made, it would probably not be possible to disentangle four-nucleon potential effects from uncertainties in the fitted parameters of V_{ijk}.

QUANTUM MONTE CARLO METHODS

The many-body problem with the full Hamiltonian described above is very difficult. We need to solve

$$\mathscr{H}\Psi(\vec{r}_1,\cdots,\vec{r}_A;s_1,\cdots,s_A;t_1,\cdots,t_A) = E\Psi(\vec{r}_1,\cdots,\vec{r}_A;s_1,\cdots,s_A;t_1,\cdots,t_A) , \tag{16}$$

where $s_i = \pm\frac{1}{2}$ are nucleon spins, and $t_i = \pm\frac{1}{2}$ are nucleon isospins (proton or neutron). Thus we need to solve the equivalent of $2^A \times \binom{A}{Z}$ complex coupled second-order equations in $3A-3$ variables (the number of isospin states can be reduced by imposing isospin conservation on the solution and time-reversal symmetry allows only half the spin states to be used for $J = 0$ nuclei). For the ground state of ^{12}C, with the above

TABLE 1. Potential coefficients for the Urbana-IX and Illinois *NNN* potentials

Model	c_π fm^{-2}	$A_{2\pi,P}$ MeV	$A_{2\pi,S}$ MeV	$A_{3\pi}$ MeV	A_R MeV	$A_{R,T=3/2}$ MeV
UIX	2.1	−0.0293	–	–	0.00480	–
IL2	2.1	−0.0370	−1.0	0.0026	0.00705	–
IL7	2.1	−0.0229	−1.0	0.0022	0.00324	0.023

symmetries, this corresponds to 270,336 coupled equations in 33 variables. Furthermore, the coupling is strong; the expectation value of the tensor component of v^π [Eq. 3] is approximately 60% of the total $\langle v_{ij}\rangle$ but it is identically zero if there are no tensor correlations. Thus we cannot perturbatively introduce the couplings.

We use two successive quantum Monte Carlo (QMC) methods to solve this problem, variational Monte Carlo (VMC) followed by Green's function Monte Carlo (GFMC). These are briefly described here; detailed descriptions are in Refs. [11, 7, 12, 9, 13].

Variational Monte Carlo

Variational Monte Carlo finds an upper bound, E_T, to an eigenenergy of H by evaluating the expectation value of H in a trial wave function, Ψ_T. The parameters in Ψ_T are varied to minimize E_T, and the lowest value is taken as the approximate energy. A good variational trial function for light nuclei has the form [7, 12]

$$|\Psi_T\rangle = \left[1 + \sum_{i<j<k} U_{ijk}^{TNI}\right]\left[\mathcal{S}\prod_{i<j}(1+U_{ij})\right]|\Psi_J\rangle\ . \tag{17}$$

The U_{ij} and U_{ijk}^{TNI} are non-commuting two- and three-nucleon correlation operators (the most important being tensor-isospin induced by v_{ij}^π); $\mathcal{S}$ indicates a symmetric sum over all possible orderings; and Ψ_J is a fully antisymmetric Jastrow wave function which determines the quantum numbers of the state being computed.

The Jastrow wave function, Ψ_J, for p-shell nuclei starts with a sum over independent-particle terms, Φ_A, that have 4 nucleons in an α-like core and $(A-4)$ nucleons in p-shell orbitals. These orbitals are coupled in a $LS[n]$ basis to obtain the desired JM value of a given state, where n specifies the spatial symmetry $[n]$ of the angular momentum coupling of the p-shell nucleons. Different possible $LS[n]$ combinations lead to multiple components in the Jastrow wave function. This independent-particle basis is acted on by products of central pair and triplet correlation functions:

$$\begin{aligned}|\Psi_J\rangle \ &= \ \mathcal{A}\left\{\left[\prod_{i<j<k} f_{ijk}^c\right]\left[\prod_{i<j\leq 4} f_{ss}(r_{ij})\right]\sum_{LS[n]}\left(\beta_{LS[n]}\left[\prod_{k\leq 4<l\leq A} f_{sp}^{LS[n]}(r_{kl})\right]\right.\right. \\ &\times \left.\left.\left[\prod_{4<l<m\leq A} f_{pp}^{LS[n]}(r_{lm})\right]|\Phi_A(LS[n]JMTT_3)_{1234:5\ldots A}\rangle\right)\right\}\ . \end{aligned} \tag{18}$$

The operator $\mathcal{A}$ indicates an antisymmetric sum over all possible partitions of the A particles into 4 s-shell and $(A-4)$ p-shell ones. The pair correlation for both particles within the s-shell, f_{ss}, is similar to that in an α-particle. The pair correlations for both particles in the p-shell, $f_{pp}^{LS[n]}$, and for mixed pairs, $f_{sp}^{LS[n]}$, are similar to f_{ss} at short distance, but their long-range structure is adjusted to give appropriate clustering behavior, and they may vary with $LS[n]$. The Φ_A, and thus Ψ_T, is translationally invariant.

Green's Function Monte Carlo

GFMC projects out the lowest-energy ground state from the VMC Ψ_T by using

$$\Psi(\tau) = e^{-(H-E_0)\tau}\Psi_T ; \tag{19}$$

$$\Psi_0 = \lim_{\tau\to\infty} \Psi(\tau). \tag{20}$$

If sufficiently large τ is reached, the eigenvalue E_0 is calculated exactly while other expectation values are generally calculated neglecting terms of order $|\Psi_0 - \Psi_T|^2$ and higher. Here I present a simplified overview of nuclear GFMC; a rather complete discussion may be found in [7, 12].

The exponential in Eq. (20) is evaluated by introducing a small time step, $\triangle\tau$, $\tau = n\triangle\tau$, (typically $\triangle\tau = 0.5$ GeV^{-1}):

$$\Psi(\tau) = \left[e^{-(H-E_0)\triangle\tau}\right]^n \Psi_T = G^n\Psi_T\,. \tag{21}$$

where G is the short-time Green's function. The $\Psi(\tau)$ is represented by a vector function of $\mathbf{R}$, and the Green's function, $G_{\alpha\beta}(\mathbf{R}',\mathbf{R})$ is a matrix function of $\mathbf{R}'$ and $\mathbf{R}$ in spin-isospin space (labeled by the subscripts α,β), defined as

$$G_{\alpha\beta}(\mathbf{R}',\mathbf{R}) = \langle \mathbf{R}',\alpha | e^{-(H-E_0)\triangle\tau} | \mathbf{R},\beta\rangle. \tag{22}$$

Omitting spin-isospin indices for brevity, $\Psi(\mathbf{R}_n,\tau)$ is given by

$$\Psi(\mathbf{R}_n,\tau) = \int G(\mathbf{R}_n,\mathbf{R}_{n-1})\cdots G(\mathbf{R}_1,\mathbf{R}_0)\Psi_T(\mathbf{R}_0)d\mathbf{R}_{n-1}\cdots d\mathbf{R}_1 d\mathbf{R}_0, \tag{23}$$

and the integration is done by Monte Carlo. We approximate the short-time propagator as a symmetrized product of exact two-body propagators and include the V_{ijk} to first-order. In benchmark calculations [14] of ^{4}He, the GFMC energy had a statistical error of only 20 keV and agreed with the other best results to this accuracy ($< 0.1\%$).

For more than four nucleons, GFMC calculations suffer significantly from the well-known Fermion sign problem. This results in exponential growth of the statistical errors as one propagates to larger τ, or as A is increased. For $A \geq 8$ the resulting limit on τ is too small to allow convergence of the energy. In the last few years we have developed and extensively tested a constrained-path algorithm for nuclear GFMC [12] to solve this problem. In this method configurations with small or negative $\Psi(\tau)^\dagger\cdot\Psi_T$ are discarded such that the average over all discarded configurations of $\Psi(\tau)^\dagger\cdot\Psi_T$ is 0. This means that, if Ψ_T were the true eigenstate, the discarded configurations would contribute nothing but noise to $\langle H\rangle$. This constrained propagation completely controls the growth of the statistical errors and in most cases produces a result that is statistically the same as unconstrained propagation (the accuracy of the comparison may be limited by the statistical errors in the unconstrained result). However we have demonstrated some cases for which constrained propagation leads to a wrong result, and in fact for which the approximate $\langle H\rangle$ is not even an upper bound to the correct eigenvalue. In all cases the correct result can be obtained by making a few (10 to 40) unconstrained

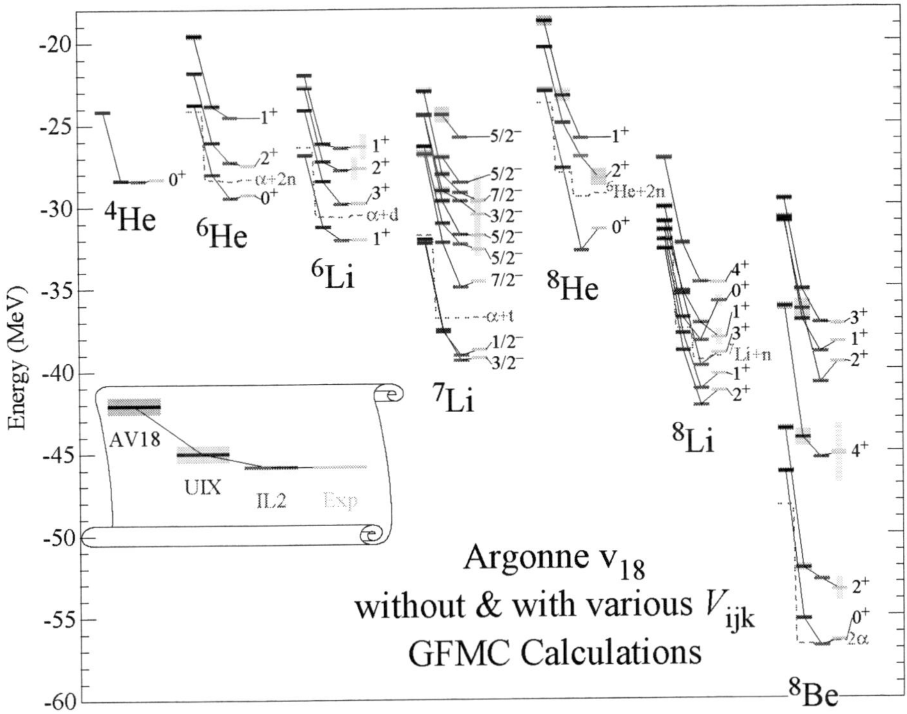

FIGURE 3. GFMC computations of energies for the AV18, AV18+UIX, and AV18+IL2 Hamiltonians compared with experiment. The light shading shows Monte Carlo statistical errors.

steps before evaluating the energy. Our calculations for $A \geq 5$ are now all made using constrained-path propagation with 10 to 40 unconstrained steps.

The number of spin-isospin components in Ψ_T grows rapidly with the number of nucleons. Thus a calculation of a state in ^{8}Be involves about 30 times more floating-point operations than one for ^{6}Li, and ^{10}Be requires 50 times more than ^{8}Be.

ENERGIES OF NUCLEAR STATES

Figure 3 shows the energies of some nuclear states computed using AV18 by itself and with either UIX or IL2. Experimental values are also shown. The strong underbinding when no *NNN* potential is used is clear. The addition of UIX produces a significant improvement but somewhat worsens the Z-dependence for a given A (note ^{8}He, ^{8}Li, and ^{8}Be). The results with IL2 improves this. Figure 4 compares computed and experimental excitation energies for a number of nuclei. Again IL2 fixes problems that arise when just a *NN* potential is used. For example, spin-orbit splittings are usually too small without the *NNN* potential (note the $\frac{1}{2}^- - \frac{3}{2}^-$ and $\frac{5}{2}^- - \frac{7}{2}^-$ splittings in ^{7}Li and the $\frac{1}{2}^- - \frac{3}{2}^-$ splitting in ^{9}Li). In the case of ^{10}B, the correct ground-state spin is obtained only by

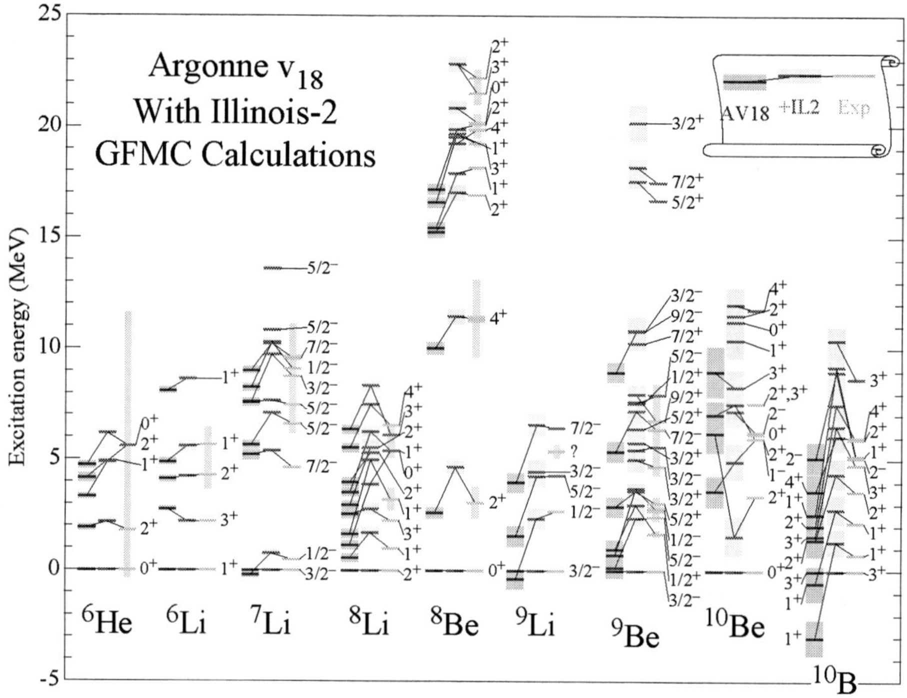

FIGURE 4. GFMC computations of excitation energies compared with experiment. The shading on the experimental energies shows the widths of resonances.

including the *NNN* potential. This figure also shows that GFMC can be used for several states with the same J^{π}. The ability of GFMC to provide such results was demonstrated in Ref. [13].

The accuracy of GFMC calculations has increased since the parameters of IL2 were determined eight years ago and modern calculations of several energy levels (most notably the ground state of ^{8}He) have shifted significantly. In addition unpublished calculations [15, 16] of nuclear and neutron matter showed that the AV18+IL2 Hamiltonian produces too much binding in dense matter. Finally Viviani and Kievsky [17] found that a mistake had been made in programming one term of the three-pion rings part of the potential. The mistake can be compensated for by changing potential parameters; with the modified parameters it appears that the reported features of the AV18+IL2 potential still apply.

A search for a new Illinois potential is now underway. To avoid the overbinding in dense matter, it appears necessary to introduce new isospin dependence in the central, repulsive, term. We do this by adding a new term:

$$V_{ijk}^{R,T=3/2} = A_{R,T=3/2} \sum_{cyclic} T^2(m_\pi r_{ij}) T^2(m_\pi r_{jk}) P_{T=3/2} \,, \tag{24}$$

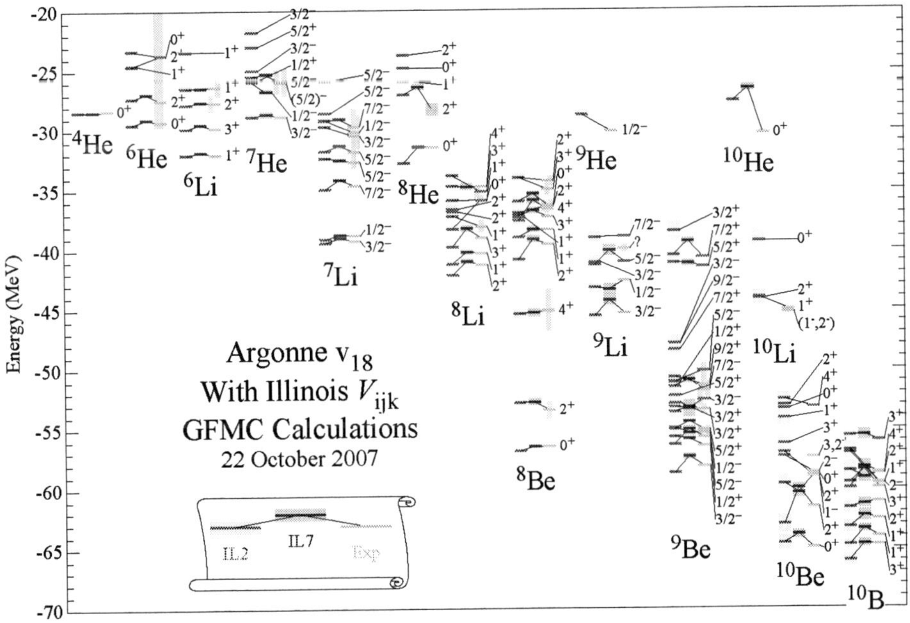

FIGURE 5. Energies for the AV18+IL2 and AV18+IL7 Hamiltonians compared with experiment.

where $P_{T=3/2}$ is a projector onto isospin-3/2 triples. We have not yet found a potential with completely satisfactory dense matter properties. We designate an intermediate result as IL7; its parameters are given in Table 1. The light-nuclei results obtained with it are significantly better than those obtained using IL2, as shown in Fig. 5.

CONCLUSIONS

Quantum Monte Carlo methods are powerful tools for studying light nuclei with realistic nuclear interactions. Calculations of $A = 6 - 12$ nuclear energies with accuracies of $1-2\%$ are possible and the Argonne with Illinois Hamiltonians reproduce nuclear levels up to A=12 well [18]. The *NNN* potential is required for overall p-shell energies and for spin-orbit splittings and several level orderings.

The QMC methods allow the computation of matrix elements of many operators of interest. Recently GFMC values of A=6,7 electromagnetic and weak transitions have also been computed; these improve on older VMC calculations and generally agree with experiment [19]. A recent VMC result for *NN* momentum distributions shows the importance of tensor correlations to an unexpected experimental result [20]. Another topic not covered here is overlap functions and the related spectroscopic factors; these are used as input to calculations (such as distorted-wave Born approximation) of nuclear reactions; recent results are presented in Refs. [21, 22]. A just-finished interesting study used VMC calculations to investigate the effects on nuclear binding energies of changes in the Hamiltonian induced by changes of the fundamental constants [23].

But perhaps the most important advance in nuclear GFMC is the computation of scattering states[24]. In these calculations the correct scattering-wave boundary condition is implemented. It was found that including the IL2 potential was necessary to achieve excellent agreement with experimental phase shifts for neutron-^{4}He scattering.

ACKNOWLEDGMENTS

The work reported in this contribution is the result of collaborations with J. Carlson, S. Gandolfi, A. Kievsky, K. M. Nollett, V. R. Pandharipande (who for many years guided and inspired our group), R. Schiavilla, M. Viviani, and R. B. Wiringa. Extensive computer resources were provided over the years by Argonne's Mathematics and Computer Science Division (most recently on the IBM Blue Gene), Argonne's Laboratory Computing Resource Center, and the U.S. Department of Energy's National Energy Research Scientific Computing Center. This work is supported by the U.S. Department of Energy, Office of Nuclear Physics, under contracts DE-AC02-06CH11357 and DE-FC02-07ER41457.

REFERENCES

1. R. B. Wiringa, V. G. J. Stoks, and R. Schiavilla, Phys. Rev. C **51**, 38 (1995).
2. J. Carlson, V. R. Pandharipande, and R. B. Wiringa, Nucl. Phys. **A401**, 59 (1983).
3. B. S. Pudliner, V. R. Pandharipande, J. Carlson, et al., Phys. Rev. Lett. **74**, 4396 (1995).
4. S. C. Pieper, V. R. Pandharipande, R. B. Wiringa, et al., Phys. Rev. C **64**, 014001 (2001).
5. J. Fujita and H. Miyazawa, Prog. Theor. Phys. **17**, 360 (1957).
6. J. Fujita and H. Miyazawa, Prog. Theor. Phys. **17**, 366 (1957).
7. B. S. Pudliner, V. R. Pandharipande, J. Carlson, et al., Phys. Rev. C **56**, 1720 (1997).
8. S. C. Pieper and R. B. Wiringa, Annu. Rev. Nucl. Part. Sci. **51**, 53 (2001).
9. S. C. Pieper, K. Varga, and R. B. Wiringa, Phys. Rev. C **66**, 044310 (2002).
10. S. A. Coon, M. D. Scadron, P. C. McNamee, et al., Nucl. Phys. **A317**, 242 (1979).
11. R. B. Wiringa, Phys. Rev. C **43**, 1585 (1991).
12. R. B. Wiringa, S. C. Pieper, J. Carlson, et al., Phys. Rev. C **62**, 014001 (2000).
13. S. C. Pieper, R. B. Wiringa, and J. Carlson, Phys. Rev. C **70**, 054325 (2004).
14. H. Kamada, A. Nogga, W. Glöckle, et al., Phys. Rev. C **64**, 044001 (2001).
15. J. Morales and D. G. Ravenhall, private communication (2007).
16. S. Gandolfi, private communication (2007).
17. M. Viviani and A. Kievsky, private communication (2007).
18. S. C. Pieper, Nucl. Phys. **A751**, 516c (2005).
19. M. Pervin, S. C. Pieper, and R. B. Wiringa, Phys. Rev. C **76**, 064319 (2007).
20. R. Schiavilla, R. B. Wiringa, S. C. Pieper, et al., Phys. Rev. Lett. **98**, 132501 (2007).
21. A. H. Wuosmaa, K. E. Rehm, J. P. Greene, et al., Phys. Rev. Lett. **94**, 082502 (2005).
22. A. H. Wuosmaa, K. E. Rehm, J. P. Greene, et al., Phys. Rev. C **72**, 061301(R) (2005).
23. V. V. Flambaum and R. B. Wiringa, Phys. Rev. C **76**, 054002 (2007).
24. K. M. Nollett, S. C. Pieper, R. B. Wiringa, et al., Phys. Rev. Lett. **99**, 022502 (2007).

The Fujita-Miyazawa 3N mechanism and neutron-rich exotic nuclei

Takaharu Otsuka*, Toshio Suzuki† and Yoshinori Akaishi**

**Department of Physics, University of Tokyo, Hongo, Bunkyo-ku, Tokyo 113-0033, Japan, Center for Nuclear Study, University of Tokyo, Hongo, Bunkyo-ku, Tokyo 113-0033, Japan, Nishina Center, RIKEN, Hirosawa, Wako-shi, Saitama, 351-0198, Japan, National Superconducting Cyclotron Laboratory, Michigan State University, East Lansing, Michigan, 48824, USA*

†Department of Physics, Nihon University, Sakurajosui, Setagaya-ku, Tokyo, 156-8550, Japan, Center for Nuclear Study, University of Tokyo, Hongo, Bunkyo-ku, Tokyo 113-0033, Japan

***Nishina Center, RIKEN, Hirosawa, Wako-shi, Saitama, 351-0198, Japan,*

Abstract. We suggest that the Fujita-Miyazawa force plays a crucial role in determining the neutron drip line for the oxygen isotopes. This is one of the consequences of the repulsive effective T=1 force produced by the Fujita-Miyazama mechanism. The same mechanism makes the effective T=1 interaction more repulsive in general.

Keywords: exotic nuclei, three-body force
PACS: 21.30.Fe, 21.60.Cs, 21.10.Pc

INTRODUCTION

The basic mechanism that makes the drip line of the oxygen isotopes lie so near (N=16) to the β stability line compared with the fluorine isotopes [1, 2, 3, 4, 5, 6, 7] has long been a puzzle in the study of the structure of exotic nuclei. In the case of the fluorine isotopes, one proton in the *sd* shell, which is sitting predominantly in the $d_{5/2}$ orbit, pulls down the neutron $d_{3/2}$ orbit as a result of strong effects from the tensor and central forces [8, 9, 10]. On the other hand, in the heavy oxygen isotopes, there are many neutrons in the $d_{5/2}$ orbit, but they do not pull down the neutron $d_{3/2}$ orbit. Otherwise, the drip line of the oxygen isotopes would be shifted to N=20.

We shall describe a possible mechanism to explain this observation. The solution arises from the Fujita-Miyazawa three-nucleon force (3NF) [11]. Other effects will also be discussed.

EFFECTS OF THE FUJITA-MIYAZAWA FORCE ON THE OXYGEN DRIP LINE

The nucleon-nucleon interaction is renormalized by a Delta excitation in the intermediate state. This makes the effective nucleon-nucleon potential much more attractive. A similar mechanism works for the renormalization of the single-particle energy through an interaction between a valence nucleon and a nucleon in the core. There is a Delta particle and a nucleon in the intermediate state. If there are other valence nucleons, some

CP1011, *New Facet of Three Nucleon Force—50 Years of Fujita Miyazawa Three Nucleon Force (FM50), The International Symposium,* edited by H. Sakai, K. Sekiguchi, and B. F. Gibson

nucleon intermediate states are Pauli blocked. This must be properly taken into account. Obviously, this is a repulsive contribution, and this mechanism can be incorporated by making the T=1 effective interaction for valence nucleons more repulsive. This mechanism is nothing but an exchange component of the Fujita-Miyazawa force.

By including this effect, the mean part (monopole component) of the T=1 interaction between $d_{5/2}$ and $d_{3/2}$ becomes repulsive. By including other effects, this repulsive mechanism is not completely canceled, and the total monopole force almost vanishes. This makes the neutron $d_{3/2}$ orbit in the heavier oxygen isotopes lie in the positive energy region, and the oxygen isotopes cease to be bound at N=16.

ADDITIONAL REPULSIVE FORCE IN THE pf SHELL

A basically identical mechanism can be seen in the pf shell. The GXPF1 [12, 13] shell model interaction has a repulsive T=1 monopole term, and this cannot be explained by a G-matrix calculation [14]. As the G-matrix formalism is rather close to optimal, the cause of this additional repulsive effect seems to be ascribed to a non-two-body interaction. In fact, we can obtain a sizable effect from the above mechanism without any adjustment. Thus, the T=1 effective interaction between valence nucleons is renormalized by the Fujita-Miyazawa mechanism, and this renormalization shifts drip lines considerably.

It is quite interesting that the Fujita-Miyazawa mechanism makes such a significant contribution to the structure and existence of exotic nuclei with a large proton-neutron asymmetry.

ACKNOWLEDGMENTS

This work has been supported in part by the RIKEN-CNS large-scale nuclear structure calculation project, and by the JSPS core-to-core project, Int. Res. Network on Exotic Femto Systems (EFES).

REFERENCES

1. M. Thoennessen, auerbachReports on Progress in Physics, 67, 1187 (2004).
2. M. Langevin et al., Phys. Lett. B **150**, 71 (1985).
3. M. Thoennessen, T. Baumann, B. A. Brown, J. Enders, N. Frank, P. G. Hansen, P. Heckman, B. A. Luther, J. Seitz, A. Stolz, and E. Tryggestad, Phys. Rev. **C** 68, 044318 (2003).
4. D. Guillemaud-Mueller, J. C. Jacmart, E. Kashy, A. Latimier, A. C. Mueller, F. Pougheon, A. Richard, Yu. E. Penionzhkevich, A. G. Artuhk, A. V. Belozyorov, S. M. Lukyanov, R. Anne, P. Bricault, C. Détraz, M. Lewitowicz, Y. Zhang, Yu. S. Lyutostansky, M. V. Zverev, D. Bazin, and W. D. Schmidt-Ott, Phys. Rev. **C**, 41, 937 (1990).
5. M. F, D. J. Morrissey, W. Benenson, B. A. Brown, M. Hellström, J. H. Kelley, R. A. Kryger, R. Pfaff, C. F. Powell, and B. M. Sherrill, Phys. Rev. **C**, 53, 647, (1996).
6. H. Sakurai, S.M. Lukyanov, M. Notani, N. Aoi, D. Beaumel, N. Fukuda, M. Hirai, E. Ideguchi, N. Imai, M. Ishihara, et al., Phys. Lett. B **448**, 180 (1999).
7. O. Tarasov, R. Allatt, JC Angélique, R. Anne, C. Borcea, Z. Dlouhy, C. Donzaud, S. Grévy, D. Guillemaud-Mueller, M. Lewitowicz, et al., Phys. Lett. B **409**, 64 (1997).

8. T. Otsuka, *et al.*, Phys. Rev. Lett. **87**, 082502 (2001).
9. T. Otsuka, T. Suzuki, R. Fujimoto, H. Grawe and Y. Akaishi, Phys. Rev. Lett. **95**, 232502 (2005).
10. T. Otsuka, T. Matsuo and D. Abe, Phys. Rev. Lett. **97**, 162501 (2006).
11. J. Fujita and H. Miyazawa, Prog. Theor. Phys. **17** No.3 (1957).
12. M. Honma, T. Otsuka, B.A. Brown and T. Mizusaki, Phys. Rev. C **65**, 061301 (2002).
13. M. Honma, T. Otsuka, B.A. Brown and T. Mizusaki, Phys. Rev. C **69**, 034335 (2004).
14. M. Hjorth-Jensen, T.T.S. Kuo and E. Osnes, Phys. Repts. **261**, 125 (1995).

Three-nucleon interactions: A frontier in nuclear structure

A. Schwenk* and J.D. Holt

TRIUMF, 4004 Wesbrook Mall, Vancouver, BC, Canada, V6T 2A3
**E-mail: schwenk@triumf.ca*

Abstract. Three-nucleon interactions are a frontier in understanding and predicting the structure of strongly-interacting matter in laboratory nuclei and in the cosmos. We present results and discuss the status of first calculations with microscopic three-nucleon interactions beyond light nuclei. This coherent effort is possible due to advances based on effective field theory and renormalization group methods in nuclear physics.

Nuclear interactions depend on a resolution scale, which we denote by a generic momentum cutoff Λ, and the Hamiltonian is always given by an effective theory for nucleon-nucleon (NN) and corresponding many-nucleon interactions (with corresponding effective operators) [1, 2, 3]:

$$H(\Lambda) = T + V_{\rm NN}(\Lambda) + V_{\rm 3N}(\Lambda) + V_{\rm 4N}(\Lambda) + \ldots . \qquad (1)$$

This scale dependence is similar to the scale dependence of parton distribution functions, and shows that the effect of 3N interactions depends on this scale and the theoretical convention that enters all parts of the Hamiltonian.

At very low momenta $Q \ll m_\pi \approx 140\,\text{MeV}$, the details of pion exchanges are not resolved and nuclear forces can be systematically expanded in contact interactions and their derivatives [1]. The corresponding pionless effective field theory (EFT) is extremely successful for capturing universal large scattering-length physics (with improvements by including effective range and higher-order terms) in loosely-bound or halo nuclei, reactions at astrophysical energies and in dilute neutron matter (see e.g., Ref. [4]).

For most nuclei, the typical Fermi momenta are $Q \sim m_\pi$ and therefore pion exchanges have to be included explicitly. In chiral EFT, nuclear interactions are then organized in an expansion in powers of $Q/\Lambda_{\rm b}$, where $\Lambda_{\rm b}$ denotes the breakdown scale, roughly $\Lambda_{\rm b} \sim m_\rho$ [1, 2]. The great advantage is that up to next-to-next-to-next-to-leading order ($\rm N^3LO$) only two new 3N couplings enter, since parts of the 3N force can be consistently determined by πN and NN couplings. The two 3N couplings, as well as the short-ranged NN couplings, depend on the resolution scale Λ and for each Λ can be fit to data.

Using the renormalization group (RG), we can evolve an initial potential to lower resolution by integrating out high momenta through discretized RG flow equations or equivalent Lee-Suzuki transformations [3, 5]. In the last two years, these methods have been refined to employ smooth regulators [6] and similarity renormalization group (SRG) transformations [7, 8], which both have technical advantages in oscillator spaces.

Changing the cutoff leaves observables unchanged by construction, but shifts contributions between the potential and the sums over intermediate states in loop integrals.

CP1011, *New Facet of Three Nucleon Force—50 Years of Fujita Miyazawa Three Nucleon Force (FM50), The International Symposium,* edited by H. Sakai, K. Sekiguchi, and B. F. Gibson

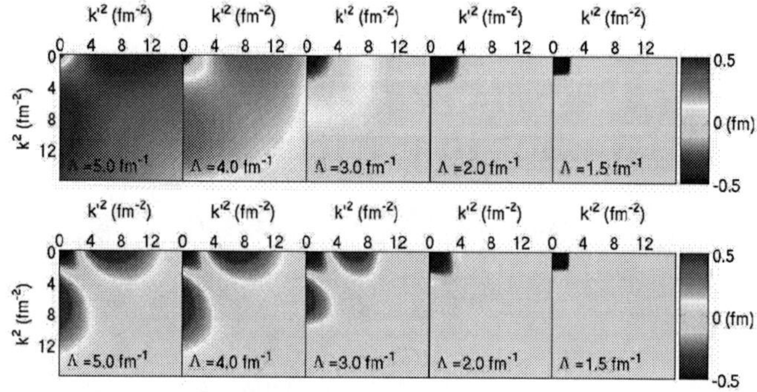

FIGURE 1. Evolution of the $^3S_1$ partial wave with a smooth $n_{\rm exp} = 8$ regulator at cutoffs $\Lambda = 5.0, 4.0, 3.0, 2.0$ and $1.5\,{\rm fm}^{-1}$ (for details see Ref. [6]). The initial potentials are Argonne v_{18} [11] (top) and a N^{3}LO chiral EFT potential from Ref. [12] (bottom). The color scale ranges from -0.5 to $0.5\,$fm.

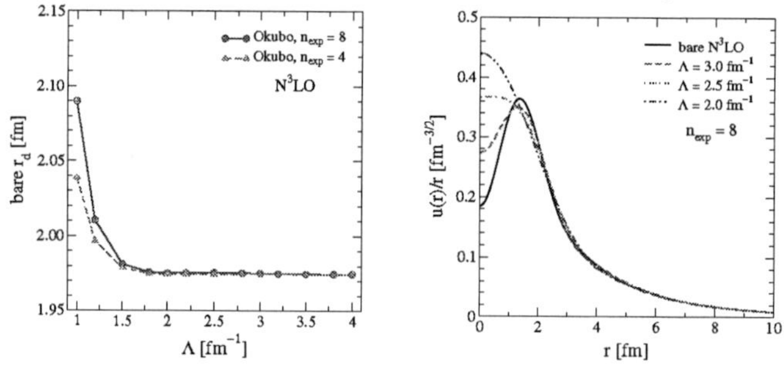

FIGURE 2. The deuteron rms radius r_d calculated with the bare operator as a function of the cutoff (left). The radius is only weakly dependent on the cutoff for $\Lambda > 1.5\,{\rm fm}^{-1}$, although the intrinsic resolution in the S-state deuteron wave function changes substantially with Λ (right). For details see Ref. [6].

Since the sums are restricted by the intrinsic resolution Λ, these shifts can weaken or largely eliminate sources of nonperturbative behavior such as strong short-range repulsion and short-range tensor forces [9, 10]. As shown in Fig. 1, the evolution leads to low-momentum interactions, generally known as "$V_{{\rm low}\,k}$", which become universal for $\Lambda \lesssim 2\,{\rm fm}^{-1}$ and have weak off-diagonal coupling. Since the RG preserves the long-range parts of nuclear interactions [6], this leads to a weak renormalization of long-range operators, for example for the deuteron rms radius in Fig. 2.

The evolution of chiral EFT interactions to lower cutoffs is beneficial for two reasons. First, the RG generates all higher-order short-range contact interactions so that observables are exactly reproduced and the theoretical uncertainty remains at the level of the truncation error in the initial potential. We illustrate this in Fig. 3 by comparing the flow of $V_{{\rm low}\,k}(k' = 0, k = 0; \Lambda)$ with the corresponding momentum-independent contact interaction $C_0(\Lambda)$ in subsequent orders of pionless EFT. Second, lower resolutions can be represented efficiently in oscillator bases, and therefore lead to direct convergence in nuclear structure calculations [6]. This is demonstrated by the very promising convergence for $N_{\rm max} \sim 10$ in NCSM calculations with SRG interactions [13] shown in Fig. 4.

$V_{{\rm low}\,k}(\Lambda)$ defines a class of NN interactions with cutoff-independent low-energy NN

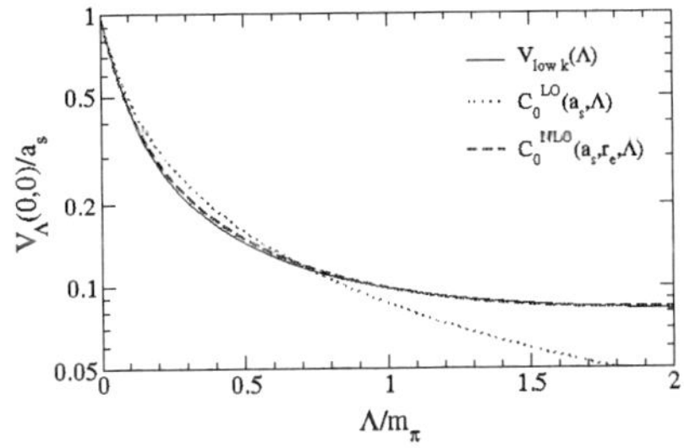

FIGURE 3. Flow of $V_{\mathrm{low}k}(k'=0,k=0;\Lambda)$ compared to the corresponding momentum-independent contact interaction $C_0(\Lambda)$ at LO and NLO, where this coupling is determined entirely from RG invariance and fits to the scattering length a_s (at LO) plus effective range r_e (at NLO).

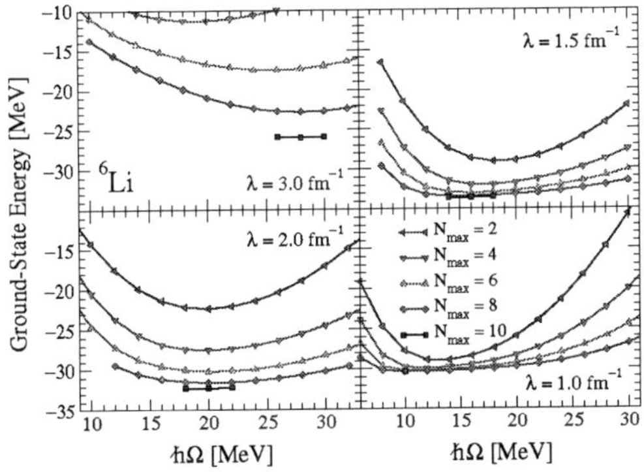

FIGURE 4. Ground-state energy of ^{6}Li as a function of the oscillator parameter $\hbar\Omega$ at four different values of SRG resolution $\lambda = 3, 2, 1.5$ and $1\,\mathrm{fm}^{-1}$. The NCSM results clearly show improved convergence with the maximum number of oscillator quanta $N_{\max}$ for lower cutoffs. For details see Ref. [13].

observables. Consequently, any cutoff variation of observables estimates the truncation errors due to neglected many-body interactions in $H(\Lambda)$ of Eq. (1). Figure 5 shows that this cutoff dependence explains the empirical Tjon line, that 3N interactions are required by renormalization and needed to break off the line to accurately describe the experimental ^{3}H and ^{4}He binding energies [14]. NN-only results also lead to Tjon lines in medium-mass nuclei, where results lie approximately on the same lines [15].

Three-nucleon interactions are crucial for many-nucleon systems. In addition to the impact on binding energies, they play a central role for spin-orbit and spin dependences, for isospin dependences of neutron- and proton-rich systems, and they drive the density dependence of nucleonic matter. The latter are pivotal for extrapolations to the extremes of astrophysics. Three-nucleon interactions are also a frontier in few-body scattering (for a critical discussion, see e.g., Ref. [16]). Since 3N contributions are amplified in nuclei, it may be necessary for controlled predictions to constrain 3N couplings with few- and many-body data. Therefore a coherent 3N effort is needed with theoretical uncertainties.

In chiral EFT without explicit Deltas, 3N interactions start at N^{2}LO or order $(Q/\Lambda_b)^3$ [1, 2] and their contributions are given diagrammatically in Fig. 6. The low-energy constants for the long-range parts relate πN, NN and 3N interactions, and the determination from πN scattering is, within errors, consistent with the extraction

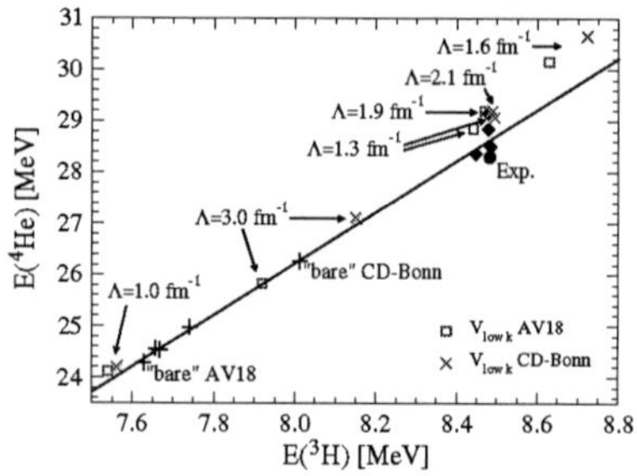

FIGURE 5. Correlation of the ^{3}H and ^{4}He binding energies. The cutoff dependence of the exact NN-only results with $V_{\mathrm{low}k}(\Lambda)$ explains the empirical (solid) Tjon line. For details see Ref. [14].

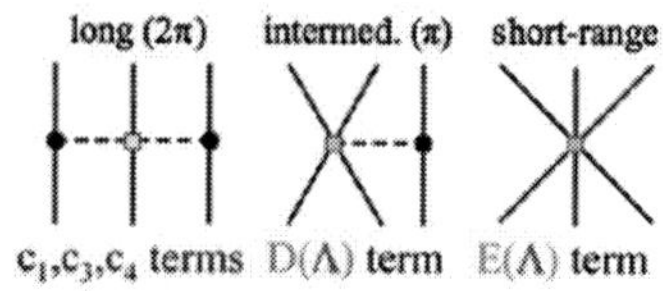

FIGURE 6. Leading (N^{2}LO) 3N interaction in chiral EFT without explicit Deltas.

from peripheral NN waves. The present constraints are $c_1 = -0.9^{+0.2}_{-0.5}$, $c_3 = -4.7^{+1.2}_{-1.0}$ and $c_4 = 3.5^{+0.5}_{-0.2}$ (all in GeV^{-1}) [17]. In particular, c_3 and c_4 are important for nuclear structure and have large uncertainties (see Fig. 10 for the impact on nucleonic matter).

For lower cutoffs, we take the corresponding 3N interactions $V_{3\mathrm{N}}(\Lambda)$ from chiral EFT by fitting the D- and E-term couplings to the ^{3}H and ^{4}He binding energies for a range of cutoffs [14]. Since chiral EFT is a complete basis, this gives the 3N force up to truncation errors. We have found that the resulting 3N interactions become perturbative for $\Lambda \lesssim 2\,\mathrm{fm}^{-1}$, while they are nonperturbative for larger cutoffs, and that the size of 3N expectation values is natural $\sim (Q/\Lambda_{\mathrm{b}})^3 \langle V_{\mathrm{low}k} \rangle$ [14].

As for the Tjon line, cutoff variation can provide lower bounds for theoretical uncertainties due to neglected many-body interactions or an incomplete many-body treatment. As shown in Fig. 7, this is a powerful and practical tool, which is important for extrapolations, in particular for matrix elements needed in fundamental symmetry tests.

Coupled-cluster (CC) theory combined with rapid convergence for low-momentum interactions pushes the limits of accurate calculations to medium-mass nuclei and sets new benchmarks for ^{16}O and ^{40}Ca [19]. In Fig. 8, first CC results with 3N forces show that low-momentum 3N interactions are accurately treated as effective 0-, 1- and 2-body terms, and that residual 3N interactions can be neglected [19]. This is promising and supports the idea that phenomenological monopole shifts in shell model interactions are due to 3N contributions [20], see also the talk by T. Otsuka and Ref. [21]. This links understanding the shell model and the drip lines to 3N forces. First investigations in this direction are presented in Fig. 9 [22]. We speculate that the $T = 0$ monopoles with large cutoff dependences receive cutoff-dependent contributions mainly from second-order NN-3N terms (expected to be attractive) and that the $T = 1$ monopoles receive cutoff-independent contributions from the 3N c_i terms (repulsive in nuclear matter). Systematic investigations of 3N forces and valence-shell interactions are in progress [22].

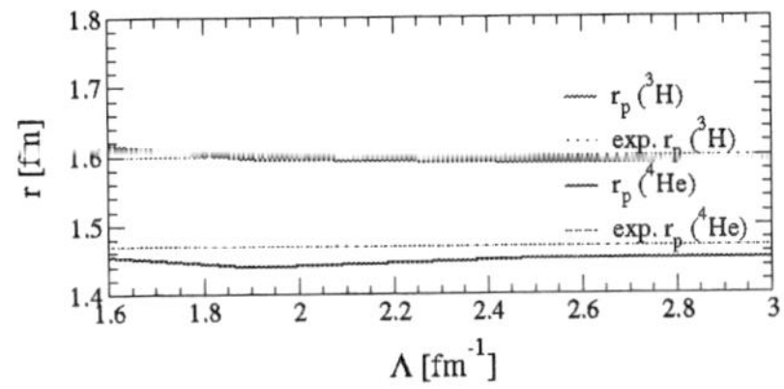

FIGURE 7. The ^{3}H and ^{4}He radii are approx. cutoff independent with NN and 3N interactions [18].

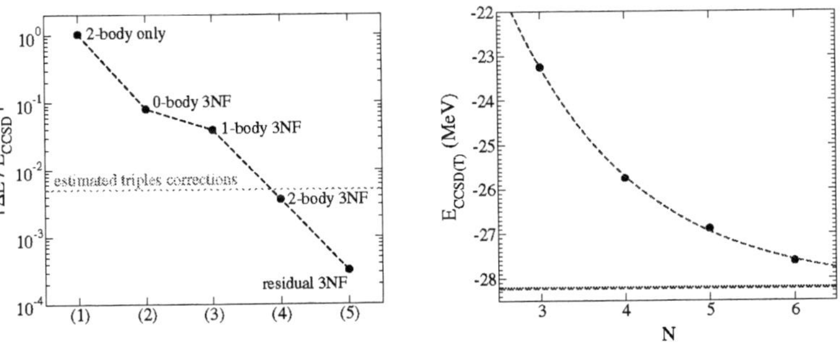

FIGURE 8. Relative contributions $|\Delta E/E|$ to the binding energy of ^{4}He at the CCSD level from $V_{\text{low}k}$ as well as normal-ordered 0-, 1-, 2-body and residual 3-body parts of the 3N interaction (left). Convergence of the corresponding CCSD(T) results with the number of oscillator shells (right). The extrapolated binding energy -28.23 MeV agrees well with the exact result $-28.20(5)$ MeV. For details see Ref. [19].

Finally, there is the possibility of perturbative nuclear matter with low-momentum NN and 3N interactions [9]. As shown in Fig. 10, with second-order contributions the cutoff dependence is very weak at low densities. 3N forces drive saturation [9]. At present the uncertainties in the c_i overwhelm other uncertainties (see the results for neutron matter) [23]. Lower values for c_3 (as expected from N^{3}LO 3N terms) add ≈ -2 MeV at saturation density in neutron matter; this also corresponds to the value from Delta resonance saturation (explaining the agreement with the results of Akmal et al.). These findings provide guidance for constructing a universal density functional for nuclei. In denser matter, 4N force contributions of $E/A \sim 1$ MeV would not be unreasonable.

This is an exciting era: with advances on many fronts, a coherent effort to understand and predict nuclear systems based on effective field theory and renormalization group interactions, where 3N forces are a frontier.

It is a pleasure to thank the organizers for a very stimulating FM50 conference and the collaborators who have contributed to these results: S. Bogner, G. Brown, D. Dean, B. Friman, R. Furnstahl, G. Hagen, T. Kuo, P. Maris, A. Nogga, T. Papenbrock, R. Perry, L. Tolos, J. Vary and A. Zuker. This work was supported by NSERC. TRIUMF receives federal funding via a contribution agreement through the NRC of Canada.

REFERENCES

1. P.F. Bedaque and U. van Kolck, *Ann. Rev. Nucl. Part. Sci.* **52**, 339 (2002).
2. E. Epelbaum, *Prog. Part. Nucl. Phys.* **57**, 654 (2006).

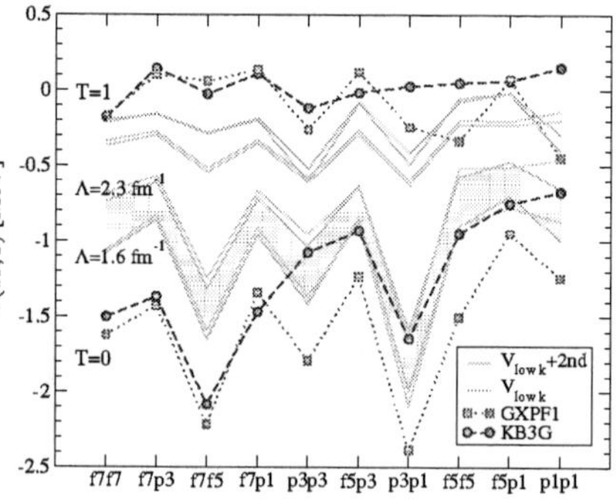

FIGURE 9. Monopole interactions in the pf shell obtained from $V_{\text{low}k}$ plus second-order many-body contributions for a range of cutoffs, compared to phenomenological matrix elements [22].

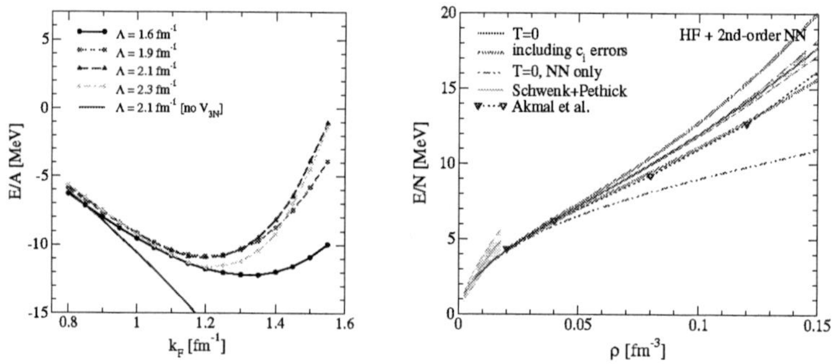

FIGURE 10. Nuclear (left) and neutron matter (right) from low-momentum NN and 3N interactions based on Hartree-Fock plus dominant second-order contributions for various cutoffs. For details see Ref. [9, 23].

3. S.K. Bogner, T.T.S. Kuo and A. Schwenk, *Phys. Rept.* **386**, 1 (2003).
4. A. Schwenk and C.J. Pethick, *Phys. Rev. Lett.* **95**, 160401 (2005).
5. S.K. Bogner, A. Schwenk, T.T.S. Kuo and G.E. Brown, nucl-th/0111042.
6. S.K. Bogner, R.J. Furnstahl, S. Ramanan and A. Schwenk, *Nucl. Phys. A* **784**, 79 (2007).
7. S.K. Bogner, R.J. Furnstahl and R.J. Perry, *Phys. Rev. C* **75**, 061001(R) (2007).
8. S.K. Bogner et al., *Phys. Rev. C* in press, arXiv:0801.1098.
9. S.K. Bogner, A. Schwenk, R.J. Furnstahl and A. Nogga, *Nucl. Phys. A* **763**, 59 (2005).
10. S.K. Bogner, R.J. Furnstahl, S. Ramanan and A. Schwenk, *Nucl. Phys. A* **773**, 203 (2006).
11. R.B. Wiringa, V.G.J. Stoks and R. Schiavilla, *Phys. Rev. C* **51**, 38 (1995).
12. D.R. Entem and R. Machleidt, *Phys. Rev. C* **68**, 041001(R) (2003).
13. S.K. Bogner et al., *Nucl. Phys. A* **801**, 21 (2008).
14. A. Nogga, S.K. Bogner and A. Schwenk, *Phys. Rev. C* **70**, 061002(R) (2004).
15. G. Hagen, D.J. Dean and A. Schwenk, in prep.
16. H.-O. Meyer, TRIUMF 3N Workshop (2007), see http://www.triumf.info/hosted/TNI/index.html.
17. U.G. Meißner, private comm. (2007).
18. A. Nogga, private comm. (2004).
19. G. Hagen et al., *Phys. Rev. C* **76**, 034302 (2007); ibid., **76**, 044305 (2007).
20. A.P. Zuker, *Phys. Rev. Lett.* **90**, 042502 (2003).
21. A. Schwenk and A.P. Zuker, *Phys. Rev. C* **74**, 061302(R) (2006).
22. J.D. Holt and A. Schwenk, in prep.
23. L. Tolos, B. Friman and A. Schwenk, arXiv:0711.3613.

Nonlocal $\alpha\alpha$ potential for ^{12}C as a 3α system

M. Orabi*, Y. Suzuki† and Y. Fujiwara**

*Graduate School of Science and Technology, Niigata University, Niigata 950-2181, Japan
Department of Physics, Faculty of Science, Cairo University, Giza 12613, Egypt
†Department of Physics, and Graduate School of Science and Technology, Niigata University, Niigata 950-2181, Japan
**Department of Physics, Kyoto University, Kyoto 606-8502, Japan

Abstract. Cluster-cluster local potentials reproducing the phase shifts of binary systems are usually adopted to describe few cluster sytems for their simplicity, but it turns out that the nonlocal nature of these composite clusters is something that we cannot discard.

Keywords: nonlocal, 3α, ^{12}C, RGM, BFW, forbidden-states, Ali-Bodmer
PACS: 21.45.-v, 21.60.Gx, 27.20.+n

INTRODUCTION

Clusters are composite particles consisting of a number of constituents. A nucleon for example consists of quarks and gluons, and an alpha particle consists of nucleons, if we forget about the structure of the nucleon itself. Therefore to construct a realistic nucleon-nucleon (NN) potential the constituents of the nucleon must be taken into account. So essentially the NN potentials are nonlocal and, in spite of this, usually local ones are used to simplify calculations. These local potentials are parametrized so as to reproduce the NN phase shifts. The same analogy can be seen when we speak about the $\alpha\alpha$ potential. The α particle is composed of four nucleons, and in order to construct a realistic $\alpha\alpha$ potential we have to take this structure into account [1]. But again this is complicated and we usually try to use simple local $\alpha\alpha$ potentials built phenomenologically to reproduce some experimental data like the $\alpha\alpha$ phase shifts. These potentials give reasonable results for the 2α system, but not for 3α (^{12}C) [2].

The ^{12}C nucleus described as a 3α cluster system has been studied by many researchers. Considering only the 0^+ states, the results for the second 0^+ state (0_2^+) are acceptable even with simple $\alpha\alpha$ potentials, cf. [2, 3]. On the other hand, the ground state (0_1^+) is not well reproduced when using such simple $\alpha\alpha$ potentials like the Buck-Friedrich-Wheatley (BFW) [4] and Ali-Bodmer (AB) [5] local potentials [2, 3, 6]. In these calculations only the two-cluster (2α) forces are taken into account. This unsatisfactory situation of using local potentials motivates us to use nonlocal ones. In the present work we use a sophisticated $\alpha\alpha$ potential constructed from a microscopic origin based on the resonating group method (RGM) [7]. The resulting potential is nonlocal and the comparison of this semi-microscopic model with a fully microscopic calculation provides us with a measure of the importance of exchange effects among the simultaneously interacting three clusters (3α) in the states studied.

CP1011, *New Facet of Three Nucleon Force—50 Years of Fujita Miyazawa Three Nucleon Force (FM50), The International Symposium,* edited by H. Sakai, K. Sekiguchi, and B. F. Gibson

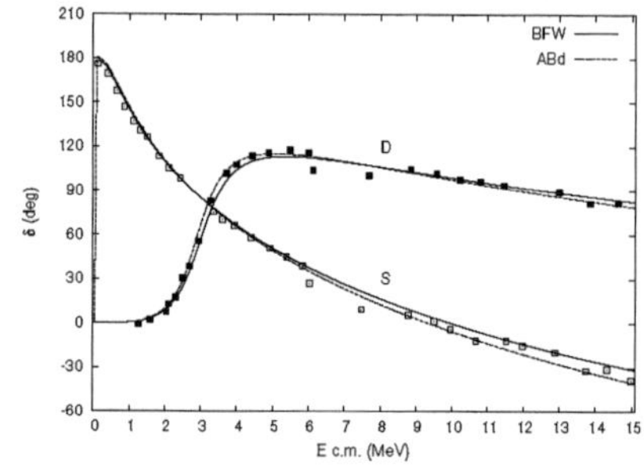

FIGURE 1. S and D partial wave $\alpha\alpha$ phase shifts using the BFW and AB local potentials in comparison with the experimental data.

LOCAL POTENTIALS

The main purpose of this paper is to illustrate the features of $\alpha\alpha$ nonlocal potentials and their results for 3α systems. So, in this section we summarize briefly some of the results which we obtain when we use local potentials, and in the next section we show the results of nonlocal potentials and compare both cases. For the local $\alpha\alpha$ potentials we use the two well-known potentials BFW and AB. The BFW potential is deep, L-independent and has three bound states, $0S$, $1S$ and $0D$. On the other hand, the AB potential is shallow, L-dependent and has no bound states. Although these two potentials are very different in their main features, both of them reproduce very well the $\alpha\alpha$ phase shifts and the ^{8}Be resonance. This is shown in Fig. 1 for both potentials in comparison with the experimental data for the S and D waves. However when we use these potentials in the 3α system we get poor results in comparison with experiment and even with microscopic calculations [2]. This motivated us to try to examine the reason behind this unacceptable behavior [2]. As mentined above, the BFW potential has three bound states. These states are unphysical, and we call them forbidden or redundant states, since it is known that ^{8}Be has no bound states, and they should be eliminated in the 3α system. They can be defined in one of two ways, either as the bound states of the BFW potential or as harmonic oscillator (HO) wave functions. In the investigation reported in Ref. [2] we adopted both definitions of the forbidden states and compared them with different methods for expressing and diagonalizing the hamiltonian. To summarize our conclusion in that paper: we found that it is safer to define the forbidden states as HO wave functions, otherwise the 3α macroscopic model loses a link to the microscopic 3α model.

NON-LOCAL POTENTIALS

The form of the hamiltonian is the same as in the case of local potentials, namely

$$\widetilde{H} = H + \lambda\Gamma, \tag{1}$$

which is called a pseudo hamiltonian. The only difference is that we replace the local potential by a nonlocal one. The projection operator Γ is defined by

$$\Gamma = \sum_{i<j=1}^{3} \Gamma_{ij}, \qquad \Gamma_{ij} = \sum_{f} \left|\chi_f^{ij}\right\rangle \left\langle \chi_f^{ij}\right|, \tag{2}$$

and the constant energy λ is selected large enough to get rid of the forbidden states. The hamiltonian H is the usual sum of the kinetic and potential energies. The nonlocal potential is derived from an effective NN interaction based on the microscopic RGM equation. So this model is semi-microscopic in the sense that exchanges involving two cluters are taken into account but not those involving three clusters together. The main motivation of this semi-microscopic model is to get more realistic results than using local potentials while avoiding at the same time the complication of a microscopic scheme. A first attempt at a 3α model with energy-dependent nonlocal $\alpha\alpha$ RGM kernels [7, 8] was developed in Ref. [9]. This interaction contains all antisymmetrization effects in the $\alpha\alpha$ system. It accurately describes the ^{8}Be ground-state resonance and the $\alpha\alpha$ phase shifts. The only model assumptions are: a simple $(0s)^4$ description of the α clusters within the translation-invariant HO model and the use of an effective nucleon-nucleon interaction [10]. Employing the resulting nonlocal $\alpha\alpha$ interaction is rendered complicated by the existence of an energy dependence and of Pauli-forbidden states. Within this semi-microscopic 3α model, the difficulties associated with the accuracy of the forbidden-state elimination disappear [11, 12]. Despite a reasonable success for the ground state, the problem now is that the 0_2^+ excited state is not obtained simultaneously. Moreover, the energy dependence of the $\alpha\alpha$ nonlocal interaction raises another difficulty. This energy is well defined in a two-body system but not in a three-body system. Choosing a definition raises another ambiguity of the model [11].

The three-body equation including clusters 1, 2 and 3 reads

$$(T_1 + T_2 + T_3 - T_{\mathrm{cm}} + V_{12} + V_{13} + V_{23})\Psi = E\Psi, \tag{3}$$

where T_i is the kinetic energy of the cluster i and the center-of-mass kinetic energy T_{cm} is subtracted. The intercluster potential V_{ij} between the ij pair can be either local in the macroscopic model or nonlocal in the semi-microscopic model. Except for shallow potentials, Eq. (3) implicitly assumes that Ψ belongs to the Pauli-allowed space of the three $\alpha\alpha$ pairs. When V_{ij} is a nonlocal RGM potential derived from an effective nucleon-nucleon interaction, the three-body equation does not contain any free parameter. This semi-microscopic model is quite different from the orthogonality-condition model [8] where the full antisymmetrization is approximately taken into account through the orthogonality with respect to the forbidden states, but where the $\alpha\alpha$ interactions are phenomenological.

Before solving Eq. (3), we briefly recall the characteristics of two variants of nonlocal RGM potentials. The RGM equation for the intercluster relative motion function χ reads

$$(T + V + \varepsilon K)\chi = \varepsilon\chi, \qquad V = V_D + V_{EX}, \tag{4}$$

where T is the intercluster kinetic energy, V_D is local (direct) potential, the nonlocal potential is $V_{EX} = K_T + K_V$, and K_T and K_V are the exchange nonlocal kernels for the kinetic and potential energies (including the Coulomb term), respectively. The energy ε is defined with respect to the two-cluster threshold. Equation (4) suggests that the microscopically founded intercluster potential is

$$V^{\varepsilon K} = V + \varepsilon K, \tag{5}$$

which is energy-dependent. The norm kernel $\mathcal{N}$ is defined by

$$\mathcal{N} = 1 - K, \tag{6}$$

where K is the overlap kernel. When applying the RGM potential Eq. (5) for V_{ij}, one must note that a particular function χ_f, called a Pauli-forbidden state (PFS), exists in general. It is defined as a solution of the eqaution, $\mathcal{N}\chi_f = 0$ or $K\chi_f = \chi_f$. The three-body wave function Ψ of Eq. (3) is constrained to be orthogonal to all the two-cluster PFSs which exist for the three intercluster pairs,

$$< \chi_f | \Psi >= 0. \tag{7}$$

A potential which has no explicit energy-dependence and still reproduces basic properties such as the phase shifts can be constructed for a macroscopic (or renormalized) relative motion function $g = \sqrt{\mathcal{N}}\chi$, by requiring the condition [13]

$$(T + V^{RGM})g = \varepsilon g \tag{8}$$

be satisfied. Comparing Eqs. (4) and (8) gives the nonlocal potential V^{RGM} as

$$V^{RGM} = \mathcal{N}^{-1/2}(T+V)\mathcal{N}^{-1/2} - T = V + W, \tag{9}$$

where the new nonlocal operator W is the difference between the renormalized RGM potential V^{RGM} and the bare RGM potential V. The relative motion function g has the useful property of preserving the orthonormality of the microscopic two-cluster wave function. In addition, the asymptotic form of g is the same as that of χ because $\mathcal{N}$ approaches unity at large distances. The phase shift determined from g using the potential V^{RGM} is exactly equal to that determined from the RGM Eq. (4) for χ.

The potential V^{RGM} is defined in the PFS-free space. To make this point more transparent, we consider the eigenvalue problem $K\psi_\kappa = \kappa\psi_\kappa$. The PFSs are nothing but the eigenfunctions with eigenvalue $\kappa = 1$. For cluster wave functions described with HO functions with a common size parameter, this eigenvalue problem can be solved. The eigenfunctions are HO wave functions $\psi_{n\ell m}$, and the corresponding eigenvalues are given by $\kappa_{n\ell} = 4(1/2)^{2n+\ell} - 3\delta_{2n+\ell,0}$ for $\alpha + \alpha$. There are thus three PFSs, ψ_{000}, ψ_{100} and ψ_{02m} for even ℓ, and all states are PFSs for odd ℓ.

The potential V^{RGM} is expanded into partial waves as

$$\begin{aligned} V_\ell^{RGM}(\mathrm{r},\mathrm{r}') &= 2\pi \mathrm{r}\mathrm{r}' \int_{-1}^{1} V^{RGM}(\mathbf{r},\mathbf{r}')P_\ell(t)\,dt \\ &= V_D(\mathrm{r})\,\delta(\mathrm{r}-\mathrm{r}') + V_{EX\ell}(\mathrm{r},\mathrm{r}') + W_\ell(\mathrm{r},\mathrm{r}'), \end{aligned} \tag{10}$$

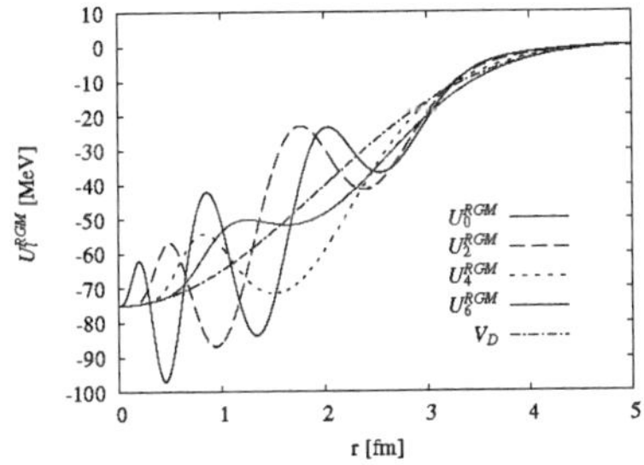

FIGURE 2. The potential function U_ℓ^{RGM} and the direct potential V_D for the $\alpha\alpha$ system.

with $t = \mathbf{r}\cdot\mathbf{r}'/\mathrm{r}\mathrm{r}'$ and P_ℓ is a Legendre polynomial. The potential V_ℓ^{RGM} acts on g_ℓ, the macroscopic wave function in partial wave ℓ. Using Eq. (10) we can calculate a so called potential function [13]. The function is not a phase-equivalent potential to the original nonlocal potential, and our purpose is just to look at its behaviour. The potential function is defined by

$$U_\ell^{RGM}(\mathrm{r}) = \int_{-2\mathrm{r}}^{2\mathrm{r}} V_\ell^{RGM}\left(\mathrm{r}+\frac{s}{2},\mathrm{r}-\frac{s}{2}\right)ds = V_D(\mathrm{r}) + U_\ell^{NL}(\mathrm{r}),$$

where the term U_ℓ^{NL} comes from the antisymmetrization effect in the nonlocal RGM kernels and from the effect of eliminating the energy-dependence of these kernels. Therefore the potential function serves as a measure of how much these effects modify V_D. This is shown in Fig. 2 where one can see the curves of U_ℓ^{RGM} oscillate around V_D and their magnitude is significant even near the tail.

RESULTS AND DISCUSSION

The oscillator parameter $\nu = 0.257$ fm^{-2} and the exchange parameter (Minnesota) $u = 0.94687$, taken from Ref. [14], reproduce the $\alpha\alpha$ phase shifts. The values $\hbar^2/m_N = 41.47$ MeV fm^2 and $m_\alpha = 4m_N$ for the nucleon and alpha masses are used in all the calculations. The Minnesota NN potential used in the microscopic calculation is the same as the one used to derive the RGM kernels. All the models reproduce the $\alpha\alpha$ phase shifts in an essentially identical way. The root-mean-square (rms) radius of the point α-particle distribution is obtained for the microscopic calculation by subtracting r_α^2 ($\mathrm{r}_\alpha = 1.479$ fm) from the $\langle \mathrm{r}^2 \rangle$ value calculated at the nucleon level. The results are summarized in Table 1 which compares the energies and rms radii (point α) of the 3α system obtained using local potentials (Local), the nonlocal $V^{\varepsilon K}$ and V^{RGM} potentials (Semi-micro) as well as a fully microscopic calculation (Micro) [15, 16]. Experimental energies are -7.27 and 0.38 MeV for the 0_1^+ and 0_2^+ states, respectively. Equation (3) is solved under different conditions depending on the potential model. In the Semi-micro model, the HO-PFSs are used for χ_f in all the cases, and in addition the self-consistency condition described in Refs. [9, 11] is put in the case of $V^{\varepsilon K}$. In this model the 0_1^+ and 0_2^+ states cannot be obtained simultaneously by changing the value of ε without losing their mutual orthogonality. For the 0_2^+ state our calculations here suggest that it is very large, around 3.5 fm.

TABLE 1. 3α results (see the text).

Model	Potential	L^π	E (MeV)	$\sqrt{\langle r^2 \rangle}$ (fm)
Local	ABd	0_1^+	−1.52	2.34
	BFW	0_1^+	−20.62	1.29
Semi-micro	$V^{\varepsilon K}$	0_1^+	−9.60	1.45
	V^{RGM}	0_1^+	−9.44	1.62
		0_2^+	0.597	
Micro		0_1^+	−11.37	1.64
		0_2^+	0.598	

CONCLUSION

We emphasize that the results of the Semi-micro model are much better than those of local potentials. One advantage for V^{RGM} over $V^{\varepsilon K}$ is that we can now calculate all the 0^+ states at the same time. Also one notices that the rms radius of V^{RGM} is in better agreement with the microscopic result than that of $V^{\varepsilon K}$. Though the Semi-micro model with V^{RGM} is qualitatively successful, its ground state energy is about 1.9 MeV higher than that for Micro. The former takes into account the binary exchanges between α clusters but not of simultaneous exchanges among the three α clusters. Such three-cluster exchanges produce attraction as observed here. This effect is rather significant in the compact ground state but negligible in the spatially extended 0_2^+ state. Extension of our work to use the V^{RGM} potential for calculating ^{16}O as a 4α system is now in progress.

REFERENCES

1. Y. Fujiwara et al., *PRC* **66**, 021001(R) (2002).
2. H. Matsumura, M. Orabi, Y. Suzuki, and Y. Fujiwara, *Nucl. Phys.* **A 776**, 1 (2006).
3. E.M. Tursunov, D. Baye, and P. Descouvemont, *Nucl. Phys.* **A 723**, 365 (2003).
4. B. Buck, H. Friedrich, and C. Wheatley, *Nucl. Phys.* **A 275**, 246 (1977).
5. S. Ali, and A. Bodmer, *Nucl. Phys.* **80**, 99 (1966).
6. Y. Suzuki et al., *Phys. Lett.* **B 659**, 160 (2008).
7. K. Widermuth, and Y.C. Tang, *A Unified Theory of the Nucleus*, Braunschweig: Vieweg (1977).
8. S. Saito, *Prog. Theor. Phys. Suppl.* **62**, 11 (1977).
9. Y. Fujiwara, H. Nemura, Y. Suzuki, K. Miyagawa, and M. Kohno, *Prog. Theor. Phys.* **107**, 745 (2002).
10. D.R. Thompson, M. LeMere, and Y.C. Tang, *Nucl. Phys.* **A 286**, 53 (1977).
11. M. Theeten, D. Baye, and P. Descouvemont, *Phys. Rev.* **C 74**, 044304 (2006).
12. Y. Fujiwara, Y. Suzuki, and M. Kohno, *PRC* **69**, 037002 (2004); *Few-Body Systems* **34**, 237 (2004).
13. W. Timm, H.R. Fiebig, and H. Friedrich, *Phys. Rev.* **C 25**, 79 (1982).
14. Y. Fujiwara et al., *PRC* **70**, 024002 (2004).
15. H. Matsumura, and Y. Suzuki, *Nucl. Phys.* **A 739**, 238 (2004).
16. S. Korennov, and P. Descouvemont, *Nucl. Phys.* **A 740**, 249 (2004).

Study of Few-Nucleon Systems at Intermediate Energies

K. Hatanaka

Research Center for Nuclear Physics, Osaka University, 10-1 Mihogaoka, Ibaraki, 567-0047 Osaka, Japan

Abstract. At the RCNP, there have been intensive experimental studies of few-nucleon systems at intermediate energies in the last decade. A variety of spin observables have been measured as well as cross sections. By comparing data to rigorous calculations, the structure of three-nucleon forces has been explored. In this talk, resent results and future perspectives are presented.

Keywords: Few-nucleon system
PACS: 21.45.-v, 21.45.Ff, 24.70.+s

INTRODUCTION

Owing to intensive theoretical and experimental effort, an often called new generation of realistic nucleon-nucleon (NN) potentials has been obtained using meson-exchange or other more phenomenological approaches, namely, AV18, CD-Bonn, Nijm I, II, and 93. They describe the rich set of experimental NN data up to 350 MeV, and the accuracy of these potentials is remarkable, giving a χ^2 per degree of freedom very close to 1. On the other hand, these realistic two-nucleon (2N) potentials fail to reproduce experimental binding energies for light nuclei where exact solutions of the Schrödinger equation are available. A 2π-exchange type three-nucleon force (3NF), Fujita-Miyazawa force [1], could eliminate the underbinding. There have been proposed several kinds of 3NFs like the Tucson-Melbourne (TM), Urbana IX, Brazil (BR), and Texas 3NF. The Hannover group have showed that similar results can be obtained by means of coupled channel calculations including Δ-isobar excitations in the nucleus.

For scattering processes, strong three-nucleon effects were hypothesized in a study of the minima of the nucleon-deuteron elastic scattering cross section at incoming nucleon energies higher than 60 MeV [2]. In the past decade, a number of studies of three nucleon (3N) systems have been performed at intermediate energies, directly comparing rigorous Faddeev calculations to precise experimental data from IUCF, KVI, RIKEN, TSL and so on. At the Research Center for Nuclear Physics (RCNP), Osaka University, there have been intensive experimental studies of three-nucleon systems as well. They cover elastic scattering, breakup reactions, and radiative capture reactions. Measurements have been performed with proton, neutron and deuteron beams at 100 - 400 MeV. Experimental data have been compared to rigorous Faddeev-type calculations to explore the structure of three-nucleon forces, Coulomb force effects, relativistic effects, etc. Recently, measurements have been performed at energies well above the pion threshold as well as for a four-nucleon system to stimulate theoretical developments. Some recent results are presented and compared with calculations.

CP1011, *New Facet of Three Nucleon Force—50 Years of Fujita Miyazawa Three Nucleon Force (FM50), The International Symposium,* edited by H. Sakai, K. Sekiguchi, and B. F. Gibson

RECENT RESULTS

Overview of the Facility

Figure 1 shows the schematic layout of the RCNP cyclotron facility. The accelerator complex consists of an injector AVF cyclotron (K=140) and a ring cyclotron (K=400). It provides ultra-high quality beams as well as moderately high intensity beams for a wide range of researches in nuclear physics, fundamental physics, interdisciplinary fields, and applications. Maximum energy of protons and heavy ions are 400 and 100A MeV, respectively. Sophisticated experimental apparatuses are equipped like a pair magnetic spectrometer (Grand Raiden, GR [3], and Large Acceptance Spectrometer, LAS), a neutron time of flight facility (NTOF) [4], a radioactive beam separator, a super-thermal Ultra Cold Neutron (UCN) source, a white neutron source and a RI production system for nuclear chemistry and biology.

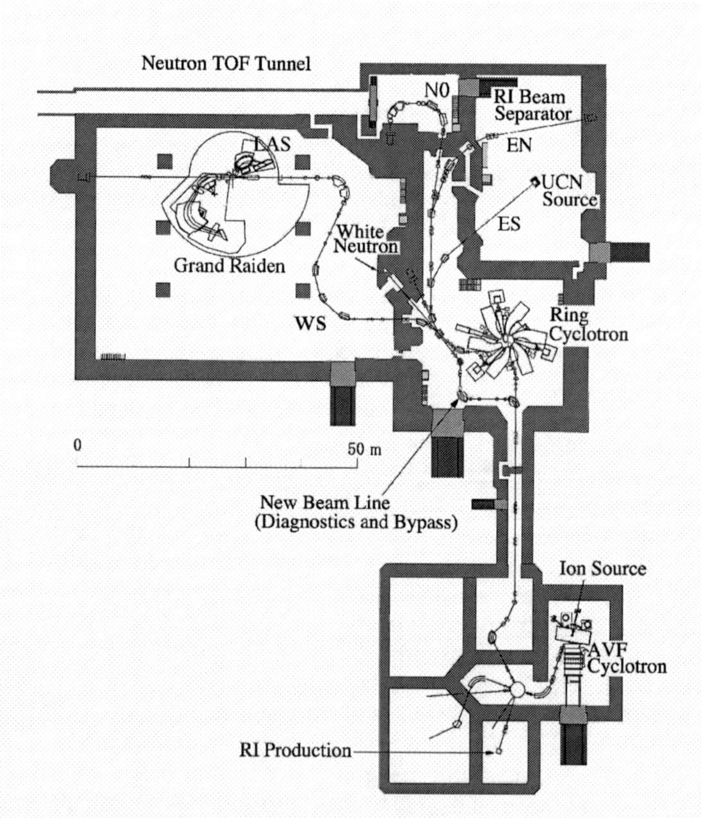

FIGURE 1. Schematic layout of the RCNP cyclotron facility.

Experimental studies of few nucleon systems are mainly performed with NTOF, GR and LAS using polarized protons and deuterons. Polarized neutrons are produced by using the ^{7}Li(p,n)^{7}Be reaction at NTOF and LAS target positions as well. It is also possible to control the beam polarization axis in horizontal plane at target positions. Both the magnetic spectrometers are equipped with focal plane polarimeters and we can measure a wide variety of spin transfer parameters as well as analyzing powers. Intensive studies have been performed for *pd* and *nd* scattering in a wide energy range to investigate the energy dependence beyond the pion production threshold, relativistic effects, and Coulomb effects, as well as for the radiative *pd* capture reaction, the inclusive and exclusive breakup reaction and the p^3He elastic backward scattering. Some results are described in the following subsections.

Three Nucleon System

The angular distributions of cross sections, proton analyzing powers and all proton polarization transfer coefficients of *pd* elastic scattering were measured at 250 MeV [5]. These are the first measurements of a complete set of proton polarization observables for *pd* elastic scattering at intermediate energies. Experimental results were compared to theoretical predictions based on exact solutions of the three-nucleon Faddeev equations and modern realistic NN potentials combined with 3NFs, namely the Tucson-Melbourne (TM) 2π-exchange model, a modification thereof closer to chiral symmetry, and the Urbana IX model. Large effects of the 3NF have been predicted at this energy. The inclusion of the 3NF gives a good description of cross sections and analyzing pow-

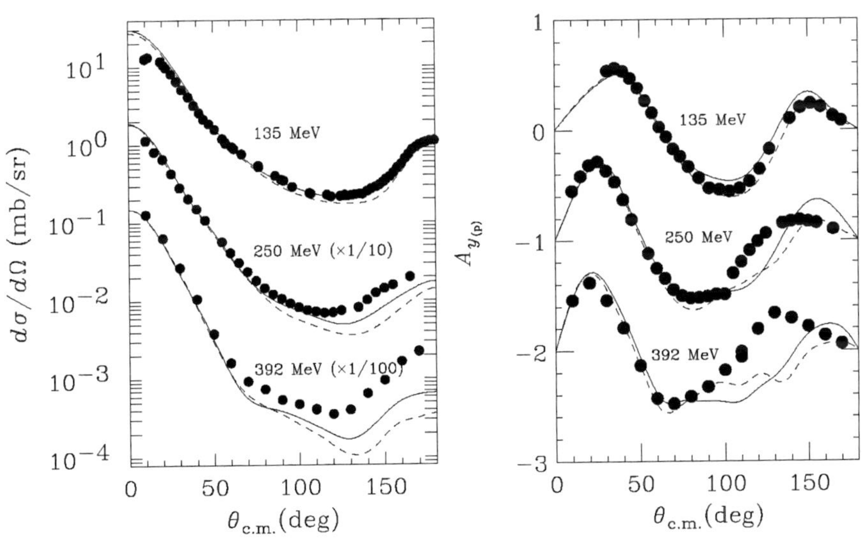

FIGURE 2. Cross sections and analyzing powers of the pd elastic scattering at 135, 250 and 392 MeV. The solid and dashed lines show the results of Faddeev calculations using CD-Bonn NN interaction including and not including TM-3NF, respectively.

ers around the cross section minimum as shown in Fig. 2. This is the similar result as those obtained at lower energies [6, 7]. However, appreciable discrepancies are observed between the data and theoretical predictions at backward angles. Polarization transfer coefficients support the Urbana IX force, but the theoretical description is still insufficient to reproduce the experimental data. These results call for a better understanding of the spin structure of the three-nucleon force as pointed out from the systematic analyses of data taken at RIKEN [6].

Relativistic corrections are expected to explain the discrepancies at backward angles between data and calculations. However they are shown to be small for both the cross sections and the analyzing powers at 250 MeV [8]. An experimental program has recently been initiated at the RCNP to measure the differential cross section and a wide variety of spin observables for *pd* elastic scattering at 392 MeV. Preliminary results of the cross sections and analyzing powers are compared with Faddeev calculations in Fig. 2. Larger discrepancies are observed between data and theoretical predictions at backward angles than those at 250 MeV. Small effects are predicted by including the relativistic corrections and the relativistic calculations cannot improve the description of data [9]. Comprehensive measurements of spin correlation parameters are in progress to clarify the spin structure of the 3NF [10].

The magnitude of the charge symmetry effects in the three-nucleon continuum was investigated by comparing the ratio of *pd* and *nd* elastic scattering cross sections to calculations. The angular dependence of the ratio at 250 MeV shows a largely oscillating structure [8]. The amplitude of the oscillation is not reproduced by calculations including Δ-isobar excitations or 3NFs when the Coulomb force is taken into account in an approximate way. Recent *nd* elastic scattering cross sections at 95 MeV [11] are compared to the *pd* scattering data at 97.5 MeV/u taken at the RCNP. Figure 3 shows the angular dependence of their ratio. The experimental result does not show such an oscillation as observed at 250 MeV. More systematic studies will be required both experimentally and theoretically in order to clarify the charge symmetry effects in the three-nucleon continuum system.

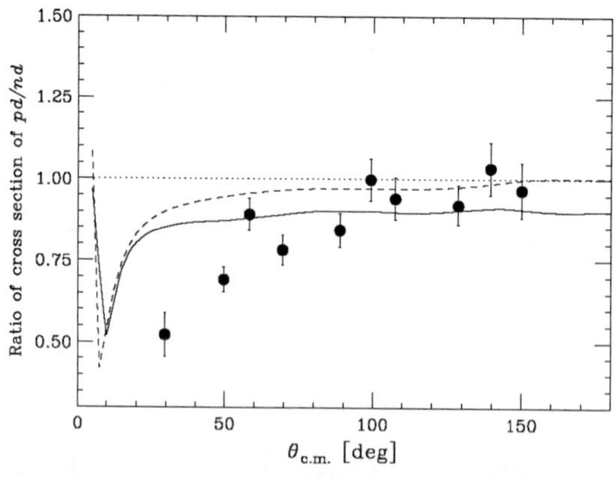

FIGURE 3. Angular dependence of ratio between the *pd* and *nd* elastic scattering cross sections. Solid and dashed lines represent the predictions based on the CD-Bonn + Δ and CD-Bonn potentials, respectively. The Coulomb force is taken into account in an approximate way.

Tensor analyzing powers, A_{xx} and A_{yy}, for the radiative *pd* capture reaction were measured for the first time at intermediate energies at RCNP [12]. Large discrepancies were observed between experimental A_{xx} data and calculations. 3NF corrections are small for this reaction to explain the discrepancies. Recently new results were reported by a KVI group. They measured tensor analyzing powers, A_{yy} and A_{zz} at incident energies of 110, 130 and 180 MeV [13]. They showed the experimental values are consistent to theoretical predictions. Since there is a relation among tensor analyzing powers like $A_{xx}+A <_{yy}+A_{zz}=0$, KVI data mean that A_{xx} of this reaction is well reproduced by calculations. There is an inconsistency among experimental results from RCNP and KVI. In order to establish experimental data, we have measured all of three kinds of tensor analyzing powers at 196 MeV and checked the consistency among them by referring the above relation. Preliminary results are presented in Fig. 4. Present results show the consistency by themselves and confirm the previous data measured at RCNP.

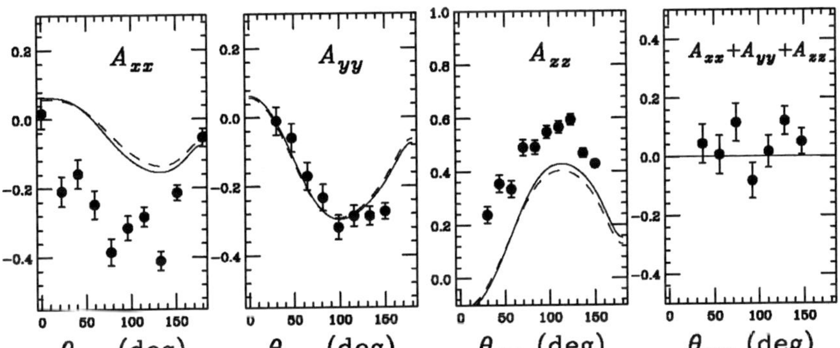

FIGURE 4. Preliminary results for *pd* capture A_{xx}, A_{yy}, A_{zz} and summation of three analyzing powers at 196 MeV (solid circles). They are compared to calculations including a 3NF (solid line) and not including a 3NF (dashed lines).

Some results on the breakup reaction are also described in these proceedings [14].

Four Nucleon System

The structure of the ^{3}He nucleus has been investigated using the ^{3}He(*p*,2*p*) and ^{3}He(*p*,*pd*) reactions at TRIUMF [15] and using spin correlations for the quasi elastic ^{3}He(*p*,*pN*) reaction at IUCF [16]. However, the structure of the ^{3}He nucleus at high momenta has not been clearly understood yet. The elastic backward scattering (EBS) may provide a tool to obtain this information for the structure of the ^{3}He nucleus because of its high momentum transfer nature. We measured the differential cross section and the spin correlation parameter C_{yy} of the p^3HeEBS at 200, 300 and 400 MeV for $\theta = 180°$ in the c.m. frame. This is the first measurement of the spin correlation parameter C_{yy} of the p^3HeEBS at intermediate energies. Experimental results were compared to theoretical calculations in order to study the mechanism of the reaction and to examine the validity of the ^{3}He wave functions based on two different realistic NN forces [17]. Although there have been remarkable developments on *ab initio* calculations of the four-nucleon continuum system, they are applicable only to low energies at present. Theoretical predictions are strongly demanded.

OUTLOOK

The RCNP AVF cyclotron has recently been upgraded to satisfy a variety of requests by users. In the upgrade project, a new beam line was installed to bypass the ring cyclotron as shown in Fig. 1 and to directly deliver low energy and high intensity beams from the AVF cyclotron to the experimental halls where sophisticated apparatuses are available. It becomes possible to use beams in such a wide energy range as from 10 MeV to 400 MeV for protons and from 2A MeV to 100A MeV for deuterons and heavier ions. Several experiments on few-nucleon systems have already been approved by the PAC and will be performed as soon as they become ready.

ACKNOWLEDGMENTS

I want to thank all of my collaborators who participated in the experiments reported in my talk. I am very grateful to theoreticians who provided many valuable comments and calculations.

REFERENCES

1. J. Fujita and H. Miyazawa, Prog. Theor. Phys. **17**, 360 (1998).
2. H. Witała *et al.*: Phys. Rev. Lett. **81**, 1183 (1998).
3. M. Fujiwara *et al.*: Nucl. Instrum. Methods A **422**, 484 (1998).
4. T. Wakasa *et al.*: Nucl. Instrum. Methods A **547**, 569 (2005).
5. K. Hatanaka *et al.*: Phys. Rev. C **66**, 044002 (2002).
6. K. Sekiguchi *et al.*: Phys. Rev. C **65**, 034003 (2002); K. Sekiguchi *et al.*: Phys. Rev. C **70**, 014001 (2004), and references therein.
7. K. Ermisch *et al.*: Phys. Rev. C **71**, 064004 (2005).
8. Y. Maeda *et al.*: Phys. Rev. C **76**, 014004 (2007).
9. H. Witala: in private communication.
10. A. Tamii *et al.*: AIP Conference Proceedings **915**, 764 (2007)
11. P. Mermod *et al.*: Phys. Rev. C **74**, 054002 (2006).
12. T. Yagita *et al.*: Mod. Phys. Lett. A **18**, 322 (2003).
13. A.A. Mehmandoost-Khajeh-Dad *et al.*: Phys. Lett. B **617**, 18 (2005).
14. S. Kuroita *et al.*: in these proceedings.
15. M.B. Epstein *et al.*: Phys. Rev. C **32**, 967, (1985), and references therein.
16. M.A. Miller *et al.*: Phys. Rev. Lett. **74**, 502 (1995).
17. Y. Shimizu *et al.*: Phys. Rev. C **76**, 044003 (2007).

Application of chiral two- and three-nucleon interactions to ^{4}He photo-disintegration

Sofia Quaglioni and Petr Navrátil

Lawrence Livermore National Laboratory, L-414, P.O. Box 808, Livermore, CA 94551, USA

Abstract. We report on an *ab initio* calculation of the ^{4}He total photo-absorption cross section using two- and three-nucleon interactions based upon chiral effective field theory. The microscopic treatment of the continuum problem is achieved using the Lorentz integral transform method, applied within the no-core shell model approach.

PACS: 25.20.Dc; 21.30.-x; 21.60.Cs, 27.10.+h

INTRODUCTION

Chiral effective filed theory (χEFT) [1, 2] represents our best opportunity to reach a consistent picture of the interaction among nucleons that is based on the underlying fundamental theory of quantum chromo-dynamics (QCD). Indeed, χEFT may be thought as an effective theory for QCD in the low energy regime relevant for describing the properties of nuclei. In the framework of χEFT the nucleon-nucleon (NN) interaction is predicted in leading order, together with a three-nucleon (NNN) interaction at the next-to-next-to-leading order or N^{2}LO [2, 3, 4], and even a four-nucleon (NNNN) interaction at the fourth order (N^{3}LO) [5]. The details of QCD are contained in parameters, the so-called low-energy constants (LECs), that are not fixed by the symmetry but can be constrained by experiment. Therefore, it is of the utmost importance to apply χEFT to nuclei in an ab initio framework.

We performed an *ab initio* calculation [6] of the ^{4}He total photo absorption cross section in the unretarded dipole approximation, using the high quality NN potential at fourth order (N^{3}LO) in the χEFT expansion of Ref. [7], and the NNN interaction at the highest order presently available (N^{2}LO) [3, 5]. The two low-energy constants that parameterize the short-range NNN interaction were selected as discussed in Ref. [8]. The microscopic treatment of the continuum problem was achieved by means of the Lorentz integral transform (LIT) method [9], applied within the *ab initio* no-core shell model (NCSM) approach [10].

THE *AB INITIO* NO-CORE SHELL MODEL

The NCSM is a technique for solution of the A-nucleon bound-state problem. Starting from a Hamiltonian containing realistic NN and NNN forces (both coordinate- and momentum-space interactions can be equally handled), the Schrrödinger equation is solved by expanding the wave functions in terms of a complete set of A-nucleon harmonic oscillator (HO) basis states up to a maximum excitation $N_{max}\hbar\Omega$ above the mini-

CP1011, *New Facet of Three Nucleon Force—50 Years of Fujita Miyazawa Three Nucleon Force (FM50), The International Symposium,* edited by H. Sakai, K. Sekiguchi, and B. F. Gibson

mum energy configuration, with Ω the HO frequency. Both Jacobi relative coordinates or Cartesian single-particle coordinates can be used. Indeed, in a complete $N_{max}\hbar\Omega$ space translational invariance is preserved even in the Slater-determinant basis. The convergence to the exact result with increasing N_{max} is accelerated by the use of an effective interaction tailored to the model-space truncation. The effective interaction is obtained using a unitary transformation in an $n-$body cluster approximation, where n is typically 2 or 3 [11].

TOTAL PHOTO-ABSORPTION CROSS SECTION VIA THE LORENTZ INTEGRAL TRANSFORM METHOD

In the long wave-length approximation, the total photo-absorption cross section

$$\sigma_\gamma(\omega) = 4\pi^2 \frac{e^2}{\hbar c} \omega R(\omega) \tag{1}$$

is proportional to the inclusive dipole response function:

$$R(\omega) = \int d\Psi_f \left| \langle \Psi_f | \hat{D} | \Psi_0 \rangle \right|^2 \delta(E_f - E_0 - \omega)\,. \tag{2}$$

This is the sum of all the transitions from the ground state $|\Psi_0\rangle$ to the various allowed final states $|\Psi_f\rangle$ induced by the dipole operator:

$$\hat{D} = \sqrt{\frac{4\pi}{3}} \sum_{i=1}^{A} \frac{\tau_i^z}{2} r_i Y_{10}(\hat{r}_i)\,. \tag{3}$$

In the above equations ground- and final-state energies are denoted by E_0 and E_f, respectively, whereas τ_i^z and $\vec{r}_i = r_i \hat{r}_i$ represent the isospin third component and center-of-mass frame coordinate of the ith nucleon. The direct calculation of such response function is extremely difficult, especially for energies beyond the three-body breakup threshold. However, it is possible to obtain the response function by following a small detour, the LIT method [12].

While in conventional approaches one usually starts from Eq. (2), in the LIT method one obtains $R(\omega)$ after the inversion of its integral transform with a Lorentzian kernel

$$L(\sigma_R, \sigma_I) = \int d\omega \frac{R(\omega)}{(\omega - \sigma_R)^2 + \sigma_I^2} = \langle \widetilde{\Psi} | \widetilde{\Psi} \rangle\,. \tag{4}$$

Indeed, in order to calculate such a transform it is sufficient to solve the in-homogeneous "Schrödinger-like" equation

$$(H - E_0 - \sigma_R + i\sigma_I) |\widetilde{\Psi}\rangle = \hat{D} |\Psi_0\rangle. \tag{5}$$

Because of the presence of an imaginary part σ_I in Eq. (5) and the fact that the right-hand side of this same equation is localized ($\langle \Psi_0 | \hat{D}^\dagger \hat{D} | \Psi_0 \rangle < \infty$), one has an asymptotic

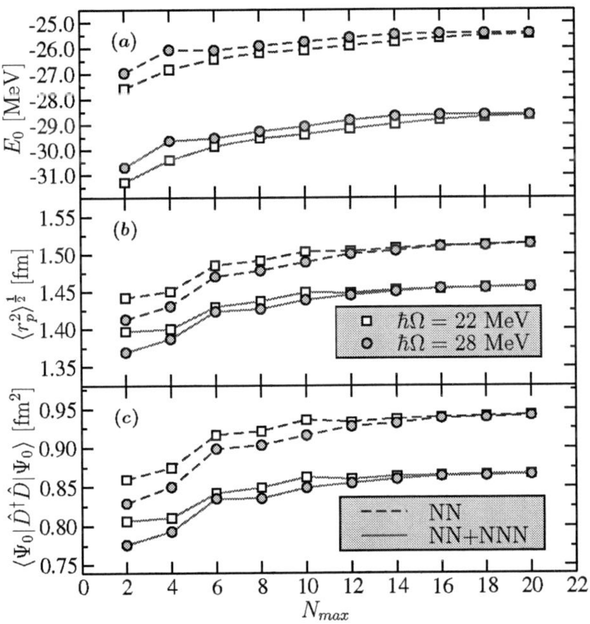

FIGURE 1. The ^{4}He ground-state (*a*) energy, (*b*) point-proton root-mean-square radius $\langle r_p^2\rangle^{1/2}$, and (*c*) total dipole strength $\langle\Psi_0|\hat{D}^\dagger\hat{D}|\Psi_0\rangle$ as functions of the model-space truncation N_{max}. Present results are for χEFT interactions with $\hbar\Omega = 22$ and 28 MeV.

boundary condition similar to a bound state. Thus, one can apply bound-state techniques for its solution, and, in particular, expansions over basis sets of localized functions such the NCSM basis. Moreover, the solution of Eq. (5) is unique. Once calculated, the response function $\langle\widetilde{\Psi}|\widetilde{\Psi}\rangle$ can be obtained by numerical inversion of the integral equation (4) [13]. During this procedure all the final-state interaction of the problem is fully taken into account.

RESULTS

We start by discussing our results for the ground state of the α particle, which enters the driving term of the LIT equation (5). In Fig. 1 we show the convergence patterns for three different ground-state observables, calculated with and without inclusion of the NNN terms of the interaction: binding energy, point-proton radius, and total dipole strength. For all of them we obtain very similar and smooth convergence patterns, using effective interactions at the three-body cluster level. In particular, already for $N_{max} = 18$ we find independence from both model space and harmonic oscillator parameter. At the ground-state level, the inclusion of the NNN force affects mostly the energy, providing 3.21 MeV additional binding, while the point-proton radius undergoes only a small reduction. The total dipole strength follows the same pattern as the radius. Indeed, these two observables

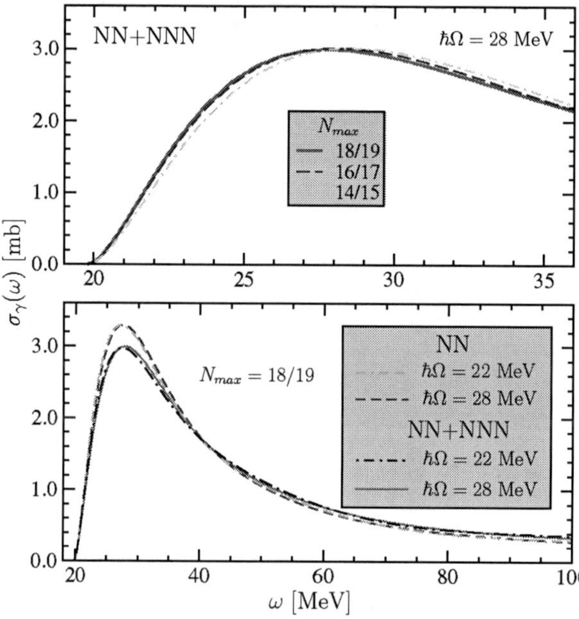

FIGURE 2. The ^{4}He total photo-absorption cross section as a function of the excitation energy ω. Present results are for χEFT interactions, and in particular: (upper panel) convergence pattern of the NN+NNN calculation with respect to the model-space truncation N_{max} for $\hbar\Omega = 28$ MeV; (lower panel) frequency dependence of the best ($N_{max} = 18/19$) results with and without inclusion of the NNN force.

are correlated [14]:

$$\langle\Psi_0|\hat{D}^\dagger\hat{D}|\Psi_0\rangle \simeq \frac{ZN}{3(A-1)}\langle r_p^2\rangle\,. \tag{6}$$

This expression, which is exact for the deuteron and triton, and for spatially symmetric systems, is violated of about 9% for ^{4}He calculated both with and without NNN interactions terms.

By applying the LIT method we have obtained the ^{4}He total photo-absorption cross section shown in Fig. 2. Also for this observable we find a stable and accurate convergence thanks to the use of three-body effective interactions. From the bottom panel of the figure we also see that for the biggest model space used, $N_{max} = 18/19$, the dependence on the HO frequency is weak. The inclusion of the NNN interaction terms induces a suppression of the peak and an enhancement of the tail of the cross section. In particular, the reduction of the low-energy cross section is related, through the inverse energy-weighted sum rule, to the the reduction found for the dipole strength

$$\int_{E_{th}}^{\infty}\frac{\sigma_\gamma(\omega)}{\omega}d\omega = 4\pi^2\frac{e^2}{\hbar c}\langle\Psi_0|\hat{D}^\dagger\hat{D}|\Psi_0\rangle\,. \tag{7}$$

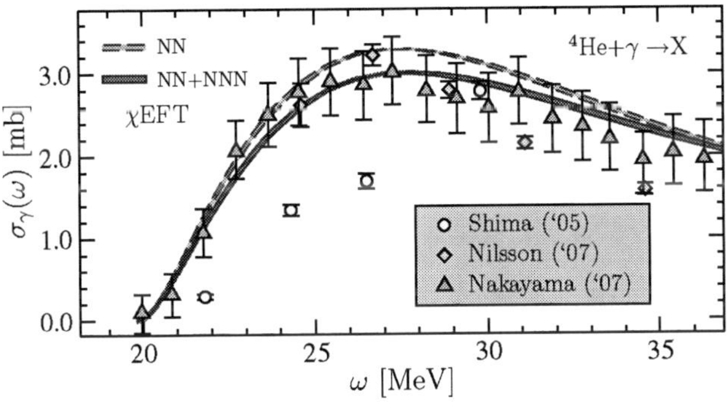

FIGURE 3. The ^{4}He photo-absorption cross section as a function of the excitation energy ω. Present results are for χEFT interactions compared to the most recent experiments [15, 16, 17].

Our final results, presented in Fig. 3 together with the most recent experiments, show a peak around the excitation energy of $\omega = 27.8$ MeV, with a peak height mildly sensitive to the NNN force. The experimental situation in the near-threshold region is controversial: two direct measurements performed using quasi-mono-energetic photons [15, 16] show discrepancies up to a factor of two with regard to the absolute magnitude of the cross-section peak. We find overall good agreement with the photo-disintegration data from bremsstrahlung photons of Nilsson *et al.* [16], while we match only the last of the experimental points of Ref. [15]. In particular, the confused experimental situation provided by these two data sets does not allow one to asses the role of the NNN force effect. Recently Nakayama *et al.* performed an indirect measurements of the α-particle total photo-absorption cross section [17] by observing its analog via the ^{4}He(^{7}Li,^{7}Be) reaction at an incident energy of 455 MeV and at forward scattering angles. Although the uncertainty in this extracted absolute cross section is 20% or more, the inclusion of NNN terms in the interaction appears to improve the agreement of the calculated cross section with this indirect measurement.

ACKNOWLEDGMENTS

This work was performed under the auspices of the U.S. Department of Energy by Lawrence Livermore National Laboratory under Contract DE-AC52-07NA27344. Support from U.S. DOE/SC/NP (Work Proposal Number SCW0498) and the Department of Energy under Grant DE-FC02-07ER41457 is acknowledged.

REFERENCES

1. S. Weinberg, *Physica A* **96**, 327 (1979); J. Gasser, and H. Leutwyler, *Ann. Phys.* **158**, 142 (1984); *Nucl. Phys.* **B250**, 465 (1985).
2. S. Weinberg, *Phys. Lett.* **B251**, 288 (1990); *Nucl. Phys.* **B363**, 3 (1991).
3. U. van Kolck, *Phys. Rev. C* **49**, 2932 (1994).

4. E. Epelbaum, A. Nogga, W. Glöckle, H. Kamada, U.-G. Meißner, and H. Witala, *Phys. Rev. C* **66**, 064001 (2002).
5. E. Epelbaum, *Phys. Lett.* **B639**, 456 (2006).
6. S. Quaglioni, and P. Navrátil, *Phys. Lett.* **B72**, 370 (2007).
7. D. R. Entem, and R. Machleidt, *Phys. Rev. C* **68**, 041001(R) (2003).
8. P. Navrátil, V. G. Gueorguiev, J. P. Vary, W. E. Ormand, and A. Nogga, *Phys. Rev. Lett.* **99**, 042501 (2007).
9. V. D. Efros, W. Leidemann, and G. Orlandini, *Phys. Lett.* **B338**, 130 (1994).
10. P. Navrátil, J. P. Vary, and B. R. Barrett, *Phys. Rev. Lett.* **84**, 5728 (2000); *Phys. Rev. C* **62**, 054311 (2000).
11. P. Navrátil, and W. E. Ormand *Phys. Rev. Lett.* **88**, 152502 (2002); *Phys. Rev. C* **68** (2003) 034305.
12. V. D. Efros, N. Barnea, W. Leidemann, and G. Orlandini, *J. Phys. G: Nucl. Part. Phys.* **34**, R459 (2007).
13. V. D. Efros, W. Leidemann, and G. Orlandini, *Few-Body Syst.* **26**, 251 (1999); D. Andreasi, W. Leidemann, C. Reiß, M. Schwamb, *Eur. Phys. J* **A24** (2005) 361.
14. L. L. Foldy, *Phys. Rev.* **107**, 1303 (1957).
15. T. Shima, S. Naito, T. Baba, K. Tamura, T. Takahashi, T. Kii, H. Ohgaki, and H. Toyokawa, *Phys. Rev. C* **72**, 044004 (2005).
16. B. Nilsson, J.-O. Adler, B.-E. Andersson, J. R. M. Annand, I. Akkurt, M. J. Boland, G. I. Crawford, K. G. Fissum, P. D. Hansen, P. Harty, and D. G. Ireland, *Phys. Lett.* **B625**, 65 (2005); *Phys. Rev. C* **75** (2007) 014007.
17. S. Nakayama, E. Matsumoto, R. Hayami, K. Fushimi, H. Kawasuso, K. Yasuda, T. Yamagata, H. Akimune, H. Ikemizu, M. Fujiwara, M. Yosoi, K. Nakanishi, K. Kawase, H. Hashimoto, T. Oota, K. Sagara, T. Kudoh, S. Asaji, T. Ishida, M. Tanaka, and M. B. Greenfield, *Phys. Rev. C* **76**, 021305 (2007).

Measurements of the ^{2}H($\vec{p}, 2p$)n breakup reaction at 250 MeV

Y. Maeda*, M. Dozono*, K. Hatanaka†, E. Ihara *, T. Kawabata **, H. Matsubara †, S. Noji ‡, K. Sagara *, S. Sakaguchi **, H. Sakai ‡, Y. Sasamoto**, M. Sasano ‡, K. Sekiguchi §, Y. Shimizu **, K. Suda†, M. Takechi §, Y. Tameshige †, A. Tamii †, T. Uesaka**, T. Wakasa * and K. Yako‡

*Department of Physics, Kyushu University, Higashi, Fukuoka 812-8581, Japan
†Research Center for Nuclear Physics, Osaka University, Ibaraki, Osaka 567-0047, Japan
**Center for Nuclear Study, The University of Tokyo, Bunkyo, Tokyo 113-0033, Japan
‡Department of Physics, The University of Tokyo, Bunkyo, Tokyo 113-0033, Japan
§RIKEN, the Institute of Physical and Chemical Research, Wako Saitama, 351-0198, Japan

Abstract. Recently it was found that the differential cross sections of elastic *Nd* scattering below 150 MeV can be well reproduced by incorporating a 3NF in Faddeev calculations based on modern nucleon-nucleon (NN) interactions. On the other hand, the differential cross sections for *nd* and *pd* at 250 MeV show large discrepancies between the data and the Faddeev calculations with a 3NF. This indicates the existence of the missing features of the three nucleon force in this energy region. For the next step in the 3NF study, we have measured the vector analyzing power A_y and the differential cross sections for the ^{2}H($\vec{p}, pp$)n breakup reaction at 250 MeV. The experiment was carried out by detecting two protons in the final state with using the double-armed spectrometer Grand Raiden (GR) and the Large Acceptance Spectrometer (LAS) at the Research Center for Nuclear Physics (RCNP). The data was compared with the results of Faddeev calculations with and without a 3NF.

Keywords: Three nucleon force, Breakup reaction
PACS: 21.30.-x, 21.45.-v, 21.45.Ff, 24.70.+s

INTRODUCTION

Study of three-nucleon force (3NF) effects in the few-nucleon system is one of the most interesting topics. Based on the meson exchange picture, the main component of the 3NF is considered to be due to 2π-exchange between three nucleons with a Δ isobar excitation as an intermediate state [1]. Firstly, study of 3NF effects was started using the bound states of few nucleon systems. For 3N and 4N systems, the correct binding energies were reproduced by introducing the TM or Urbana IX 3NFs, which showed clearly the need for 3NFs in the nucleus. The study of the 3N continuum for nucleon-deuteron (Nd) scattering allows us to treat not only the cross sections but also the spin observables, which are sensitive to more details of 3NF properties. Recent progress in computational resources has made it possible to obtain rigorous numerical results for Faddeev-type calculations in 3N scattering processes using realistic NN and 3N forces.

Recently, highly precise measurements of dp elastic scattering for deuteron energies of E_d^{lab} = 140, 200, and 270 MeV were performed [2, 3]. The calculations with only NN

CP1011, *New Facet of Three Nucleon Force—50 Years of Fujita Miyazawa Three Nucleon Force (FM50), The International Symposium,* edited by H. Sakai, K. Sekiguchi, and B. F. Gibson

forces fail to reproduce the data for cross sections and vector analyzing powers in the angular range where the cross section reaches a minimum. These discrepancies are filled by adding the 2π-exchange 3NFs. These results support the prediction that the *Nd* elastic scattering at intermediate energies is a good probe to study 3NF effects. However, the calculations including 3NFs do not improve enough to reproduce the data of the deuteron tensor analyzing powers. This clearly indicates that we are just beginning to study the details of the 3NF.

Theoretically, 3NF effects are expected to be larger at higher energy. At RCNP, we measured the *nd*[4] and *pd*[5] elastic scattering data at 250 MeV. From comparison between the differential cross section data and the theoretical predictions, we see that the Faddeev calculations with a 3NF still underestimate the data. This is quite different from the results of the comparisons between the *dp* data and the theoretical predictions at 135 MeV/A. Recently, relativistic calculations were reported and compared with data at 250 MeV [4, 6]. However, the discrepancies between the data and the theory can not be explained by relativistic corrections. It indicates the presence of missing features of the three nucleon force in this energy region, for instance $\pi-\rho$- or $\rho-\rho$- exchange type 3NF effects which are not included in the present calculations.

For a next step in the study of 3NF effects in the higher energy region, the breakup reactions are expected to be more important, because the total cross sections of the breakup reactions are predicted to become larger than those of the elastic scattering. However, the three-nucleon final state in the breakup reaction is kinematically much more complicated than that of the elastic reaction. Recently, theoretical survey for the configurations which exhibit large 3NF effects in some observables was carried out[7]. In Ref.[7], we can see that the data for A_y in the configurations for $\theta_1 = \theta_2$ and $\phi_{12} = 0$, the so-called final state interaction (FSI) geometry, is a good probe for not only the 3NF effects but also the difference between 3NF models. According to this theoretical prediction, we measured A_y and the differential cross sections of the following reaction

$$\vec{p}+d \rightarrow (p_1 n)_{FSI} + p_2 \tag{1}$$

at E_p = 250 MeV. We chose the configuration in which the *np* pair is around the FSI region, because it is difficult to include the Coulomb force in Faddeev calculations for the *pp* pair in the FSI region.

EXPERIMENTAL PROCEDURE

A schematic view of the experimental setup is shown in the left panel of Fig.1. We accelerated the polarized proton beam up to E_p = 250 MeV and transported it to the west experimental hall at RCNP. The typical beam polarization during the experiment was 0.7. We used the liquid deuteron target with a thickness of about 30mg/cm^2, which had windows of aramid foils with a thickness of 12 μm. During the experiment, we monitored the thickness of the liquid target by measuring *pd* elastic scattering. For this purpose, we installed a pair of plastic scintillation counters in the scattering chamber at $\theta_{lab} = 50.8°$ in both up and down directions and detected scattered protons and recoil deuterons in coincidence. We calibrated the absolute value of the liquid target thickness

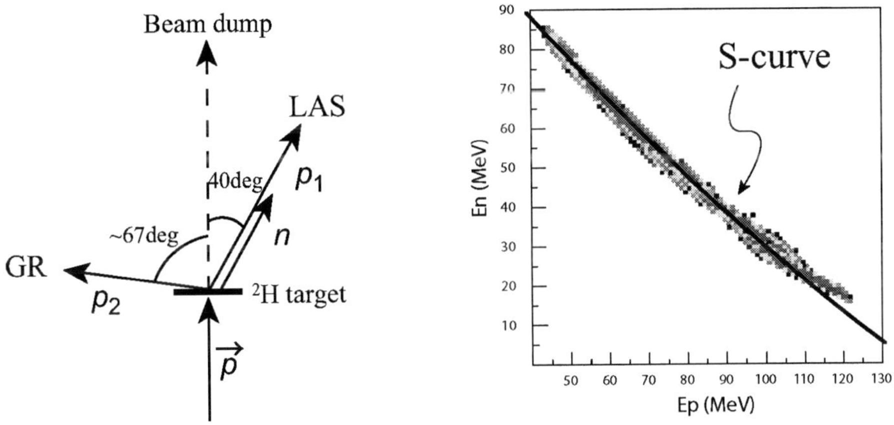

FIGURE 1. Left panel shows the schematic view of the experimental setup for the breakup measurement. Momentum of p_1 in FSI are measured by using LAS and that of another proton p_2 is measured by using GR. Right panel shows the kinematical correlation of the measured energies of E_{p1} and E_n. The solid line shows the prediction under the restrictions of energy and momentum conservation (S curve) for the configuration of $\theta_{p1} = \theta_n = 40°$, $\phi_{p1n} = 0°$.

by measuring *pd* elastic scattering using a deuterated polyethylene (CD_2) target[8] of which the thickness was known to be 44 mg/cm^2.

We determined the kinematic configuration of the final state by detecting two protons and measuring the scattered angles and their momenta. The recoil proton p_1 in FSI was detected using the magnetic spectrometer LAS at $\theta_{\mathrm{LAB}} = 40°$. Another proton p_2 which scattered in the direction of around $\theta_{\mathrm{LAB}} = 67°$, which depends of the relative energy between proton p_1 and the neutron, was detected using the magnetic spectrometer GR in coincidence with p_1. By using the momentum vectors of p_1 and p_2, we obtained the missing mass spectrum of neutrons with an energy resolutions of 1.2 MeV. Right panel of Fig. 1 shows the energy correlation (S curve) between E_{p1} and E_n in the final state. The number of breakup events are obtained by subtracting the contribution from the windows of the liquid target and from the accidental coincidence events. For this subtraction, we also carried out the measurements using an empty target cell.

RESULTS

We obtained the preliminary results for the differential cross sections and the vector analyzing powers with an S-curve energy bin of 7 MeV. The results are plotted in Fig. 2 by solid circles as a function of the energy of p_1. The statistical errors of the data are less than 1% and 0.02 for the differential cross sections and the vector analyzing powers, respectively. The peak corresponding to the FSI can be seen around $E_p = 65$ MeV.

The solid (dashed) lines represent the results of the Faddeev calculations [9] with (without) the Tucson-Melbourne99 (TM99) 3NF and the CDBonn potential. The dot-dashed (dotted) lines are the same predictions with solid (dashed) lines but folded

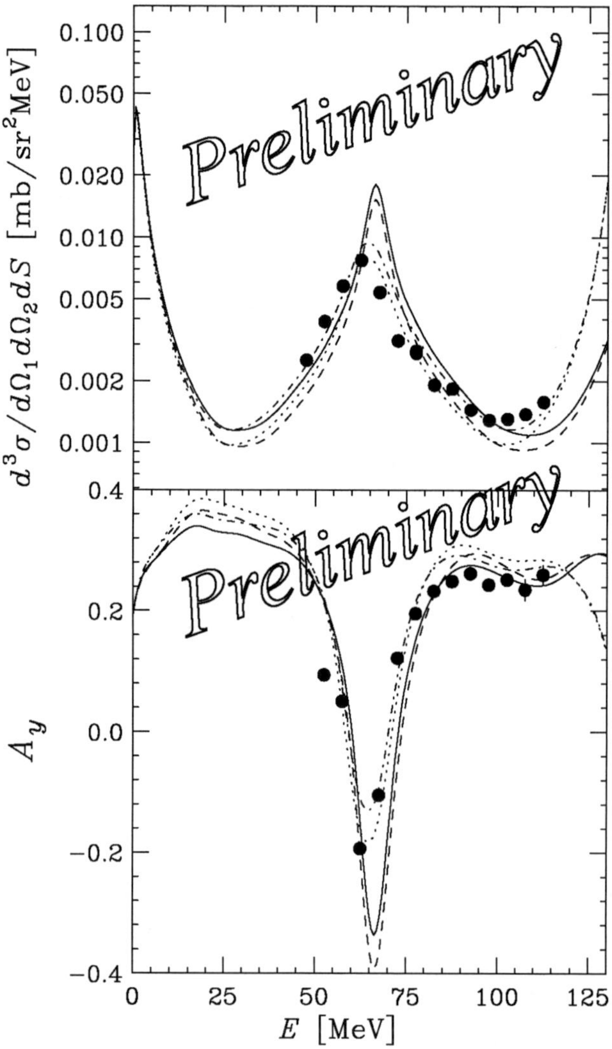

FIGURE 2. The upper and the lower panel show the results of the differential cross sections and the vector analyzing powers for the *pd* break reactions at 250 MeV for $\theta_p = \theta_n = 40°$ in the laboratory system (solid circles). Only statistical errors are represented. The predictions with CDBonn and TM99-3NF (solid line) and that with CDBonn only (dashed line) are also shown. The dot-dashed (dotted) lines are the theoretical results with (without) TM99-3NF which was folded with the angular acceptance of the measurements.

with the experimental solid angles. From Fig. 2, we can see that the experimental results are well reproduced by the calculations folded with the angular acceptance of the measurements. Concerning the differential cross sections, the data around the FSI region is well reproduced by the calculations with CDBonn only but the data around the cross section minimum is well reproduced by that including TM99-3NF. The case of the vector analyzing powers is similar to that for the differential cross sections; the data around the FSI peak and the cross section minimum are well reproduced by the calculations without and with TM99-3NF, respectively. However, the vector analyzing powers still exhibit some discrepancies between the data and the predictions around the cross section minimum. It would be interesting to compare the data with theoretical predictions which include relativistic corrections[6] or Coulomb interaction effects[10].

SUMMARY

We have carried out a measurement of the differential cross sections and the vector analyzing powers for the *pd* breakup reaction at 250 MeV at RCNP. The kinematic configuration $\theta_p = \theta_n = 40°$ and $\phi_{pn} = 0°$ was measured by momentum analyzing the two protons using the GR and LAS spectrometers. The results were compared with Faddeev calculations performed with and without the TM99 3NF. The data is well reproduced by the theoretical predicions which took into account the experimental solid angles. In addition to that, the data around the FSI region is well reproduced by predictions with CDBonn only but the data at distance from the FSI region is well reproduced only by including the TM99 3NF. On the other hand, in the case of *pd* elastic scattering at $\theta^d_{\rm LAB} = 40°$, for which the momentum transfer is almost equal to that of this work, the data is well reproduced by the predictions with 3NFs. It differs from this work in which the data partially prefer calculations with only an NN force. This indicates that we may be able to explore the details of 3NF effects, which cannot be deduced from elastic scattering, using breakup reactions.

ACKNOWLEDGMENTS

We acknowledge the outstanding work of the RCNP Accelerator group for delivering excellent proton beam. This work was supported financially in part by Research Fellowships of the Japan Society for the Promotion of Science (JSPS) for Young Scientists.

REFERENCES

1. J. Fujita, and H. Miyazawa, *Prog. Theor. Phys.* **17**, 360 (1957).
2. H. Sakai, et al., *Phys. Rev. Lett.* **84**, 5288 (2000).
3. K. Sekiguchi, et al., *Phys. Rev. C* **65**, 034003 (2002).
4. Y. Maeda, et al., *Phys. Rev. C* **76**, 014004 (2007).
5. K. Hatanaka, et al., *Phys. Rev. C* **66**, 044002 (2002).
6. H. Witała, J. Golak, W. Glöckle, and H. Kamada, *Phys. Rev. C* **71**, 054001 (2005).
7. J. Kuroś-Żołnierczuk, et al., *Phys. Rev. C* **66**, 024003 (2002).
8. Y. Maeda, et al., *Nucl. Instrum. Meth. A* **490**, 518 (2002).
9. H. Kamada, private communication.
10. A. Deltuva, A. C. Fonseca and P. U. Sauer, *Phys. Rev. C* **71**, 054005 (2005).

Many-Body Interactions in Condensed-Matters

Seiji MIYASHITA

Department of Physics, The University of Tokyo, 7-3-1 Hongo, Bunkyo-ku, Tokyo, 113-0033, Japan
CREST, JST, 4-1-8 Honcho Kawaguchi, Saitama, 332-0012, Japan

Abstract.
In commemoration of the 50 years anniversary of Fujita-Miyaszawa theory on the 3-body interaction among nucleons, we would like to review on the phenomena inherent to many-body interactions in condensed matters: (1) 3-body non-additive interaction among atoms which causes significant effect on the third order virial coefficient, (2) the many-body (ring) exchange in the solid helium 3 which plays important role, and (3) effective long range interaction induced by the elastic interaction in spin-crossover materials which causes a peculiar phase transition of the mean-field universality class.

Keywords: Non-addtitive interaction, Virial expansion, Ring exchange in He, Spin-Crossover

INTRODUCTION

If many particles interact via a common variable, there appear interactions among them. There the interaction is not necessarily sum of two-body interaction, but they are usually non-additive multiple-body interaction. Here, in commemoration of the 50 years anniversary of Fujita-Miyaszawa theory on the 3-body interaction among nucleons, we would like to review on the phenomena inherent to many-body interactions in condensed matters. As phenomena inherent to many-body interactions in condensed matters, the following systems will be picked up. First, we study the virial expansion of the equation of state of non-polar gases. About 50 years ago, Taro Kihara pointed out that the three-body interaction among atoms (called non-additive force) causes significant correction to the third order virial coefficient. Since then, various experiments confirmed this correction. The history of the discoveries of the three-body interaction is also introduced. Second, we study the magnetic properties of He3. In the solid helium 3, multiple-exchange of atoms causes important effects. In particular, the magnetic properties due to the many-body (ring) exchange have been studied extensively. Low temperature susceptibility of two-dimensional (surface) helium He3 adsorbed on graphite surface has been studied experimentally. There, in the first layer, the effect of 3 spin permutation revealed to be significant and it causes ferromagnetic exchange interaction although the bare 2-spin exchange is antiferromagnetic. Moreover, in the second layer He3, the higher order permutation effect plays important role and it causes a spin liquid state although the effective two body exchange is ferromagnetic due to the 3 spin permutation effect. It has been turn out that the spin liquid has no energy gap, and various new types of phases for gapless spin liquid state have been proposed. Finally, we study the phase transitions of spin-crossover materials. Recently, it has been pointed out that the elastic interaction plays important role for the interaction among spin transition between the so-call high

CP1011, *New Facet of Three Nucleon Force—50 Years of Fujita Miyazawa Three Nucleon Force (FM50), The International Symposium,* edited by H. Sakai, K. Sekiguchi, and B. F. Gibson

spin state and low spin state. There, the elastic interaction causes an effective long range interaction and that a phase transition of the mean-field universality class takes place, where domain structure in the ordering process is suppressed. The system exhibits a peculiar spin correlation which can not attribute simple two body interaction.

EFFECT OF THE 3-BODY INTERACTION IN THE VIRIAL EXPANSION

About 50 yeas ago, the studies on the virial coefficients were done very extensively.[1] If the interaction between gas atom is weak, the equation of state of the gas is expressed by the Boyle-Charles law. With increasing the interaction, we need corrections to the equation of motion. The correction can be expressed in the following form

$$\frac{PV}{RT} = 1 + \frac{B(T)}{V} + \frac{C(T)}{V^2} + \cdots,. \tag{1}$$

Here, the first term represents the Boyle-Charles law. The coefficient $B(T)$ is called the second virial coefficient and $C(T)$ the third virial coefficient.

The three-body interaction of non-polar gases

The original interaction between atoms is the electric dipole interaction. Because the non-polar gases such as Ar, Ne, etc. do not have a permanent dipole moment, there is no direct force due to the dipole interaction. However, the atoms may have virtual excitations to the excited state. Thus, we express the situation by the Hamiltonian

$$H = H_0 + V, \quad V = \sum_{i<j} p_i T(ij) p_j \tag{2}$$

where H_0 is the unperturbed Hamiltonian of the three independent atoms, p_i is the electric dipole of the i-th atom, and T is the dipole interaction

$$T(ij) = \frac{1 - 3e_{ij}e_{ij}}{r_{ij}^3}, \quad e_{ij} = \frac{r_i - r_j}{r_{ij}}. \tag{3}$$

Let E_n be the eigenvalue of the unperturbed Hamiltonian

$$H_0|n\rangle = E_n|n\rangle \tag{4}$$

By standard perturbation, the energy W of the ground state is given by

$$W = E_0 + W_2 + W_3, \tag{5}$$

where

$$W_2 = \sum_{n \neq 0} \frac{\langle 0|V|n\rangle\langle n|V|0\rangle}{E_0 - E_n}, \tag{6}$$

and

$$W_3 = \sum_{n\neq 0}\sum_{m\neq 0} \frac{\langle 0|V|n\rangle\langle n|V|m\rangle\langle m|V|0\rangle}{(E_0-E_n)(E_0-E_m)}. \tag{7}$$

The second order term is the van der Waals force.[2] The potential of this force is given by

$$W_2(r_{12}) = -\frac{\mu_{12}}{r_{12}^6}, \tag{8}$$

where $r_{12} = |r_1 - r_2|$. The third term has the following form:

$$W_{123}(r_1,r_2,r_3)^{\mathrm{non-add}} = v\frac{3\cos\theta_1\cos\theta_2\cos\theta_3+1}{r_{12}^3 r_{23}^3 r_{31}^3}, \tag{9}$$

where θ_i is the angle at the corner of the i-th atom of the triangle consisting of the three atoms 1,2, and 3. As we showed above, the three-body interaction among non-polar molecules was obtained by a perturbation theory of the electric dipole interaction among excited dipole moments of atoms. Axilroad and Teller studied the case of a particular atomic model in 1943.[3] At the same year, Muto also published the almost the same result for the harmonic oscillator model in "Nippon Sugaku-Buturigakkai-si".[4] Mizuno and Kihara made a generalization to any spherical symmetric atoms.[5]

Effect of the 3-Body interaction in the virial expansion

The second virial coefficients were well explain quantitatively by this force. On the other hand, the third virial coefficients C_3^{add} calculated using the sum of the two-body interactions

$$W_{123}(r_1,r_2,r_3)^{\mathrm{add}} = W_2(r_{12}) + W_2(r_{23}) + W_2(r_{31}) \tag{10}$$

does not agree with the experimental data. Then, Mizuno and Kihara pointed out the importance of the contribution of the non-additive three-body interaction, $C_3^{\mathrm{non-add}}$. [5] After this work, extensive studied on this problem have been done and the importance of the three body interaction has been established.[6, 7, 8] In Fig. 1(b), an example of the contribution of $C_3^{\mathrm{non-add}}$ given in the paper of Graben and Present[7] is shown. The lower curve shows contribution of C_3^{add}, and the upper curve $C_3^{\mathrm{add}} + C_3^{\mathrm{non-add}}$. We find the non-additive term gives significant effect at low temperature.

RING EXCHANGE INTERACTION IN HE3 MAGNETISM

He3 is a fermion with nuclear spin 1/2. The magnetic interaction among spins has been studied extensively.[9, 10] Heisenberg pointed out the magnetic interaction between spins come from a quantum mechanical effect among indistinguishable fermi particles, and derived the Heisenberg exchange interaction between the electron spins from the spin-independent Coulomb interaction. This mechanism works also in the He3

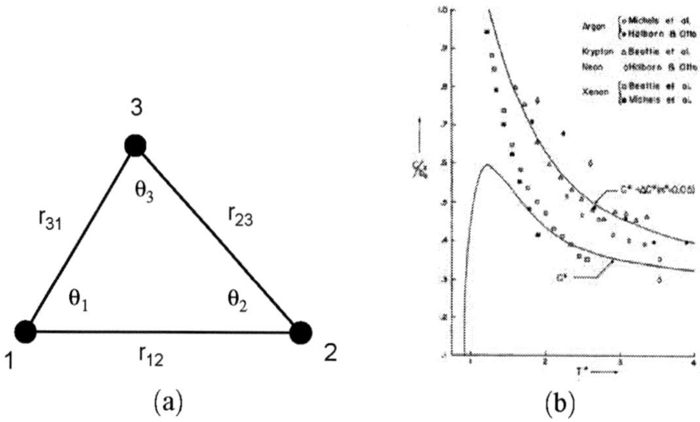

FIGURE 1. (a)Three-body interaction and (b) its contribution to the third virial coefficient (from *Phys. Rev. Lett.* **9** 247 (1962))

system. Extending the idea, Thouless obtained multiple exchange interaction among particles[11]:

$$\mathscr{H} = \sum_n (-1)^n J_n P_n, \tag{11}$$

where P_n denotes the n particle permutation operation, and all $J_n \geq 0$. For example, the 2-particle exchange is

$$P_{12} = \frac{1}{2}(1 + \sigma_1 \cdot \sigma_2), \tag{12}$$

and the 3-particle permutation is

$$P_{123} = \frac{1}{4}(1 + \sigma_1 \cdot \sigma_2)(1 + \sigma_1 \cdot \sigma_3). \tag{13}$$

This means that the bare exchange interaction is antiferromagnetic. Because σ_i is the Pauli matrix for $spin 1/2$, the three-body term can be written by the sum of two-body exchanges

$$P_{123} = P_{12} + P_{23} + P_{31} - 1, \tag{14}$$

and thus the three-body interaction gives modification of the exchange interaction

$$J_2^{\text{eff}} = J_2 - 2J_3. \tag{15}$$

Indeed, high density He3 which forms the hcp lattice has been known to show a ferromagnetic ground state.[12] Here we pick up experiments on the magnetic properties of the adsorbed He3 atoms on the graphite surface. The adsorbed atoms form layers which are triangular lattice with different densities. In the first layer, in a density region, the adsorbed He3 are trapped among graphite atoms and form a triangle lattice called "$\sqrt{3} \times \sqrt{3}$ commensurate phase" which is depicted in Fig. 2(a) In this region, effective exchange interaction was estimated to be ferromagnetic. This indicates that the 3-body interaction dominates there.[13] (see Fig. 2(b)) In the figure $J_\chi = -J_2^{\text{eff}}$.

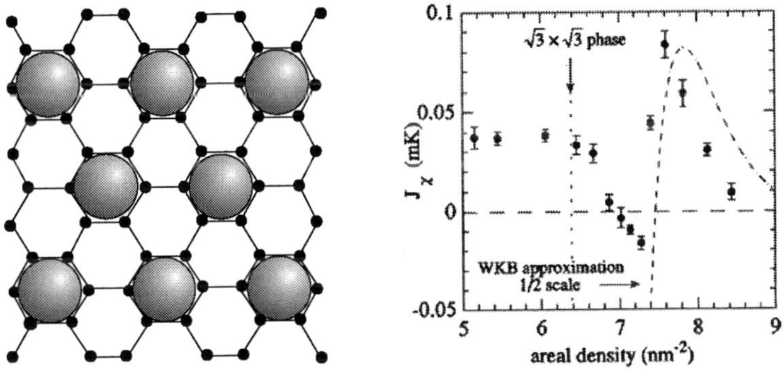

(a) (b)

FIGURE 2. (a)He3 adosorped on the graphite surface, (b) estimated value of $J_2^{\rm eff}$ from Ikegami, et al. (*Phys. Rev. Lett.* **81** 2478 (1998).)

On the second layer, the He3 atoms form more complicated structure. At a density, they form a commensurate structure called "$\sqrt{7}\times\sqrt{7}$ phase". There, the two-body exchange interaction is estimated to be ferromagnetic.[14] However, the magnetic state is expected to be the spin liquid state due to the higher order permutation effects. The 4th order interaction is given by

$$P_{1234} = J_4[\sum_{i<j} \sigma_i \cdot \sigma_j +$$

$$(1+\sigma_1\cdot\sigma_2)(1+\sigma_3\cdot\sigma_4)+(1+\sigma_1\cdot\sigma_4)(1+\sigma_2\cdot\sigma_3)-(1+\sigma_1\cdot\sigma_3)(1+\sigma_2\cdot\sigma_4).] \tag{16}$$

This can not be reduced to be a sum of two-body interactions and thus we expect that it causes a new type effect on the ordered state. A new type of phase due to the 4-th interaction (chiral ordered state) has been proposed.[15]

The effect of higher exchange interaction was studied theoretically by using exact diagonalization method.[16] They estimated a phase diagram in the space $(J_2^{\rm eff}/J_4, J_5/J_4)$, and they predicted that the case of the $\sqrt{7}\times\sqrt{7}$ phase belonged to a kind of spin liquid phase, i.e., there is no long range spin structure and there is a energy gap between non-magnetic ground state and the first magnetic excited state. However, the most recent experimental data indicate no gap.[17, 18, 19] In Fig. 3, we show the magnetic susceptibility down to 10^{-5}K achieved by the nuclear-spin adiabatic demagnetization method.[10, 19] There the susceptibility does not vanish at low temperature. This indicates that there is no energy gap. The dotted line is an estimation from a theory which predicted an energy gap.

In order to realize a spin liquid state without energy gap, there have been proposed various new types of order such as the nematic order, and octupolar order, etc. for which the 4-th interaction plays important roles.[20]

It is also interesting to study the effect of higher permutation effects. The fact suggests much more higher permutation may be need to be taken into account. If very many

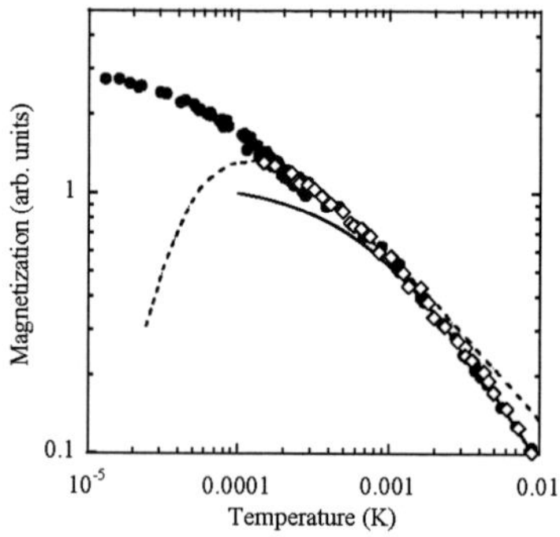

FIGURE 3. The magnetic susceptibility at very low temperatures from the paper of Masutomi, et al. (*Phys. Rev. Lett.* **92** 025301 (2004).)

higher terms are necessary, the spin Hamiltonian of the form (11) may not be appropriate to study the phenomena, and we must study original model of motion of Fermi particle. Indeed a possibility of gapless spin liquid state has been pointed out in numerical study on the Hubbard model on the triangular lattice.[21]

EFFECTIVE MANY-BODY INTERACTION IN SPIN-CROSSOVER MATERIALS

In some transition metals such as Fe and Co, the spin state has two different electric states.: the high spin (HS) state where the spins of ion aligns due to the Hund rule, and the low spin (LS) state where the crystal field causes a split of the degenerate d states and the electrons occupy the low energy states with antiparallel spins. Generally, HS state has more degeneracy than LS state and HS state, and it is more favorable at high temperature, while HS has higher energy and favorable at low temperature. The one site energy is expressed by the Hamiltonian

$$H = (D - k_{\rm B}T\ln g)\sigma, \tag{17}$$

where $\sigma = 1$ and -1 denotes HS state and LS state, respectively. Parameters D is the energy difference of HS and LS state, and g is the ratio of degeneracy of HS and LS state. T denotes the temperature and $k_{\rm B}$ is the Boltzmann constant. The change between HS and LS causes various interesting responses to external perturbations, e.g., changes of the temperature, pressure and photo-irradiation. And thus this type of materials attract much interest as a possible candidates of new devices with novel functionalities.[22]

Cooperative phenomena among them have been studied extensively, and various properties of phase transitions have been clarified. There we need interaction among spins for the cooerativity. So far a short range Ising type interaction has been adopted.[23]

$$H = -J\sum_{<ij>}\sigma_i\sigma_j + \sum_i (D - k_{\rm B}T\ln g)\sigma_i, \tag{18}$$

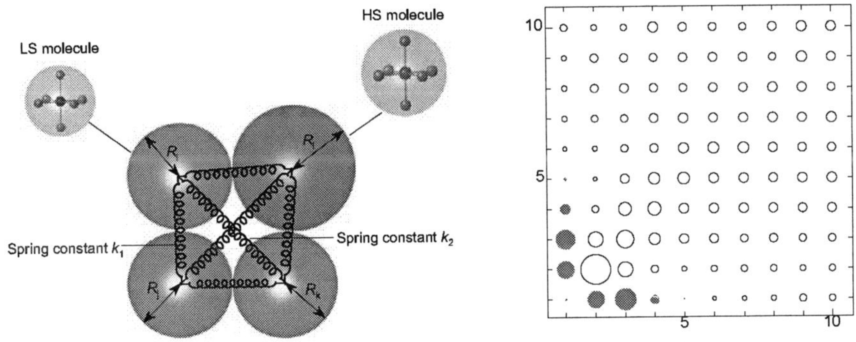

(a) (b)

FIGURE 4. (a) Distortion of the lattice due to the change of molecular size, and (b) short range spin correlation functions.

where the sum $< ij >$ is taken over all nearest neighbor pairs. Depending on the parameter, the transition between HS and LS states changes from a smooth crossover to a discontinuous first-order phase transition. The general feature of this change was studied in detail.[24]

Recently, we point out that the elastic interaction among lattice distortion due to the change of the size of molecules (Fig. 4(a)) could be an important origin of the interaction of the spin state. We have demonstrated that the spin-state independent elastic interaction

$$\mathscr{H}_{\text{elastic}} = \sum_{<ij>} \frac{1}{2} k \left(|r_i - r_j| - R_i - R_j \right)^2 \tag{19}$$

can induce phase transitions and reproduce a general structure of ordering processes.[25, 26] Here r_j is the position of the i-th molecule, and R_i is its radius.

We have point out that the phase transition of the elastic interaction belongs to the mean-field universality class.[27] The properties of the mean-field universality class[28, 29, 30, 31, 32, 33] are clearly demonstrated in the model. This fact comes from that the elastic interaction causes multiple and long range interaction among spin state which is similar to the interaction between charges via the gauge field. In this elastic interaction model, we find that formation of large cluster is strongly suppressed, and the critical opalescence near the critical point also the nucleation breakdown of the metastability would not appear in this phase transition. In Fig. 4(b), the correlation functions $\langle \sigma_{0,0} \sigma_{i,j} \rangle$ near the critical point are shown by the circles. The radius denotes the strength of the correlation, the open circles denote the ferromagnetic correlation and the closed ones antiferromagnetic. We find antiferromagnetic correlation only in the x and y direction, which should be attributed a peculiar multi-spin interactions due to the elestic interaction.

ACKNOWLEDGMENTS

The author would like to thank Professor Hideyuki Sakai for this opportunity to present this article. He also thank Professor Tutomu Momoi, Professor Hiroshi Fukuyama and Dr. Ryuichi Masutomi for valuable informations.

REFERENCES

1. T. Kihara, *Rev. Mod. Phys.* **27** 412 (1955).
2. F. London, *Z. Phys.* **63** 245 (1939).
3. B. M. Axilrod and E. Teller, *J. Chem. Phys.* 11 299 (1943).
4. Y. Muto, *Nippon Sugaku-Buturigakukai-si* **17** 629 (1943) in Japanese.
5. Y. Midzuno and T. Kihara, *J. Phys. Soc. Jpn.* **11** 1045 (1956).
6. S. Koba, S. Kaneko and T. Kihara, *J. Phys. Soc. Jpn.* **11** 1050 (1956).
7. H. W. Graben and R. D. Present, *Phys. Rev. Lett.* **9** 247 (1962).
8. A. E. Sherwood and J. M. Prausnitz, *J. Chem. Phys.* **41** 429 (1964).
9. M. Roger, J. H. Hetherrington and J. M. Delriew, *Rev. Mod. Phys.* **55** (1983) 1. M. Roger: *Phys. Rev. Lett.***64** 298 (1990).
10. R. Masutomi, PhD thesis (Univ. Tokyo).
11. D. J. Thouless, *Proc. Phys. Soc. (London)* **86** (1965) 893.
12. T. Okamoto, H. Fukuyama, H. Akimoto, H. Ishimoto and S. Ogawa, *Phys. Rev. Lett.* **72** 868 (1994).
13. H. Ikegami, K. Obara, D. Ito and H. Ishimoto, *Phys. Rev. Lett.* **81** 2478 (1998).
14. M. Roger, C. Bäuerle, Yu. M. Bunkov, A.-S. Chen, and H. Godfrin, *Phys. Rev. Lett.* **80** (1998) 1308.
15. K. Kubo and T. Momoi, *Z. Phys.* B**103** 485 (1997). T. Momoi, K. Kubo and K. Nikuni, *Phys. Rev. Lett.* **79** 2081 (1997). T. Momoi, H. Sakamoto and K. Kubo, *Phys. Rev.* B**59** 9491 (1999).
16. G. Misguich, B. Bernu, C. Lhuillie, and C. Baldmann: *Phys. Rev. Lett.* **81** 1098 (1998).
17. K. Ishida, M. Morishita, K. Yawata, and H. Fukuyama, *Phys. Rev. Lett.* **79** (1997) 3451.
18. H. Ikegami, R. Masutomi, K. Obara, and H. Ishimoto, *Phys. Rev. Lett.* **85** (2000) 5146.
19. R. Masutomi, Y. Karaki, and H. Ishimoto, *Phys. Rev. Lett.* **92** 025301 (2004).
20. N. Shannon, T. Momoi, and P. Sindzingre, *Phys. Rev. Lett.* **96** 02713 (2006). T. Momoi, P. Sindzingre and N. Shannon, *Phys. Rev. Lett.* **97** 257204 (2006).
21. T. Koretsune and Ogata, *Phys. Rev. Lett.* **89** 116401 (2002), and *J. Phys. Soc. Jpn.* **72** 2437 (2003). T. Mizusaki and M. Imada, *Phys. Rev. B* **74** 014421 (2006).
22. P. Gütlich, A. Hauser and H. Spiering, *Angew. Chem. Int. Ed. Engl.* **33** 2024 (1994) and references therein.
23. J. Wajnflasz and R. Pick, *J. Phys. (Paris), Colloq.* **32** C1-91 (1971).
24. S. Miyashita, Y. Konishi, H. Tokoro, M. Nishino, K. Boukheddaden, and F. Varret, *Prog. Theor. Phys.* **1**14, 719 (2005).
25. M. Nishino, K. Boukhedadden, Y. Konishi, and S. Miyashita, *Phys. Rev. Lett.***98** (2007) 247203.
26. Y. Konishi. H. Tokoro, M. Nishino and S. Miyashita, *Phys. Rev. Lett.* (2008) in press, and arXiv:0707.0352.
27. S. Miyashita, Y. Konishi, M. Nishino, H. Tokoro and P. A. Rikvold, *Phys. Rev. B* (2008) in press, and arXiv:0710.0921.
28. M. E. Fisher, *Phys. Rev.* **176**, 257 (1968).
29. K. Binder, *Phys. Rev. Lett.* **47**, 693 (1981)
30. V. Privman and M. E. Fisher, *Phys. Rev. B* **30**, 322 (1984).
31. E. Brezin and J. Zinn-Justin, *Nucl. Phys. B* **257**, 867 (1985).
32. B. Dünweg and D. P. Landau, Phys. Rev. B **48**, 14182 (1993). M. Laradji, D. P. Landau, and B. Dünweg, *Phys. Rev. B* **51**, 4894 (1995). X. Zhu, F. Tavazza, and D. P. Landau, *Phys. Rev. B* **72**, 104102 (2005).
33. E. Luijten and H. W. J. Blöte, *Phys. Rev. B* **56**, 8945 (1997), and references therein.

Gravitational Many-Body Problem

J. Makino

Center for Computational Astrophysics, National Astronomical Observatory, Osawa, Mitaka, Tokyo 181, Japan

Abstract. In this paper, we briefly review some aspects of the gravitational many-body problem, which is one of the oldest problems in the modern mathematical science. Then we review our GRAPE project to design computers specialized to this problem.

Keywords: Stellar dynamics and kinematics
PACS: 98.10.+z

INTRODUCTION

The gravitational many-body (or more usually "N-body") problem is expressed by the following system of second-order ordinary differential equations:

$$\frac{d^2 x_i}{dt^2} = -\sum_{j \neq i} G m_j \frac{x_i - x_j}{|x_i - x_j|^3}, \tag{1}$$

where x_i and m_i are the position and mass of the particle with index i, t is the time, and G is the gravitational constant. The summation is taken over all particles (typically stars) in the system under study. In this paper, We'll first describe in what kind of behavior of this kind of systems we are interested in, and then briefly describe our GRAPE project to develop specialized computers for this problem.

GRAVITATIONAL MANY-BODY PROBLEM

The two-body problem and planetary systems

The case of $N = 2$ was formulated and solved by Newton for the motion of planets around the Sun, and that marked the beginning of the modern mathematical science. In the case of $N = 2$, the equation of motion can be reduced to the motion of the center of mass of two particles, and the relative motion of the two particles. The center of mass does not feel any acceleration and just moves on a straight line with a constant speed.

Newton showed that the shape of the orbit is expressed as a conic curve. If the energy of the relative motion of two particles is negative, the orbit is an ellipse with one of the two foci located at the origin. If the energy is positive, the solution is a hyperbola again with one focus at the origin. The elliptic orbit was very good approximation of the orbit of Mars. However, the effect of gravitational interaction between planets, in particular the effects of massive planets like Jupiter and Saturn, cannot be neglected if we want to

CP1011, *New Facet of Three Nucleon Force—50 Years of Fujita Miyazawa Three Nucleon Force (FM50), The International Symposium,* edited by H. Sakai, K. Sekiguchi, and B. F. Gibson

predict the exact motions of planets. Throughout the nineteenth century, the main tool for the study of the effect of the interaction between planets was perturbation theory, where the interaction between planets was expanded in terms of the ratio between the strength of the solar gravity and that of the interaction between planets, and the effect of higher-order terms were ignored.

Whether or not the series expansion converges depends, roughly speaking, on whether the system is integrable. Modern celestial mechanics began with Poincare's finding that the general three-body problem is not integrable. Our current understanding of the solar system is such that it is probably not stable, but the instability timescale, at least for the eight main planets, is much longer than the age of the solar system.

High-accuracy numerical integration of the solar systems by Wisdom and his collaborators [1] and more recent works by [2] and [3] have played important roles, since we cannot rely on expansion techniques.

Stellar systems

The dynamics of stellar systems, which consist of a large number of stars, is rather different from that of the planets. In the case of planetary systems, the central star dominates the total mass of the system. Typically more than 99% of the total mass is in the central star. On the other hand, a typical globular cluster, which is a relatively small stellar system, consists of around one million stars.

Thus, in the first approximation, stars in a stellar system move in a smooth gravitational potential field, which is an overlay of gravitational potentials of all stars in the system. In the case of a star cluster or a galaxy, though the individual stars move, the system as a whole is in dynamical equilibrium.

We need to define the meaning of the term equilibrium. We start with the distribution function $f(\mathbf{x},\mathbf{v})$, which is defined in six-dimensional phase space and gives the number of stars per unit volume. Here, we assume that all stars are of the same mass. This is of course not a very good approximation, but is okay for many applications. The dynamical equilibrium is defined simply as $\frac{\partial}{\partial t} f(\mathbf{x},\mathbf{v}) = 0$. Of course, with a finite number of stars, the distribution function fluctuates and we do not have an exact dynamical equilibrium. However, for theoretical understanding it is a very important concept.

There are two rather different effects of the fact that real stellar systems have only a finite number of stars. The first one is, as we stated above, the distribution function shows some fluctuations. Another effect is that, because of these fluctuations, the gravitational potential field also fluctuates, resulting in random changes of orbits for all stars.

The average amplitude of the fluctuation would be proportional to $1/\sqrt{N}$, where N is the number of particles. Thus, very roughly, we can expect that the energy of a particle would change by $1/\sqrt{N}$ per orbital timescale, and that changes are random. In about N orbital timescales, the cumulative change would become of the order of the energy itself.

The actual rate of the change is somewhat more complex to estimate. It is relatively straightforward to show that particles with different distances contribute to the energy change of one particle logarithmically weighted. Thus, the timescale is given by

$$t_r \sim \frac{N}{\log N} t_d, \tag{2}$$

where t_d is the dynamical timescale of the system (the time in which a typical star travels through the system), and t_r is the relaxation timescale defined as the timescale in which the energy change of a typical particle becomes comparable to its energy.

Very roughly speaking, the relaxation timescale is much longer than the age of the universe for galaxies, but significantly shorter for star clusters and clusters of galaxies. In the case of star clusters the relaxation time is short because the number of stars is relatively small ($\sim 10^6$) and the dynamical timescale is short ($\sim 10^6$ years). In the case of clusters of galaxies, though the dynamical timescale is long ($> 10^8$ years), the number of galaxies in a cluster of galaxies is small (typically less than 10^4). Thus, the evolution through relaxation is important in these systems, while it is not very important in determining the overall structure of a typical galaxy. That is why galaxies have wide variation in their morphologies.

In a sense, this change in the energy of particles has the same effect as the change in motion of gas molecules due to collisions, and leads to the thermal relaxation of the system. However, the thermal relaxation process of a self-gravitating N-body system is different from that of a gaseous system in two ways. First, the "mean free path" of particles in an N-body system is usually much larger than the system size, while the mean free path of the molecules in a gaseous system is a small fraction of the system size.

In a self-gravitating system, the meaning of the term "mean free path" is somewhat unclear, since a particle is, in a sense, always having gravitational encounters with all other particles in the system. However, as we have seen, the effect of one particle is small, and we need to discuss the cumulative effect of many encounters.

In common thermodynamic systems such as gas molecules in a box, the fact that the mean-free path is smaller than the system size is the reason why we can apply the assumption of Local Thermal Equilibrium (LTE). In other words, in the gravitational N-body system, the very assumption of local thermal equilibrium, which makes it possible to use thermodynamical concepts like temperature or pressure, is totally broken.

To make the matter even more complex, a self-gravitating system (either made of point-mass particles or gas) has no thermal equilibrium.

Let us first consider an N-body system in isolation. We can assume that the system is in dynamical equilibrium. If the system is also in thermal equilibrium, the velocity distribution function must be Maxwellian with the same temperature everywhere. This Maxwellian velocity profile, however, cannot be realized in a stellar system. If a stellar system is in thermal equilibrium, it must have a spherically symmetric density distribution which in the limit of $r \to \infty$ given by $\rho \propto 1/r^2$, where r is the distance from the center. This means the system is infinitely large both in spatial size and mass.

Thus, any stellar system with a finite mass is not in thermal equilibrium. As a result of thermal relaxation, however, it tries to approach the isothermal, Maxwellian distribution function. In other words, the high-velocity tail of the distribution function, which is non-existent because only stars with negative energy are bound to the system, is constantly generated. These stars will naturally escape from the system. Thus, a stellar system spontaneously creates escaping particles.

This evolution of the system through the formation of escaping stars is important. However, in real star clusters the main driving force of the evolution is the structure formation within the cluster. The reason is that the local relaxation time is proportional to the density. Thus, the central high-density region can evolve much faster than the rest of the system does. To give a clear understanding of the system, it will be helpful to consider the system in a spherical adiabatic wall [4, 5, 6].

Antonov [4] found that the system has neutral stability at $D = 709$, where D is the ratio between the central density and density just inside the spherical wall. Hachisu and Sugimoto [6] performed a stability analysis of the system by means of the perturbation equation, and found that the system is unstable against the redistribution of the heat if $D > 709$. In other words, if $D > 709$, the isothermal equilibrium state is unstable and would spontaneously develop a spatial structure if there is any small perturbation. When the central region starts to contract, it will continue to do so since the central temperature becomes higher and higher. Eventually, the central density and temperature reaches infinity. This process is called gravothermal catastrophe. This concept of gravothermal catastrophe can only be applied to gas systems. However, it has been strikingly successful in providing qualitative (and in many cases, quantitative) descriptions of the behavior of more sophisticated models without the assumption of LTE, such as the direct N-body calculation, orbit-averaged Monte-Carlo and direct Fokker–Planck calculations.

In a real N-body system, the contraction cannot continue indefinitely since there are only a finite number of stars. Clearly, the central mass density cannot go higher than that of individual stars without making stars collide with each other. Also, the chance that binary stars are formed through triplet interactions becomes higher, since the formation rate should be proportional to the density squared.

From the point of view of thermodynamics, the formation of binaries in a gravitational N-body system has a rather special meaning. In the case of normal matter, molecules and other bound stars are preferred in a low-temperature state, because the bound state can increase the kinetic energy by reducing the number of degrees of freedoms. Usual molecules have zero-energy states, and there is a limit to the amount of the energy it produces. However, gravitationally bound binary stars have no such limit, and binaries can always increase the total entropy of the system.

Spontaneous formation of binary systems is rare in "normal" environments, where the number density of stars is low. However, as the density increases as the result of the gravothermal catastrophe, the central number density of stars can go arbitrarily high, to the point that the formation rate of binaries is sufficient to compensate for the energy loss through the outward heat transfer. At this point, the central region starts to expand.

Since the system is thermally unstable and because the region outside the high-density core is very close to isothermal, this expansion is driven again by thermal instability, and the central density shows oscillation. This oscillation was first found by Sugimoto and Bettwieser [7] using the fluid approximation, but eventually it was confirmed using Fokker-Planck approximation [8] and a direct N-body calculation [9].

Need for direct N-body simulation

In real star clusters, there are many other effects which can play important roles, such as the effect of the physical collisions between stars, the external gravitational field of the parent galaxy, the effect of stellar evolution and formation of black holes. In clusters of galaxies, close interaction and merging of galaxies play critical roles. These additional effects are very difficult to handle using approximate models such as the fluid approximation or Fokker-Plank model, but it is straightforward to take into account these effects in an N-body simulation. On the other hand, the major problem with direct N-body simulation is its calculational cost. Thus, efforts have been made to increase the effective speed of N-body simulation, through improvement of algorithms, parallelization, and hardware development.

In the rest of this manuscript, we briefly describe our approach of designing specialized hardware for an N-body simulation.

SPECIAL-PURPOSE COMPUTING

Astrophysical N-body simulation

The idea of building special-purpose computers for specific numerical problems is certainly not new. In fact, the very first digital electronic computers (ABC and ENIAC) were designed to solve specific problems. However, the designers of early computers found that their machines could be used for a much wider range of problems, and thus the evolution of the programmable general-purpose computer started.

At the time when even the largest supercomputers did not have a fully parallel multiplier, it did not make sense to design a special-purpose computer. Any computer, either special- or general-purpose, consists of arithmetic units (mainly multipliers), control logic and memory. A special-purpose computer might have simpler control logic or smaller memory than general-purpose computer. However, if the cost of the arithmetic unit itself is a fair fraction of the total cost of the machine, we cannot gain much by trying to reduce the cost of the rest of the system.

However, with present microprocessors, very small fraction of transistors are used for floating-point units. In fact, a typical microprocessor of 2008 has about 1,000 times more transistors compared to Cray-1, but number of floating-point multipliers is larger by only a factor of 8.

There are several reasons why it is difficult to use large number of floating-point units in a single microprocessor. However, if we are to design a computer specialized for N-body simulation, or rather, the calculation of the gravitational interaction between particles, we can use almost all transistors for floating-point units without much problem.

Figure 1 shows the basic "processor" for our N-body computer. Calculation of the gravitational force from one particle to another requires some 40 floating point operations, depending on how you count division and square-root operations. If we assign 10 operations to each of them, we end up with about 40 operations per interaction. Instead of using one arithmetic unit to perform these 40 operations, we connect 40 floating point arithmetic units in the form of a pipeline, which calculates and accumulate the force

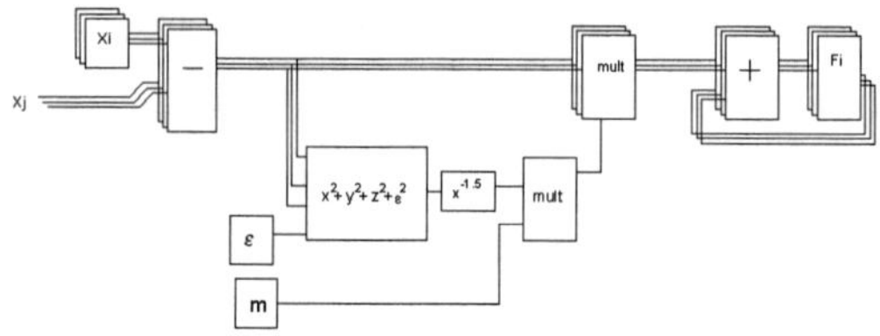

FIGURE 1. The force calculation pipeline

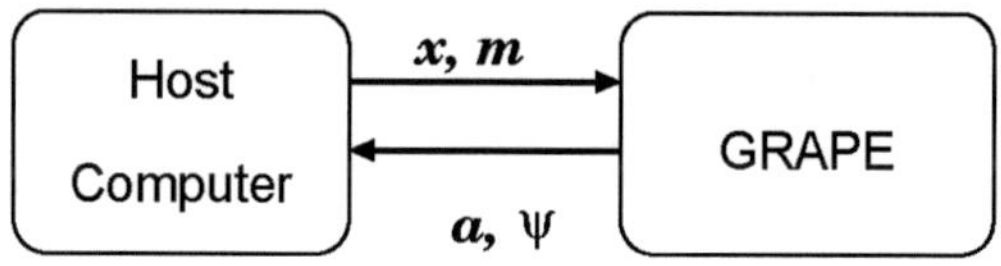

FIGURE 2. The basic architecture of GRAPE

from one particle at each clock cycle.

Our GRAPE (GRAvity PipE) project is based on this idea of integrating multiple (both physical and virtual) pipelines specialized for gravitational force calculation into one chip (or one board, when the number of available transistors was not as large as what is now). To give an idea, the GRAPE-6 system which was completed in March 2002 integrates 2048 pipeline chips, each with 6 pipeline processors which calculate the gravitational force and its first time derivative. As a result, single GRAPE-6 processor chip integrate about 360 floating point arithmetic units.

Since the GRAPE hardware performs only the force calculation, all other operations, such as the time integration of the orbits of stars and analysis of the calculated results must be done somewhere else. Therefore we connect the GRAPE hardware and general-purpose computer by some communication link. This is the basic structure of our GRAPE systems, shown in figure 2.

GRAPE Project

We started GRAPE Project in 1988. The first machine we completed, the GRAPE-1 [10], was a single-board unit on which around 100 IC and LSI chips were mounted and wire-wrapped. We used commercially available IC and LSI chips to implement force calculation pipeline. This choice was natural consequence of the fact that we lacked both money and experience to design custom LSI chips. In fact, none of the original design and development team of GRAPE-1 had the knowledge of electronic circuit more than what was learned in basic undergraduate course for physics students.

GRAPE-2 is similar to GRAPE-1 since it was also based on commercial LSI chips. The difference is in the numerical accuracy. For GRAPE-1, we used an unusually short word format, to make the hardware as simple as possible. The input coordinates are expressed in 15-bit fixed point format. After subtraction, the result is converted to 8-bit logarithmic format, in which we use just 3 bits for "fractional" part. This format is used until we obtain $1/r^3$. The final accumulation was done in 48-bit fixed point, to avoid overflow and underflow. The advantage of the short format like 8-bit logarithmic format is that we could use ROM chips to implement complex functions that require two inputs. Any function of two 8-bit words can be implemented by one ROM chip with 16-bit address input. Thus, all operations other than the initial subtraction of the coordinates and final accumulation of the force were implemented by ROM chips.

GRAPE-3 was our first machine with custom LSI chip. The number format was again the combination of the fixed point and logarithmic format similar to what were used in GRAPE-1. The chip was fabricated using 1μm design rule by National Semiconductor. The number of transistors on chip was 110K. The chip operated at 20MHz clock speed, offering the speed of about 0.8 Gflops. We designed a printed-circuit board with 8 chips, for the speed of 6.4 Gflops per board. Thus, GRAPE-3 was also our first trial to integrate multiple pipelines into a system. Also, GRAPE-3 was the first GRAPE machine to be manufactured and sold by a commercial company. Nearly 100 copies of GRAPE-3 have been sold to more than 30 institutes (more than 20 outside Japan).

With GRAPE-4, we finally integrated a high-accuracy pipeline into one chip. Also, with this chip we added the pipeline to calculate the first time derivative of the force, so that we can implement high-order time integration algorithms in a simple and efficient way [11]. We could not fit a full single pipeline to a chip with the technology available at that time. So we designed a "1/3" version of the pipeline which processes x, y and z components of coordinates serially in three consecutive clock cycles. The chip was fabricated using 1μm design rule by LSI Logic. Total transistor count was about 400K.

We built a large system consisted of 1728 pipeline chips (36 PCB boards each with 48 pipeline chips). GRAPE-4 system operates on 32 MHz clock, delivering the speed of 1.1 Tflops. Completed in 1995, GRAPE-4 was the first computer for scientific calculation to achieve the peak speed higher than 1 Tflops. Also, for 1995 and 1996 it was awarded the Gordon Bell Prize for peak performance, which is given to a real scientific calculation on parallel computer with the highest performance. Technical details of machines from GRAPE-1 through GRAPE-4 can be found in our book[12] and reference therein.

GRAPE-5[13] was an improvement over GRAPE-3. It integrated two full pipelines which operate on 80 MHz clock. Thus, single GRAPE-5 chip offered the speed 8 times more than that of the GRAPE-3 chip, or the same speed as that of a 8-chip GRAPE-3 board. GRAPE-5 was awarded the 1999 Gordon Bell Prize for price-performance. The GRAPE-5 chip was fabricated with 0.35μm design rule by NEC.

GRAPE-6 described above is similarly the improvement over GRAPE-4.

Table 1 summarizes the history of GRAPE project. Figure 3 shows the evolution of GRAPE systems and general-purpose parallel computers. One can see that evolution of GRAPE is faster than that of general-purpose computers.

These GRAPE hardwares, including GRAPE-6, have been applied to a number of astrophysical problems both by our group and by other researchers worldwide. Since there are too many of interesting results, we do not try to list them. Some of the results are summarized in the proceedings of IAU Symposium 208 "Astrophysical Supercomputing Using Particles", held in July 2001. Some more informations can be found at `http://www.astrogrape.org`.

TABLE 1. History of GRAPE project

GRAPE-1	(89/4 — 89/10)	120 Mflops, low accuracy
GRAPE-2	(89/8 — 90/5)	40 Mflops, high accuracy(32bit/64bit)
GRAPE-1A	(90/4 — 90/10)	240 Mflops, low accuracy
GRAPE-3	(90/9 — 91/9)	14 Gflops, high accuracy
GRAPE-2A	(91/7 — 92/5)	180 Mflops, high accuracy
HARP-1	(92/7 — 93/3)	180 Mflops, high accuracy Hermite scheme
GRAPE-3A	(92/1 — 93/7)	6 Gflops/board some 80 copies are used all over the world
GRAPE-4	(92/7 — 95/7)	1 Tflops, high accuracy Some 10 copies of small machines
MD-GRAPE	(94/7 — 95/4)	1Gflops/chip, high accuracy programmable interaction
GRAPE-5	(96/4 — 99/8)	5Gflops/chip, low accuracy
GRAPE-6	(97/8 — 02/3)	64 Tflops, high accuracy

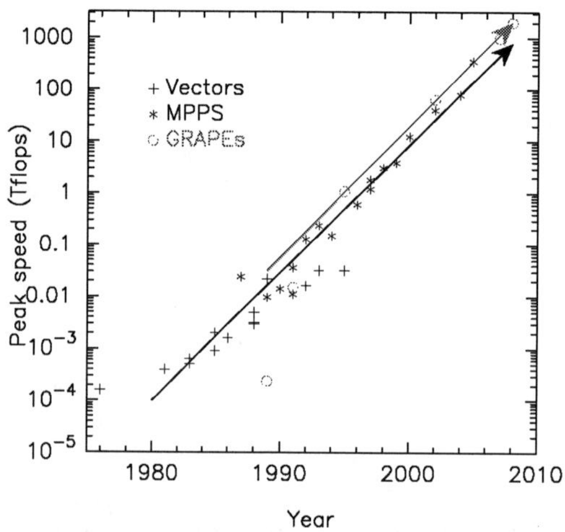

FIGURE 3. The evolution of GRAPE and general-purpose parallel computers. The peak speed is plotted against the year of delivery. Open circles, crosses and stars denote GRAPEs, vector processors, and parallel processors, respectively.

GRAPE-DR

In July 2004, we were awarded the grant to develop the next-generation GRAPE system, which we call GRAPE-DR. This grant was, however, not for a special-purpose computer for astrophysical N-body simulation, but for a programmable massively-parallel processor. In the following we overview what is GRAPE-DR and why we chose to develop such a system.

I believe GRAPE hardwares have been pretty successful. One main problem with the GRAPE project is the initial cost to develop custom chips, which has been increasing exponentially. The initial sample chips for GRAPE-4 cost around 200K USD, while that

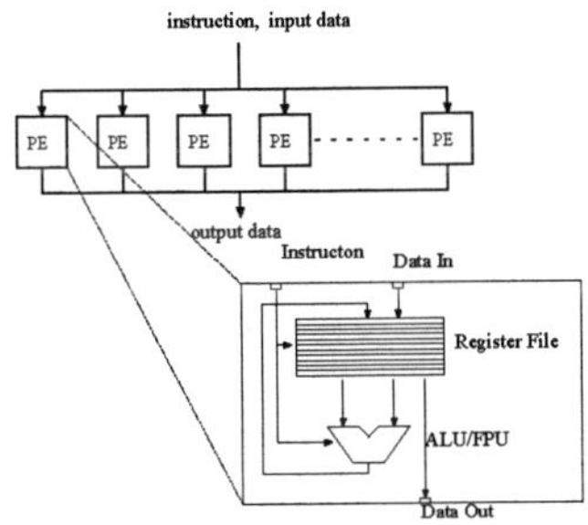

FIGURE 4. Basic structure of an SIMD processor

for GRAPE-6 more than 1M USD, and for our current GRAPE-DR project we paid more than 4 M USD. We need the total budget 3-4 times more than the initial chip cost, and a project to develop a parallel machine with special chip could be done with 1 M USD in 1992, 4M USD in 1996, but 15M USD in 2004. It is clearly not easy to raise a 15M USD grant for a machine with relatively narrow range of applications. In other words, we need to extend the application range.

The GRAPE-DR architecture

Figure 4 shows the basic structure of the new programmable GRAPE. It consists of a number of processing elements (PEs), each of which consists of an FPU and a register file. They all receive the same instruction from outside the chip, and perform the same operation.

Compared to the classic SIMD architecture such as that of TMC CM-2, the main difference are the followings.

a) PEs do not have large local memories.

b) There is no communication network between PEs.

We introduced these two simplifications so that a large number of PEs could be integrated into a single chip. If we want to have a large memory connected to each PE, the only economical way is to attach DRAM chips. However, once we decide to use external memory chips, it becomes very difficult to integrate large number of processors into a chip, since it is practically impossible to add an external memory with sufficient bandwidth.

A communication network is not very expensive, as far as it is limited to processors in a single chip. A two-dimensional mesh network would be quite natural, for physically two-dimensional array of PEs on a single silicon chip. However, such a two-dimensional network poses a very hard problem, if we try to extend it to multi-chip systems. Again, external wires would be too costly.

If we eliminate the inter-PE communication network right from the beginning, we have no problem in constructing multi-chip systems, since PEs in different chips need not be connected.

Thus, this simple architecture has two advantages. First, we can integrate a very large number of PEs into a single chip. Second, it is easy to construct a system with multiple chips. As a result, we can construct a system with very high peak performance.

We made a few improvements on this basic architecture, so that it can offer good performance on at least a few selected applications such as the gravitational force calculation and dense matrix multiplication.

The processor chip with 512 simple PEs was completed in 2006, and we are currently working on the processor board with 4 chips, which will offer the peak speed of 2 Tflops (single precision)/1 Tflops (double precision).

ACKNOWLEDGMENTS

I would like to thank Daiichiro Sugimoto, Toshikazu Ebisuzaki, Makoto Taiji, Tomoyoshi Ito, Toshiyuki Fukushige and many others who were involved in the development of the six generations of GRAPE hardware, and Kei Hiraki and Mary Inaba and many others involved in the GRAPE-DR project. I also thank Yoko Funato, Simon Portegies Zwart, Steve McMillan, Piet Hut and again many others for discussions and collaboration in the software development and scientific work on the GRAPE hardware. This research is partially supported by the Special Coordination Fund for Promoting Science and Technology (GRAPE-DR project), Ministry of Education, Culture, Sports, Science and Technology, Japan.

REFERENCES

1. G. J. Sussman, and J. Wisdom, *Science* **241**, 433–437 (1988).
2. H. Kinoshita, and H. Nakai, *Earth Moon and Planets* **72**, 165–173 (1996).
3. T. Ito, and K. Tanikawa, *Monthly Notices of Royal Astronomical Society* **336**, 483–500 (2002).
4. V. A. Antonov, *Vest. Leningrad Univ.* **7**, 135 (1962).
5. D. Lynden-Bell, and R. Wood, *Monthly Notices of Royal Astronomical Society* **138**, 495–525 (1968).
6. I. Hachisu, and D. Sugimoto, *Progess of Theoretical Physics* **60**, 123–135 (1978).
7. D. Sugimoto, and E. Bettwieser, *Monthly Notices of Royal Astronomical Society* **204**, 19P–22P (1983).
8. H. Cohn, P. Hut, and M. Wise, *The Astrophysical Journal* **342**, 814–822 (1989).
9. J. Makino, *The Astrophysical Journal* **471**, 796+ (1996).
10. T. Ito, J. Makino, T. Ebisuzaki, and D. Sugimoto, *Computer Physics Communications* **60**, 187–194 (1990).
11. J. Makino, and S. J. Aarseth, *Publications of the Astronomical Society of Japan* **44**, 141–151 (1992).
12. J. Makino, and M. Taiji, *Scientific Simulations with Special-Purpose Computers — The GRAPE Systems*, John Wiley and Sons, Chichester, 1998.
13. A. Kawai, T. Fukushige, J. Makino, and M. Taiji, *Publications of the Astronomical Society of Japan* **52**, 659–676 (2000).

Universal three-body repulsion suggested by neutron stars

T. TAKATSUKA*, S. NISHIZAKI* and R. TAMAGAKI†

*Faculty of Humanities and Social Sciences, Iwate University,
Morioka 020-8550, Japan
†Kamitakano Maeda-Cho 26-5, Kyoto 606-0097, Japan

Abstract. Because of a serious inconsistency between theory and observation for the mass of hyperon-mixed neutron stars, it is suggested that some "extra repulsion" is needed in hypernuclear systems. A 3-body force repulsion is tested for two cases, a 2π-exchange via Δ-excitation type ($2\pi\Delta$) and a string-junction model (SJM) for the quark structure of baryons. It is found that the $2\pi\Delta$ model generates an increasing repulsion with increasing density but cannot produce the "extra repulsion"in hyperon-mixed neutron stars because it does not work on a Λ due to the lack of $\Lambda\Lambda\pi$ direct coupling. It is pointed out that the "extra repulsion" should act universally, *i.e.*, independent of baryon species. The SJM 3-body repulsion can satisfy the condition of universality because of the flavor-independence and the $\{2\pi\Delta + \mathrm{SJM}\}$ scheme is shown to be a promising candidate for the "extra repulsion", maintaining consistency with the empirical saturation property of nuclear matter.

Keywords: hyperon-mixed neutron stars, three-body repulsion
PACS: 26.60.Kp

INTRODUCTION

According to the increase of density ρ toward the central region in neutron stars (NSs), hyperons (Y), such as Λ and Σ^-, inevitabley appear as new components and develop the population with ρ, becoming important constituents comparable to nucleons (N). The most striking aspect of Y-mixed NSs is that the equation of state (EOS) is extremely softened and as a result conflicts with observations[1∼3]. That is, the maximum mass M_{max} for Y-mixed NSs fails to satisfy the condition that M_{max} should exceed M_{obs}(PSR1913+16)=$1.44M_{\odot}$, minimally requested from recent NS mass observations. This serious inconsistency between theory and observation cannot be resolved by enhancing the NN repulsion and strongly suggests the necessity of some "extra repulsion" in hypernuclear systems, i.e., in the YN and YY interaction parts[1∼3]. In our preceding works[3], we introduced a repulsion from phenomenological three nucleon interaction (TNI) "universally", i.e., equally to YN and YY parts as well as to NN part, and found that such universal 3-body repulsion is a solution to overcome the confronting problem.

The aim of the present paper is to make a research for the origin of such a 3-body repulsion. We discuss two cases. The one is the 2π-exchange 3-body force via isobar Δ excitation (abbreviated to $2\pi\Delta$) which has been initiated by Fujita-Miyazawa[4] and whose importance is well recognized in nuclear physics. We extend this $2\pi\Delta$ into $\{Y+N\}$ system in order to be applicable for the present case. The other is the universal short-range 3-body repulsion from a viewpoint of the string-junction quark model of

CP1011, *New Facet of Three Nucleon Force—50 Years of Fujita Miyazawa Three Nucleon Force (FM50), The International Symposium*, edited by H. Sakai, K. Sekiguchi, and B. F. Gibson

baryons (SJM) proposed very recently by one of the present authors (R. T.)[5]. By applying these $2\pi\Delta$ and SJM 3-body repulsions to the Y-mixed NS matter, we address to what extent these 3-body repulsion can substitue the "extra repulsion" demanded for Y-mixed NSs.

OUTLINE OF APPROACH AND RESULTS IN PREVIOUS WORKS

Here we briefly recapitulate our way of approach to the Y-mixing problem in NS matter and show main points of our previous works[3], remarking a dramatic softening of the EOS for Y-mixed NS matter.

Effective interaction approach

The short-range correlation (s.r.c.) between baryons is of prime importance in the energy calculations of dense baryon matter. To take account of this effect, we adopt a G-matrix-based effective interaction approach. We construct ρ-dependently the YN and YY effective local potentials $\tilde{V}_{YN}$ and $\tilde{V}_{YY}$, from the G-martix calculations with the Nijmegen model D hard core potential[6] performed for $\{n+Y\}$ matter with the mixing ratio $y_Y(\equiv \rho_Y/\rho\ ;\ \rho = \rho_Y + \rho_n)$. For simplicity, we neglect the contamination of Ξ^- with larger rest mass ($m_{\Xi^-} \simeq 1321$ MeV, compared to $m_\Lambda \simeq 1116$ MeV and $m_{\Sigma^-} \simeq 1192$ MeV) and treat the Y-mixed NS matter as composed of n, p, Λ, Σ^-, e^- and μ^-. Concerning the NN interactions, we use the ρ- and $y_p(\equiv \rho_p/\rho\ ;\ \rho = \rho_n + \rho_p)$-dependent effective local potential ($\tilde{V}_{\rm RSC}$) previously constructed from the G-matrix in asymmetric nuclear matter with y_p and with the Reid-soft-core NN potential[7]. Our local effective potential $\tilde{V}_{BB}$ ($B \equiv n$, p, Λ, Σ^-) are expressed by a superposition of Gaussian functions and parameters are reffered to Refs.3 and 8.

In practice, $\tilde{V}_{NN} = \tilde{V}_{\rm RSC}$ is supplemented by the phenomenological three nucleon interaction $\tilde{U}_{\rm TNI}$ of Lagaris-Pandharipande type[9, 10], so as to satisfy the empirical saturation property of symmetric nuclear matter, i.e., $\tilde{V}_{NN} \to \tilde{V}_{\rm RSC} + \tilde{U}_{\rm TNI}$. The ρ-dependent $\tilde{U}_{\rm TNI}$ is given in a form of 2-body potential and consists of two parts ($\tilde{U}_{\rm TNI} = \tilde{U}_{\rm TNA} + \tilde{U}_{\rm TNR}$), the attractive $\tilde{U}_{\rm TNA}$ and the repulsive $\tilde{U}_{\rm TNR}$ parts. As ρ gets higher, $\tilde{U}_{\rm TNA}$ becomes minor compared to $\tilde{U}_{\rm TNR}$ which brings about a large repulsive contribution with ρ. The parameters inherent in $\tilde{U}_{\rm TNI}$ are determined ao as to reproduce the binding energy $E_B = -16$ MeV, saturation density $\rho_s = \rho_0$ ($\equiv 0.17$ nucleons/fm^3; nuclear density) and the nuclear incompressibility κ which is an important measure for the stiffness of the nuclear-part (N-part) EOS (details are referred to Ref.11). Thus our EOS (energy per baryon versus ρ) is specified by κ and we consider three cases, TNI2($\kappa = 250$MeV), TNI6($\kappa = 280$MeV) and TNI3($\kappa = 300$MeV). When applied to neutron matter, TNI6 generates the EOS very close to that by Wiringa et al.[12] which is currently taken to be most realistic, and TNI2 (TNI3) is introduced as a softer (stiffer) case compared to TNI6.

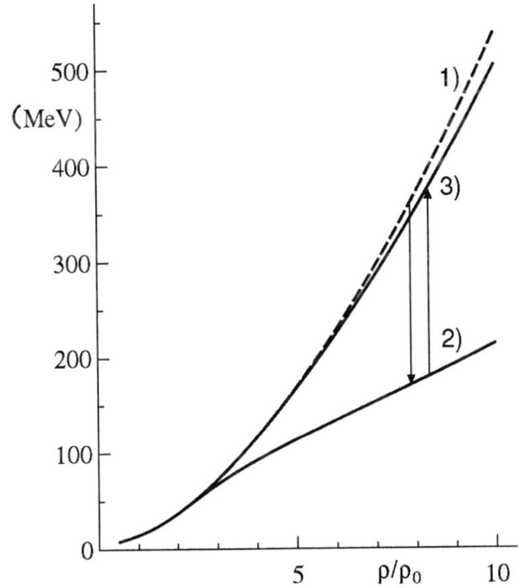

FIGURE 1. Energy per baryon versus density ρ with ρ_0 being the nuclear density. A dashed curve with 1) and a solid curve with 2) correspond to the case without Y-mixing and with Y-mixing, respectively, for the inclusion of phenomenological three nucleon interaction (TNI3). A curve with 3) is for the case with Y-mixing by including of TNI3 (precisely, the repulsive part TNR3) universally in YN and YY parts as well as NN one, denoted as TNI3u in the text. Downward (upward) arrow indicates a dramatic softening effect due to Y-mixing (moderate softening recovered by TNI3u).

Dramatically softened EOS and demand for "Extra Repulsion"

Under the conditions of charge neutrality, chemical equilibrium and baryon number conservation, we calculate the fractions $y_i(\equiv \rho_i/\rho;\ i \equiv n,\ p,\ \Lambda,\ \Sigma^-,\ e^-,\ \mu^-)$ and the EOS for Y-mixed NS matter. In Fig.1, the EOS without and with Y, the lines attached by 1) and 2), respectively, are compared for the TNI3 case. We observe that the Y-mixing effect causes a dramatic softening of the EOS. Other cases, TNI2- and TNI6-EOSs are quite similar. By this strong softening, we encounter a serious problem that the gravitational mass M of NSs with Y-mixing leads to the inconsistency between theory and observations, i.e., cannot satisfy the minimum condition that at least the maximum mass $M_{\rm max}$ from theory should exceed $M_{\rm obs} = 1.44 M_\odot$ from observation for the PSR1913+16 NS. For TNI3 case as an example, $M_{\rm max} \simeq 1.10 M_\odot$ which is by far smaller than $M_{\rm obs} = 1.44 M_\odot$. A significant point is that this inconsistency cannot be resolved by introducing a stronger NN repulsion, namely, stiffer N-part EOS as $M_{\rm max} \simeq 1.08 M_\odot \to 1.09 M_\odot \to 1.10 M_\odot$ for TNI2→TNI6→TNI3 with increasing κ. This is because, when the N-part EOS gets stiffer, the Y-mixed phase develops from lower densities and consequently the softening effect becomes stronger, making the enhanced NN repulsion ineffective.

In our preceding works, as one possibility for such a repulsion, we introduced a repulsion from the 3-body force $\tilde{U}_{\rm TNR}$ in TNI also to the YN nad YY parts, quite similarly to NN part (called the "universal inclusion of TNI" and denoted by TNIu with suffix "u"), considering that the 3-body force should not be restricted to NN part when extended to $\{N+Y\}$ system. That is , we made a treatment

$$\tilde{V}_{YN} \to \tilde{V}_{YN} + \tilde{U}_{\rm TNR}, \qquad \tilde{V}_{YY} \to \tilde{V}_{YY} + \tilde{U}_{\rm TNR} \tag{1}$$

as well as $\tilde{V}_{NN} \to \tilde{V}_{\rm RSC} + \tilde{U}_{\rm TNI}$. By this treatment, we found that the Y-mixing effect leads to a moderate softening of the EOS, not a dramatic one as before (see the line with 3) in Fig.1 for example), and the condition $M_{\rm max} > 1.44 M_\odot$ is nicely satisfied; $M_{\rm max} \simeq 1.52$, 1.71 and $1.83 M_\odot$ for TNI2u, TNI6u and TNI3u, respectively.

Here we wish to note that the dramatic softening of the Y-mixed EOS is not a result peculiar to our way of approach. Other approaches such as the works by M. Baldo et al.[13] and I. Vidaña et al.[14] encounter the same problem, i.e., $M_{\rm max} < 1.44 M_\odot$ for Y-mixed NSs, although in these works the point that the inconsistency should be taken as the necessity of "extra repulsion" was not recognized. We remark that too strong softening effect contradicting observation is common to Y-mixed NSs, as far as a realistic treatment of baryon-baryon interaction and s.r.c. are duely taken into account, and is nothing but a significant message as to a demand for "extra repulsion in hypernuclear systems" called for by NSs.

THREE-BODY FORCE FROM 2π-EXCHANGE VIA Δ EXCITATION

In our previous works[1~3], we have introduced the 3-body force $\tilde{U}_{\rm TNR}$ as an "extra repulsion" and found that $\tilde{U}_{\rm TNR}$ is a solution to overcome a serious inconsistency between theory and observation for Y-mixed NSs. The 3-body force, however, is introduced phenomenologically and in this section, as one of the origin of such a repulsion, we discuss the 3-body force from 2π-exchange via isobar Δ excitation ($2\pi\Delta$). It has been found in nuclear matter calculations by Hokkaido group[15] that the $2\pi\Delta$ process including s.r.c. function $f(r)$ generates a small attraction at low ρ but gives rise to an increasing repulsion with increasing ρ. In order to be applicable to the present case, we extend the $2\pi\Delta$ process to $\{N+Y\}$-space as shown in Fig.2, where B denotes a baryon and B^* includes the excitation of Σ^* (strangeness $S=-1$) in addition to usual $\Delta(S=0)$ excitation[16]. The coupling constants $g_{BB\pi}$ for vertexes are estimated by the quark model of baryons and s.r.c. is taken into account for each baryon pairs (Fig.2), but the mass difference of baryons in the energy denominator is ignored for the sake of simplicity. We have

$$g_{N\Delta\pi} : g_{\Lambda\Sigma^{*0}\pi} : g_{\Sigma^-\Sigma^{*-}\pi} = 2 : \sqrt{3} : 1. \tag{2}$$

On the other hand, from the OBE (one-boson-exchange) with SU(3) symmetry hypothesis, we have

$$g_{\Sigma^-\Sigma^-\pi} = 2\alpha g_{NN\pi} \simeq g_{NN\pi} \tag{3}$$

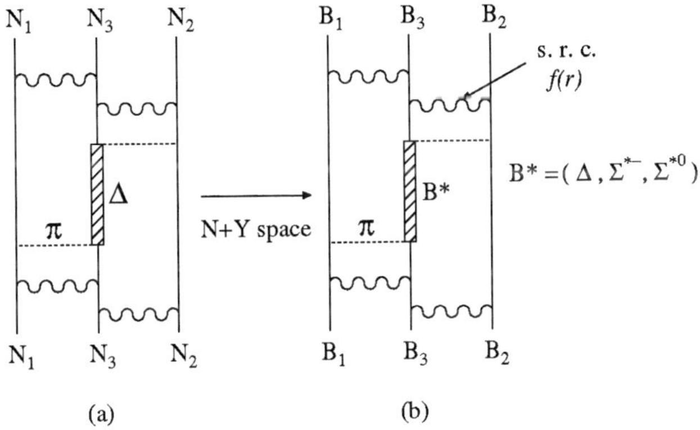

FIGURE 2. Extension of 3-body force from 2π-exchange via Δ excitation type ($2\pi\Delta$) in N-space (a) into $\{N+Y\}$ space (b), where B^* stands for Δ, Σ^{*-}, Σ^{*0} and $f(r)$ is the short-range correlation function.

with α the socalled F-D ratio. The α is given as $\alpha = 0.49, 0.52$ and 0.464 by Funabashi-Gifu[17], Ehime[18] and Nijmegen[6] BB potentials, respectively, and can be taken model-independently as $\alpha \simeq 0.5$ leading to $g_{\Sigma^-\Sigma^-\pi} \simeq g_{NN\pi}$ in Eq.(3).

After summing up the contribution of the third baryon (B_3) intervening between the first (B_1) and the second one (B_2) and also taking into account the s.r.c. for B_1B_3 and B_2B_3 pairs, we have the 3-body force $U_{2\pi\Delta}(B_1B_2)$ expressed in a form of 2-body potential between B_1 and B_2, as has been done in Ref.15 for the case of nuclear matter. Eq.(3) means $U_{2\pi\Delta}(NN) \simeq U_{2\pi\Delta}(\Sigma^-N) \simeq U_{2\pi\Delta}(\Sigma^-\Sigma^-)(= U_{2\pi\Delta})$, that is, the $2\pi\Delta$ effects are almost equal for $B_1B_2 \equiv NN$, $N\Sigma^-$ and $\Sigma^-\Sigma^-$ interactions. It should be noted that the $2\pi\Delta$ process is missing for ΛB parts; $U_{2\pi\Delta}(\Lambda N) = U_{2\pi\Delta}(\Lambda\Sigma^-) = U_{2\pi\Delta}(\Lambda\Lambda) = 0$ because of the lack of direct $\Lambda\Lambda\pi$ coupling. The repulsion of $U_{2\pi\Delta}$ comes from its central part, whereas the contribution from its tensor part, though attractive, is relatively small at high densities, compared to the central force contributions. Finally the effective 3-body interaction from $2\pi\Delta$-central force, expressed in a form of 2-body one, $\tilde{U}_{2\pi\Delta\text{-C}}(B_1B_2)$, is given by

$$\tilde{U}_{2\pi\Delta\text{-C}}(NN) \simeq \tilde{U}_{2\pi\Delta\text{-C}}(\Sigma^-N) \simeq \tilde{U}_{2\pi\Delta\text{-C}}(\Sigma^-\Sigma^-) = \tilde{U}_{2\pi\Delta\text{-C}}, \tag{4}$$

$$\tilde{U}_{2\pi\Delta\text{-C}} = f(r)[U_{2\pi\Delta\text{-C}}(\rho_N) + \frac{3}{4}U_{2\pi\Delta\text{-C}}(\rho_\Lambda) + \frac{1}{4}U_{2\pi\Delta\text{-C}}(\rho_{\Sigma^-})], \tag{5}$$

$$U_{2\pi\Delta\text{-C}}(\rho_B) = 110\rho_B(1+0.055\tilde{\rho})e^{-(r/0.6)^2} - 8.4\tilde{\rho}_B e^{-(r/1.4)^2} \tag{6}$$

with $U_{2\pi\Delta\text{-C}}$ in MeV, $\tilde{\rho}_B = \rho_B/\rho_0$ and $\rho_B = y_B\rho$ the density of the third baryon B_3 ($B = N$, Λ, Σ^-, $\rho = \rho_N + \rho_\Lambda + \rho_{\Sigma^-}$). The s.r.c. $f(r)$ is taken from G-matrix calculations and is given as, neglecting the weak ρ-dependence,

$$f(r) = 1. - 1.2e^{-(r/0.6)^2} + 0.4e^{-(r/1.2)^2} \quad \text{for } l = \text{even state}, \tag{7}$$

$$= 1. - 0.55e^{-(r/0.7)^2} + 0.06e^{-(r/1.6)^2} \quad \text{for } l = \text{odd state} \tag{8}$$

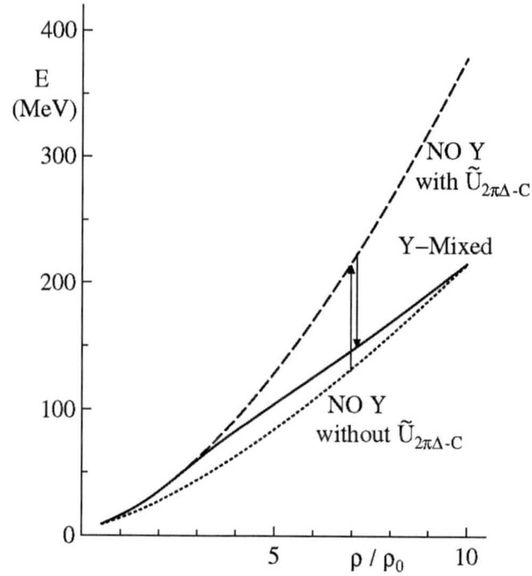

FIGURE 3. Energy per baryon versus ρ in the scheme of 2-body force ($\tilde{V}_{BB}$) plus 3-body force ($\tilde{U}_{2\pi\Delta\text{-C}}$) from $2\pi\Delta$-central part. Dotted, dashed and solid lines are for the case without Y and without $\tilde{U}_{2\pi\Delta\text{-C}}$, the case without Y and with $\tilde{U}_{2\pi\Delta\text{-C}}$ and with both Y and $\tilde{U}_{2\pi\Delta\text{-C}}$, respectively. Dramatic softening of EOS due to the Y-mixing (downward arrow) is inevitable even for the inclusion of $2\pi\Delta$-central repulsion.

with l the orbital angular momentum.

Our prime interst is to what extent $2\pi\Delta$ 3-body force can play the role of "extra repulsion". We calculate the EOS of Y-mixed matter by concentrating the effect of the central force (i.e., $\tilde{V}_{BB} \to \tilde{V}_{BB} + \tilde{U}_{2\pi\Delta\text{-C}}$). Main points of our results are as follows. The effect of $\tilde{U}_{2\pi\Delta\text{-C}}$ surely brings about a remarkable repulsion for usual neutron star matter (without Y) as shown in Fig.3 and gives $M_{\max} \simeq 1.66 M_{\odot}$ for normal NSs, in comparison with $M_{\max} \simeq 1.4 M_{\odot}$ without $\tilde{U}_{2\pi\Delta\text{-C}}$. With the Y-mixing degrees of freedom, however, the stiffening effect due to $\tilde{U}_{2\pi\Delta\text{-C}}$ is overwhelmed by the softening effect due to Y-mixing and consequently contradicts M_{obs}, namely, $M_{\max} \simeq 1.2 M_{\odot} < M_{\text{obs}} = 1.44 M_{\odot}$. This is because the extended $2\pi\Delta$ cannot act on Λ due to the absence of $\Lambda\Lambda\pi$ direct coupling, in different from N and Σ^- cases. It shows obviously the importance of universality nature of "extra repulsion" like TNIu treatment.

THREE-BODY FORCE FROM STRING-JUNCTION MODEL

Motivated by the universal nature demanded for the "extra repulsion", one of the present author (R.T.) recently proposed a new idea for the universal three-baryon repulsive interaction, from a viewpoint of the quark confinement mechanism described in a string-junction model (SJM)[5]. In the SJM, the color flux-tubes embodying the confinement mechanism are expressed by strings with orientation and an ordinary baryon is drawn by the Y-shaped net with three strings and a junction. The junction plays a role to neutralize

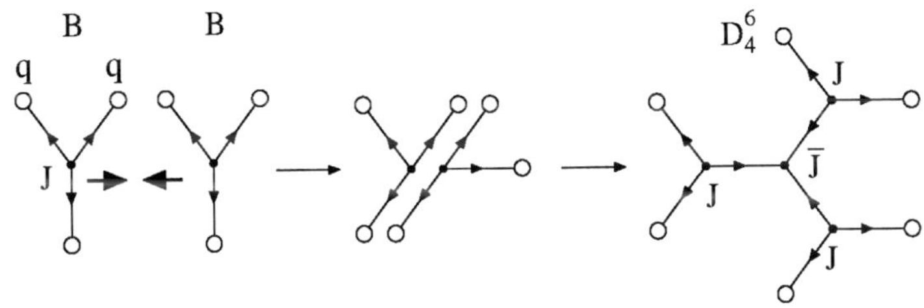

FIGURE 4. Schematic picture that two baryons with *Y*-shaped quark structure fuse into a six-quark state, from the viewpoint of the string-junction quark model (SJM) explained in the text.

the color and the strings stem from the junction and end at the quarks. This Y-shaped structure representing the confinement mechanism in the single baryon has been founded from the lattice QCD calculations performed recently by Suganuma et al.[19]. The key concept to the baryon-baryon short-range repulsion from the viewpoint of SJM is that the repulsive core preventing two baryons from close approach appears as the latent effect of the high excitation energy needed for the system to fully overlap.

As two baryons come near (Fig.4(a)), they make registance to the fusion, keeping their identity by deforming the Y-shaped quark-string structure (Fig.4(b)), but eventually fuse into a six-quark state by recombining their string-junction when they fully overlap (Fig.4(c)). Then the energy cost from Fig.4(a) to Fig.4(c), estimated as about 2GeV from SJM mass formula, can be taken as the repulsive core height of BB interaction. The range of repulsive core (η_c) would be $(0.45-0.5)$fm from a quark bag size of a single baryon. Thus the short-range repulsion is expressed by taking the Gaussian form as

$$V_{SJM}(r) = V_0 exp[-(r/\eta_c)^2], \;\; V_0 \simeq 2\text{GeV} \;\text{and}\; \eta_c \simeq (0.45-0.5)\;\text{fm}. \qquad (9)$$

Quite similarly to the $2\pi\Delta$ case, the 3-body force from SJM, $U_{SJM}(B_1B_2)$ expressed in a form of 2-body potential acting between B_1 and B_2, is derived along the line of Hokkaido group[15], by taking into account the s.r.c. between B_1B_3 and B_2B_3 and taking the summation over the effects from B_3. We consider two cases, SJM1 and SJM2 corresponding to $\lambda = (1/\eta_c)^2 = 4.5$ and $4.0\ \text{fm}^{-2}$, respectively. The repulsion from U_{SJM} grows almost linearly with ρ and is well approximated by

$$\begin{aligned} U_{SJM} &= 70\rho(1+0.025\tilde{\rho})e^{-1.5r^2} \quad \text{for SJM1}, & (10)\\ &= 95\rho(1+0.024\tilde{\rho})e^{-1.35r^2} \quad \text{for SJM2} & (11) \end{aligned}$$

with U_{SJM} in MeV, ρ in fm^{-3} and $\tilde{\rho} = \rho/\rho_0$. The effective force $\tilde{U}_{SJM}$ to take account of the s.r.c. between B_1 and B_2, is given by

$$\tilde{U}_{SJM} = f(r)U_{SJM} \qquad (12)$$

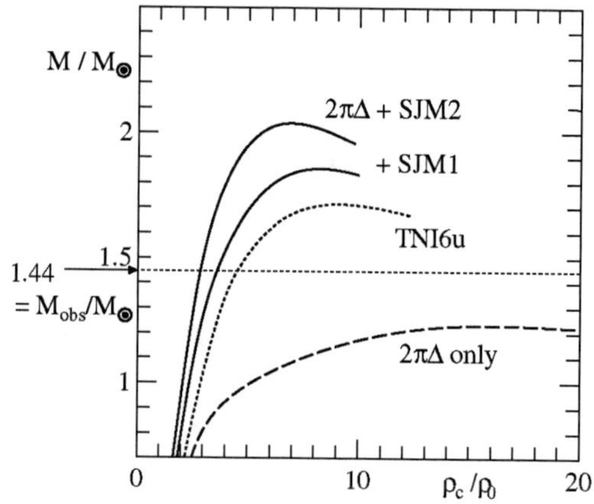

FIGURE 5. Comparison of neutron star models (gravitational mass M versus central density ρ_c) for various use of 3-body repulsion in the scheme of 2-body force ($\tilde{V}_{BB}$) plus 3-body force. Notations are explained in the text.

with $f(r)$ given by Eqs.(7) and (8). Since $U_{\rm SJM}(B_1B_2)$ originates from the color degrees of freedom, it is independent of quark flavor, i.e., independent of baryon species and nicely meets the condition of universally acting repulsion. Actually, the calculations with the introduction of $\tilde{U}_{\rm SJM}$, i.e., $\tilde{V}_{BB} \to \tilde{V}_{BB} + \tilde{U}_{\rm SJM1}$, gives the EOS of Y-mixed NS matter comparable to that for TNI6u case, playing the expected role for "extra repulsion".

In Fig.5, NS models are compared for several EOSs corresponding to the choice of 3-body repulsion. As already mentioned, $M_{\rm max} \simeq 1.2M_\odot$ for the case only with $2\pi\Delta$ (a dashed line), but $M_{\rm max}$ is made larger by including $\tilde{U}_{\rm SJM}$, i.e., $\tilde{V}_{BB} \to \tilde{V}_{BB} + \tilde{U}_{2\pi\Delta\text{-C}} + \tilde{U}_{\rm SJM}$; $M_{\rm max} \simeq 1.86M_\odot(2.04M_\odot)$ for $2\pi\Delta$+SJM1(SJM2). These $M_{\rm max}$ is larger than that for TNI6u case (a dotted line) and can surely be consistent with observation ($M_{\rm max} > M_{\rm obs} = 1.44M_\odot$), satisfying the nature of "extra repulsion" acting universally on baryons.

CONCLUDING REMARKS

Dramatically softened EOS of NSs due to the effects of Y-mixing, in conjunction with NS mass observations, strongly suggests the necessity of some "extra repulsion"acting on YN and YY interaction parts. As such a repulsion, 3-body force from two mechanisms, the one from the 2π-exchange via Δ excitation process ($2\pi\Delta$) extended to $\{N+Y\}$ system and the other from the latent effect for 2-baryons to fully overlap in the string-junction quark model (SJM). The former, associated with the one-boson-exchange process, is mainly concerned with the 3-body force effect at intermediate and long distances between two baryons and the latter, relevant to the quark confinement mechanism, is concerned with that for short distance.

From the numerical calculations for Y-mixed NS matter with these 3-body repulsions, following points are noted. For normal NS matter without Y-mixing, the 3-body

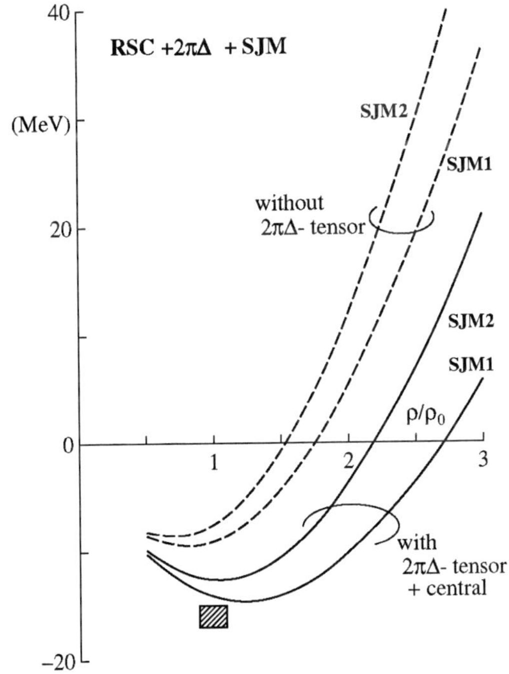

FIGURE 6. Energy per nucleon for symmetric nuclear matter as functions of ρ in the scheme of {2-body force (RSC)+$2\pi\Delta$ 3-body force + SJM 3-body force} explained in the text. SJM1 (SJM2) is for the case with the range of repulsive core $\eta_c = 1/\sqrt{4.5}$ fm $(1/\sqrt{4.0}$ fm) in the original SJM 2-body repulsion (Eq.(9)). A hatched box indicates the region corresponding to the empirical saturation property.

repulsion coming from $2\pi\Delta$ central force remarkably makes the EOS stiffer than that without $2\pi\Delta$, but, when applied to the Y-mixed NS matter, its stiffening effect turns out to be ineffective and consequently generates a softened EOS contradicting observations ($M_{\rm max} < M_{\rm obs} = 1.44 M_\odot$). This is because the $2\pi\Delta$ 3-body force does not work on Λ due to the lack of $\Lambda\Lambda\pi$ direct coupling and clearly shows the importance of universality nature of "extra repulsion", namely, working for any baryons. In this respect, the SJM 3-body repulsion can meet the condition due to the flavor independence. In fact, the 3-body repulsion from $2\pi\Delta$+SJM scheme can avoid a dramatic softening effect on the EOS due to the Y-mixing and assure the consistency between theory and observation with respect to NS masses.

Here it is worthwhile to note whether this {$2\pi\Delta$+SJM} scheme satisfies another experimental condition, i.e., empirical saturation property of symmetric nuclear matter. Results are shown in Fig.6. It is observed that the inclusion of central and tensor parts from $2\pi\Delta$ largely satisfies the condition, although only with central part the attraction remains to be insufficient. Thus we can say that the repulsion from {$2\pi\Delta$+SJM} scheme is a promising candidate for "extra repulsion", keeping the consistency with both NS mass observations and empirical saturation property of nuclear matter.

ACKNOWLEDGMENTS

The authors wish to thank Y. Yamamoto for collaborations in the series of studies on Y-mixed NSs. They also thank S. Tsuruta, T. Tatsumi and T. Muto for discussions and interests in the work.

REFERENCES

1. As review articles, T. Takatsuka, Prog. Theor. Phys. Suppl. No.**156**, 84 (2000).
2. T. Takatsuka, S. Nishizaki and Y. Yamamoto, Eur. Phys. J. **A13**, 213 (2002).
3. S. Nishizaki, Y. Yamamoto and T. Takatsuka, Prog. Theor. Phys. **105**, 607 (2001); **108**, 703 (2002).
4. J. Fujita and H. Miyazawa, Prog. Theor. Phys. **17**, 360 (1957).
5. R. Tamagaki, Soryushiron Kenkyu (kyoto) **115**, 27 (2007); poster presentation (QT-131) in International Nuclear Physics Conference (INPC2007), Tokyo, June 3-8 (2007); arXiv:0801.2289v1 [nucl-th] 15 Jan 2008.
6. M.M. Nagels, T.A. Rijken and J.T. deSwart, Phys. Rev. **D12**, 744 (1975); **D15**, 2547 (1977); **D20**, 1633 (1979).
7. R.V. Reid, Ann. of Phys. **50**, 411 (1965).
8. S. Nishizaki, T. Takatsuka, N. Yahagi and J. Hiura, Prog. Theor. Phys. **86**, 853 (1991).
9. L.E. Lagaris and V.R. Pandharipande, Nucl. Phys. **A369**, 470 (1981).
10. B. Friedman and V.R. Pandharipande, Nucl. Phys. **A361**, 502 (1981).
11. S. Nishizaki, T. Takatsuka and J. Hiura, Prog. Theor. Phys. **92**, 93 (1994).
12. R.B. Wiringa, V. Fiks and A. Fabrocini, Phys. Rev. **C38**, 1010 (1988).
13. M. Baldo, G.F. Burgio and H.-J. Shulze, Phys. Rev. **C61**, 055801 (2000).
14. L. Vidaña, A. Polls, A. Ramos, L. Engvik and M. Hjorth-Jensen, Phys. Rev. **C62**, 035801 (2000).
15. T. Kasahara, Y. Akaishi and H. Tanaka, Prog. Theor. Phys. Suppl. No.**56**, 96 (1974).
16. T. Takatsuka, S. Nishizaki and R. Tamagaki, poster presentation (QT-223) in International Nuclear Physics Conference (INPC2007), Tokyo, June 3-8 (2007).
17. I. Arisaka, K. Nakagawa, S. Shinmura and M. Wada, Prog. Theor. Phys. **104**, 995 (2000).
18. T. Ueda, K. Tominaga, M. Yamamoto, N. Kijima, D. Okamoto, K. Miyanaga and T. Yamada, Prog. Theor. Phys. **99**, 891 (1998).
19. H. Suganuma, F. Okihara, T. Takahashi and H. Ichie, Nucl. Phys. **A755**, 399c (2005).

Three-Nucleon Force, Fifty Years Ago

Hironari Miyazawa

University of Tokyo

Abstract. The three-body nuclear force was calculated fifty years ago. This talk will describe how and why it was introduced and what the status of pion theory was at that time.

Keywords: three-nucleon force
PACS: 21.45.Ff, 13.75.Gx

Introduction

First, I wish to thank you for coming to this meeting. I am really pleased and honored to have this symposium at the fiftieth anniversary of our work.

It was just fifty years ago that Jun-ichi Fujita and I performed the calculation of the three-nucleon force. In this talk I will describe why and how it was introduced, and also discuss the status of pion physics at that time. Yesterday the first speaker, Dr. Robilotta, gave a nice review of the history of three-nucleon forces, including our work. My talk necessarily overlaps with his.

Pion physics, 50 years ago

The pi-meson was discovered in 1947, and shortly thereafter was artificially produced in accelerators. Electron synchrotrons produced photo-pions, and proton synchrocyclotrons produced pion beams. Hence, by 1950, the properties of pions were almost known: A pion's mass was about 140MeV, its spin was zero, and the parity was negative. Reactions were charge independent, *i. e.*, the isospin was nearly conserved, and pion's isospin was 1. However, the theory of pions was completely unsuccessful.

At that time quantum electrodynamics was very successful. It correctly explained the Lamb shift, the electron's anomalous magnetic moment, etc.. Naturally we employed the same technique, *i. e.*, perturbation theory, to pion processes such as photopion production, pion-nucleon scattering, etc.. We found that the results were in complete disagreement with experiment. It was anticipated that the perturbation expansion would not work, because the pion-nucleon interaction was strong. The presence of a resonance was proposed, but few believed in it. Hence we did not know how to deal with pion interactions. We could not do anything except for qualitative discussion of pion processes.

However, in 1955, pion theory succeeded in two cases. One was the dispersion relations for forward pion-nucleon scattering, and the other was the two-nucleon interaction at a large separation.

CP1011, *New Facet of Three Nucleon Force—50 Years of Fujita Miyazawa Three Nucleon Force (FM50), The International Symposium,* edited by H. Sakai, K. Sekiguchi, and B. F. Gibson

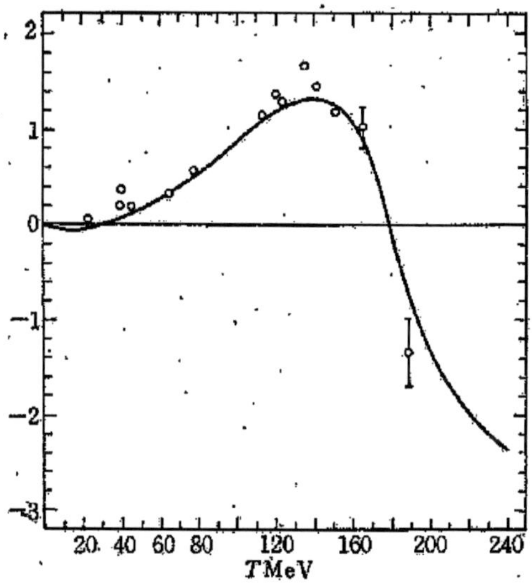

FIGURE 1. Experimental fit of the dispersion relation for $F_+ + F_-$

The dispersion relation expresses the pion-nucleon forward scattering amplitude as an analytic function of the incident energy. The following equation is an example:

$$\mathrm{Re}[F_-(k_0) - F_+(k_0)] = \frac{f^2}{2\pi}\frac{1}{k_0} + \frac{k_0}{4\pi^2}\int_m^\infty \frac{\sigma_-(k_0') - \sigma_+(k_0')}{k_0'^2 - k_0^2} k_0' \mathrm{d}k_0'. \tag{1}$$

The left-hand side is the difference of the positive and negative pion amplitudes, and the right-hand side is expressed in terms of the total cross sections and the Yukawa coupling constant, f. These equations were compared with experiments. Figure 1 shows the comparison for another example, the sum of positive and negative pion amplitudes. The fit was OK, and the coupling constant was determined to be

$$\frac{f^2}{4\pi} = 0.08. \tag{2}$$

This was the first numerical fit of pion dynamics with experimental data.

Another success was the two-nucleon interaction problem for a large inter-nucleon distance, r. I discuss this briefly in detail, because this is closely connected with my main topic.

Two nucleons interact by exchanging pions, and a method utilized in molecular physics can be followed. In the Hamiltonian of the pion-nucleon system,

$$\sum_{1,2} \frac{p_i^2}{2M} + \sum \omega_\alpha N_\alpha + \frac{f}{m}\sum_{1,2} \tau^i(\boldsymbol{\sigma}^i \nabla)\phi(\boldsymbol{x}_i), \tag{3}$$

one initially neglects the nucleon kinetic energy term. For $r > r_0 = 1/m$ (m: pion mass), the momentum of an exchanged pion $k = |\boldsymbol{k}| < m$ is much less than the nucleon mass, M, and the recoil energy of the nucleon is small. By neglecting this kinetic energy the position, $\boldsymbol{x}_i$, of the nucleon is a c-number, *i. e.*, the nucleon is fixed at $\boldsymbol{x}_i$. Thus, one obtains the potential energy of a two-nucleon system. This is the nuclear potential. Finally one reinstates the nucleon kinetic energy, and solves the Schrödinger equation.

The potential energy is given by

$$V(r) = -\frac{f^2}{4\pi m^2}(\boldsymbol{\tau\tau})(\boldsymbol{\sigma}\nabla)(\boldsymbol{\sigma}\nabla)\, Y(1,2), \quad Y(1,2) = Y(r) = \frac{\mathrm{e}^{-mr}}{r}. \tag{4}$$

This is called the Yukawa potential. At large distances, the two-pion exchange potential is smaller than the one-pion exchange by a factor of

$$\frac{f^2}{4\pi}\frac{\mathrm{e}^{-mr}}{\sqrt{mr}}. \tag{5}$$

This expression is by no means exact, but important factors are apparent. For $r \gg r_0$ multi-pion exchange contributions can be safely neglected.

A deuteron is a loosely bound nucleus, which extends into the region of $r \approx 3r_0$. Here the adiabatic potential can be used and multi-pion exchanges can be safely neglected. Tamagaki and others employed the one-pion exchange adiabatic potential (Yukawa potential) and, from the deuteron quadrupole monent, they obtained the value of the coupling constant as

$$\frac{f^2}{4\pi} = 0.08. \tag{6}$$

This value completely agrees with that in Eq. (2), indicating that the adiabatic or static approximation, *i. e.*, neglecting the nucleon recoil energy in low energy processes, is experimentally justified.

Static pion theory

At this point, we had two reliable tools in pion dynamics: the dispersion formula and the static approximation. These two combined provided a powerful means of calculating pion processes. While calculating the processes a la Feynman, we encountered a part that corresponded to pion-nucleon scattering (Figure 2). The shaded area contained a number of pion emission and reabsorption corrections which could not be calculated perturbationally. I tried to express the pion-nucleon part exactly in terms of experimentally observed quantities.

P-wave pion-nucleon scattering has four independent amplitudes, which can be written as

$$F_{ij}(\boldsymbol{k},k_0;\boldsymbol{q},k_0) = A(\boldsymbol{k},\boldsymbol{q},k_0)\tau_i\tau_j\boldsymbol{\sigma k\sigma q} + B_1(\,)\tau_i\tau_j\boldsymbol{\sigma q\sigma k} + \\ + B_2(\,)\tau_j\tau_i\boldsymbol{\sigma k\sigma q} + C(\,)\tau_j\tau_i\boldsymbol{\sigma q\sigma k}. \tag{7}$$

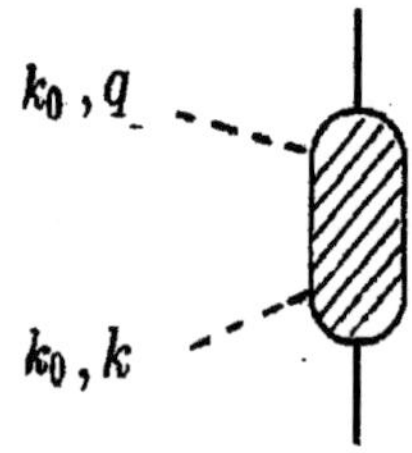

FIGURE 2. Scattering part in Feynman graph

The amplitudes $A,\ldots$ depend on four-momenta k_μ and q_μ. The static approximation simplifies the matter. Because the nucleon kinetic energy can be neglected, pion energy is conserved: $q_0 = k_0$. Moreover, the quantity A does not depend on the three-momenta $\boldsymbol{k}$ or $\boldsymbol{q}$. The heavy nucleon does not move when it absorbs a pion of momentum $\boldsymbol{k}$, so that the information $\boldsymbol{k}$ is lost except for the factor $\boldsymbol{\sigma k}$ due to P-wave absorption.

In this way A, B and C are functions of only k_0 . Furthermore, there is a so-called crossing symmetry which means that a positive pion of energy k_0 coming in is the same as a negative pion of energy $-k_0$ going out. From this one knows that

$$A(k_0) = C(-k_0), \quad B_1(k_0) = B_2(-k_0). \tag{8}$$

Thus, the Feynman amplitude for the scattering can be written as

$$F_{ij}(\boldsymbol{k}, k_0; \boldsymbol{q}, k_0) = A(k_0)\boldsymbol{\tau}_i\boldsymbol{\tau}_j\boldsymbol{\sigma k \sigma q} + B(k_0)\boldsymbol{\tau}_i\boldsymbol{\tau}_j\boldsymbol{\sigma q \sigma k} + \\ + B(-k_0)\boldsymbol{\tau}_j\boldsymbol{\tau}_i\boldsymbol{\sigma k \sigma q} + A(-k_0)\boldsymbol{\tau}_j\boldsymbol{\tau}_i\boldsymbol{\sigma q \sigma k}. \tag{9}$$

The two functions A and B can be determined by the forward (and near forward) dispersion relations.

In addition to P-wave scattering, there is small but finite S-wave scattering. This can be treated in a similar way as the P-wave.

Once the Feynman amplitude for pion-nucleon scattering was known, we could calculate a number of processes that contained the scattering part. For the one-nucleon problem, it gave the properties of a pion cloud around the nucleon (Fig. 3a). For the two-nucleon problem the two-pion exchange potential could be calculated (Fig. 3b). For the three-nucleon case it gave the three-nucleon force (Fig. 3c)! For the many-nucleon case, the amplitude gave the behavior of pion propagation in nuclear matter (Fig. 3d).

Three-body force

The graph in Fig. 3c gives the three-body scattering amplitude, but also includes contributions from two-body forces. After subtracting these latter contributions we had

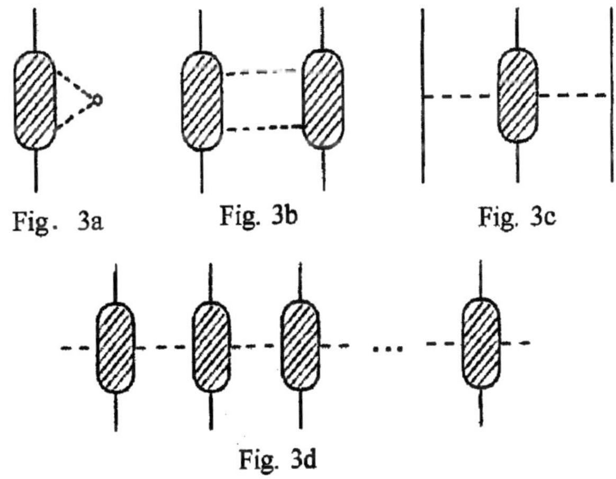

FIGURE 3. Graphs containin scattering parts

the three-body nuclear potential given by

$$V = V(123) + V(231) + V(312), \tag{10}$$

$$\begin{aligned} V(123) = [\{A\,(\boldsymbol{\tau}^1\boldsymbol{\tau}^2)(\boldsymbol{\tau}^2\boldsymbol{\tau}^3) + B(\boldsymbol{\tau}^2\boldsymbol{\tau}^3)(\boldsymbol{\tau}^2\boldsymbol{\tau}^1)\}(\boldsymbol{\sigma}^1\nabla^1)(\boldsymbol{\sigma}^2\nabla^1)(\boldsymbol{\sigma}^2\nabla^3)(\boldsymbol{\sigma}^3\nabla^3) \\ +\{B\,(\boldsymbol{\tau}^1\boldsymbol{\tau}^2)(\boldsymbol{\tau}^2\boldsymbol{\tau}^3) + A(\boldsymbol{\tau}^2\boldsymbol{\tau}^3)(\boldsymbol{\tau}^2\boldsymbol{\tau}^1)\}(\boldsymbol{\sigma}^3\nabla^1)(\boldsymbol{\sigma}^2\nabla^1)(\boldsymbol{\sigma}^2\nabla^3)(\boldsymbol{\sigma}^1\nabla^3) \\ +2D(\boldsymbol{\tau}^1\boldsymbol{\tau}^3)(\boldsymbol{\sigma}^1\nabla^1)(\boldsymbol{\sigma}^3\nabla^3)]\;Y(12)Y(23), \end{aligned} \tag{11}$$

$$Y(ij) = \mathrm{e}^{-m|\boldsymbol{x}_i - \boldsymbol{x}_j|}/|\boldsymbol{x}_i - \boldsymbol{x}_j|. \tag{12}$$

The third line in the expression of $V(123)$, the D-term, was due to the S-wave scattering which was not explicitly written in the amplitude. The constants A and B could be written in terms of $A(0)$ and $B(0)$, obtained in the last section, and the Yukawa coupling constant, f. The quantity D could be written in terms of the S-wave pion-nucleon scattering lengths .

As Dr. Robilotta pointed out in his talk, Eq. (11) is the general form of the three-nucleon potential, and is valid so far as the adiabatic approximation is valid and multi-pion exchange is negligible. That is, Eq. (11) is the correct form of the three-nucleon force for all $r_{ij} > r_0$, in the same sense that the two-body potential is given by the Yukawa potential.

In the numerical calculation of the constants A and B, Fujita and I took the resonance dominance approximation and considered only the delta-resonance in pion-nucleon scattering. However, it is not correct to say that we took only the delta-excitation from the beginning. We wanted to apply a new and powerful method, the static pion theory, to any pion processes whenever possible. The three-nucleon force was one of these examples. All pion corrections were accurately considered and were expressed in terms

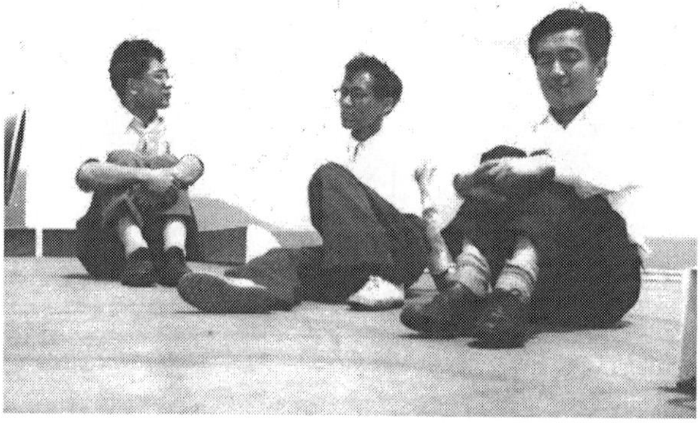

FIGURE 4. Picture of Jun-ichi Fujita. From right to left, Fujita, Miyazawa and Konuma

of experimental entities. Our aim was not to propose or predict the existence of the three-body force, but to calculate the potential numerically accurately.

That the three-body force exsists can be seen as follows. In general, a particle has discrete variables T like spin, isospin, *etc.* In a three-particle system, A, B, and C, the third particle C turns over T variables of A and B. This naturally affects the force between A and B, resulting in a three-body force. That is, in a system of particles that depend on the non-commuting variable T and interact via the two-body force of the form,

$$V_2 = \sum T_i T_j V(x_i, x_j), \tag{13}$$

there appears a many-body potential. In the nucleon case, T includes not only the spin and isospin matrices but also the operator that excites a nucleon into a delta-resonance. The nucleon-delta transition provides a dominant contribution to the many-nucleon force.

Figure 4 is a picture of Jun-ichi Fujita, around 1957, when he was a graduate student. Although he later became a superb nuclear theorist, he sadly passed away thirty years ago.

Conclusion

The three-nucleon force was calculated as described. Although it was not clearly written in our paper, our aim was not to propose or predict that a three-body force must exist, but to calculate the force with pion corrections accurately taken into account.

Recently, I learned that the presence of a three-body force has been experimentally verified. I am very pleased with this news and I hope, in the future, that our results will be quantitatively verified.

Thank you very much!

COULOMB EFFECTS IN THREE- AND FOUR-NUCLEON SCATTERING: A MASK OF THE 3N FORCE

A. Deltuva*, A. C. Fonseca* and P.U. Sauer†

*Centro de Física Nuclear da Universidade de Lisboa,
Av. Prof. Gama Pinto 2, P-1649-003 Lisboa, Portugal
†Institut für Theoretische Physik, Leibniz Universität Hannover, D-30167 Hannover, Germany

Abstract. Recent progress on the solution of *ab initio* three- and four-nucleon scattering equations in momentum space that include the correct treatment of the Coulomb interaction is reviewed; results for specific observables in reactions initiated by $p+d$, $p+{}^3$He and $n+{}^3$He indicate that the inclusion of the Coulomb interaction is paramount to validate two- and three-nucleon force models. Three- and four-nucleon force effects in ^{4}He binding energy and four-nucleon scattering observables are studied for the first time. The effect of the four-nucleon force is found to be much smaller than the effect of the three-nucleon force.

Keywords: Three-nucleon, four-nucleon, Coulomb effect, three-nucleon forces
PACS: 21.30-.x, 21.45.+v, 24.70.+s, 25.10.+s

INTRODUCTION

In recent years considerable progress has been achieved in the study of nuclei up to $A \leq 12$ through *ab initio* structure calculations based on the underlying concept that nucleons interact through pairwise forces fitted to the two-nucleon ($2N$) scattering up to the pion production threshold plus a three-nucleon force [1, 2] when necessary. The same concept applies to *ab initio* scattering calculations but here restricted to three- [3, 4] or four-nucleon [5, 6, 7] reactions given the greater difficulty involved in the solution of many-body scattering equations, particularly in the presence of the repulsive Coulomb force between protons.

In the present review we account for the changes that have taken place since 2005, leading to fully converged *ab initio* calculations of $p+d$ elastic scattering and breakup at energies up to the pion production threshold and all possible four-nucleon reactions initiated by $n+{}^3$He, $p+{}^3$H, $d+d$, $n+{}^3$H and $p+{}^3$He below three-body breakup threshold. In addition to the correct treatment of Coulomb, the effect of the three-nucleon ($3N$) and four-nucleon ($4N$) forces is studied in $3N$ and $4N$ scattering observables.

THREE-NUCLEON SCATTERING WITH TWO PROTONS

The treatment of the Coulomb interaction is based on the ideas proposed in Ref. [8] for the scattering of two charged particles and extended in Ref. [9] for three-particle scattering. The Coulomb potential is screened, standard scattering theory for short-range

CP1011, *New Facet of Three Nucleon Force—50 Years of Fujita Miyazawa Three Nucleon Force (FM50), The International Symposium*, edited by H. Sakai, K. Sekiguchi, and B. F. Gibson

potentials is used and the renormalization procedure [8, 9] is applied to obtain results in the unscreened limit. Although the theory was well established, no reliable calculations existed due to the shortcomings of the practical implementation. In Refs. [10, 11] we have demonstrated how one obtains fully converged calculations for $p+d$ elastic scattering and breakup, $p+d$ capture into $^3\mathrm{He}+\gamma$, and two- and three-body photo and electro disintegration using realistic nucleon-nucleon interactions plus the Coulomb potential between protons.

One major difference relative to previous work is that we use a Coulomb potential w_R, screened around the separation $r = R$ between two charged baryons which, in configuration space, reads

$$w_R(r) = w(r)\, e^{-(r/R)^n}, \tag{1}$$

with the true Coulomb potential $w(r) = \alpha_e/r$, where $\alpha_e \approx 1/137$ is the fine structure constant and n controls the smoothness of the screening. We find that values $3 \leq n \leq 8$ provide a sufficiently smooth, but at the same time a sufficiently rapid screening around $r = R$.

The starting point in our approach is the full three-body Alt, Grassberger and Sandhas (AGS) equation [12] for the transition operator $U^{(R)}_{\beta\alpha}(Z)$ that depends parametrically on the screening radius R

$$U^{(R)}_{\beta\alpha}(Z) = \bar{\delta}_{\beta\alpha} G_0^{-1}(Z) + \sum_{\sigma} \bar{\delta}_{\beta\sigma} T^{(R)}_{\sigma}(Z) G_0(Z) U^{(R)}_{\sigma\alpha}(Z), \tag{2a}$$

where the two-particle transition matrix $T^{(R)}_{\sigma}$ is derived from the full channel interaction $v_\alpha + w_{\alpha R}$

$$T^{(R)}_{\alpha}(Z) = (v_\alpha + w_{\alpha R}) + (v_\alpha + w_{\alpha R}) G_0(Z) T^{(R)}_{\alpha}(Z), \tag{2b}$$

$G_0(Z) = (Z - H_0)^{-1}$ is the free resolvent and $\bar{\delta}_{\beta\alpha} = 1 - \delta_{\alpha\alpha}$. Of course, the full multichannel transition matrix $U^{(R)}_{\alpha\alpha}(Z)$ must contain the pure Coulomb transition matrix $T^{\mathrm{c.m.}}_{\alpha R}(Z)$ derived from the screened Coulomb potential $W^{\mathrm{c.m.}}_{\alpha R}$ between the spectator proton and the center of mass (c.m.) of the remaining neutron-proton pair in channel α, that is,

$$T^{\mathrm{c.m.}}_{\alpha R}(Z) = W^{\mathrm{c.m.}}_{\alpha R} + W^{\mathrm{c.m.}}_{\alpha R} G^{(R)}_{\alpha}(Z) T^{\mathrm{c.m.}}_{\alpha R}(Z), \tag{3a}$$

$$G^{(R)}_{\alpha}(Z) = (Z - H_0 - v_\alpha - w_{\alpha R})^{-1}, \tag{3b}$$

with the pd channel being one of those channels α. The same screening function is used for both Coulomb potentials $w_{\alpha R}$ and $W^{\mathrm{c.m.}}_{\alpha R}$. As shown in [10] $[U^{(R)}_{\beta\alpha}(Z) - \delta_{\beta\alpha} T^{\mathrm{c.m.}}_{\alpha R}(Z)]$ is a short range operator which after renormalization [9] has a well defined $R \to \infty$ limit. Therefore, in the $R \to \infty$ limit

$$\begin{aligned} \langle \mathbf{q}_\beta | U_{\beta\alpha} | \mathbf{q}_\alpha \rangle &= \delta_{\beta\alpha} \langle \mathbf{q}_\beta | T^{\mathrm{c.m.}}_{\alpha C} | \mathbf{q}_\alpha \rangle \\ &+ \lim_{R\to\infty} \left\{ \mathscr{Z}^{-\frac{1}{2}}_{\beta R}(q_\beta) \langle \mathbf{q}_\beta | [U^{(R)}_{\beta\alpha}(Z) - \delta_{\beta\alpha}\, T^{\mathrm{c.m.}}_{\alpha R}(Z)] | \mathbf{q}_\alpha \rangle \mathscr{Z}^{-\frac{1}{2}}_{\alpha R}(q_\alpha) \right\}, \end{aligned} \tag{4}$$

where $T^{\text{c.m.}}_{\alpha C}$ is the pure Coulomb amplitude and

$$\mathscr{Z}_{\alpha R}(q) = e^{-2i(\sigma_l^\alpha(q) - \eta_{lR}^\alpha(q))}, \tag{5}$$

with the diverging screened Coulomb pd phase shift $\eta_{lR}^\alpha(q)$ corresponding to standard boundary conditions and the proper Coulomb one $\sigma_l^\alpha(q)$ referring to the logarithmically distorted proper Coulomb boundary conditions; l is the pd relative orbital angular momentum. Likewise for breakup we show [10] that

$$\langle \mathbf{p}_f \mathbf{q}_f | U_{0\alpha} | \mathbf{q}_\alpha \rangle = \lim_{R\to\infty} \{ z_R^{-\frac{1}{2}}(p_f) \langle \mathbf{p}_f \mathbf{q}_f | U_{0\alpha}^{(R)}(Z) | \mathbf{q}_\alpha \rangle \mathscr{Z}_{\alpha R}^{-\frac{1}{2}}(q_\alpha) \}, \tag{6}$$

$$U_{0\alpha}^{(R)}(Z) = G_0^{-1}(Z) + \sum_\sigma T_\sigma^{(R)}(Z) G_0(Z) U_{\sigma\alpha}^{(R)}(Z), \tag{7}$$

where $\mathscr{Z}_{\alpha R}(q_\alpha)$ and $z_R(p_f)$ are the corresponding pd and pp renormalizations factors.

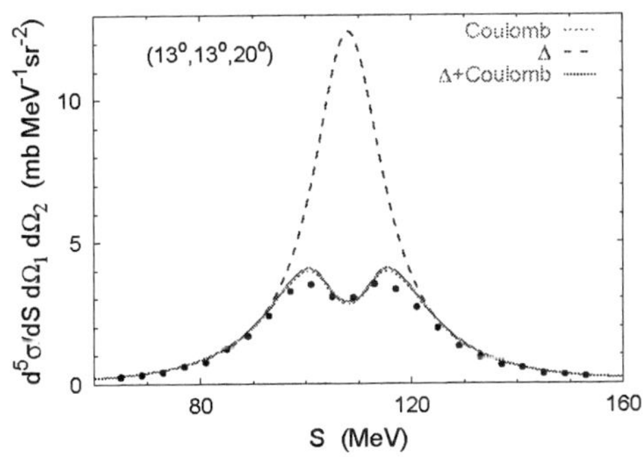

FIGURE 1. Fivefold differential cross section for dp breakup at $E_d = 130$ MeV close to pp−FSI. The solid line corresponds to calculations with CD Bonn + Δ + Coulomb, the dotted line to CD Bonn + Coulomb and the dashed line to CD Bonn + Δ. The dots are data from Ref. [15].

Equations (2a), (3a) and (7) are solved independently for each R and the corresponding on-shell amplitudes for nuclear plus Coulomb calculated in the $R \to \infty$ limit through (4) and (6). In effect the convergence with R is obtained at small finite R as demonstrated in Refs. [10, 11] for elastic scattering and breakup. All configurations converge at $R = 20$ fm except the pp-FSI configuration where a pp-FSI peak in the absence of Coulomb, becomes a pp-FSI depression with zero cross section at $E_{pp} = 0$ MeV when Coulomb is added. Although further details may be found in Refs. [10, 11] we show in Figs. 1-3 three examples on how Coulomb effects in pd breakup and three-body photo disintegration of ^{3}He are paramount to describe the data. The calculations are based on the purely nucleonic charge-dependent CD-Bonn potential [13] and its coupled-channel extension, CD-Bonn + Δ [14], allowing for a single virtual Δ-isobar excitation and fitted to the experimental data with the same degree of accuracy as the CD-Bonn itself. In the three-nucleon system the Δ isobar mediates an effective $3N$ force and effective $2N$ and $3N$ currents, both consistent with the underlying $2N$ force.

In all cases we show, Δ effects are much smaller than Coulomb effects. Thus, Coulomb masks the $3N$ force whose effect we really want to study. Calculations without the

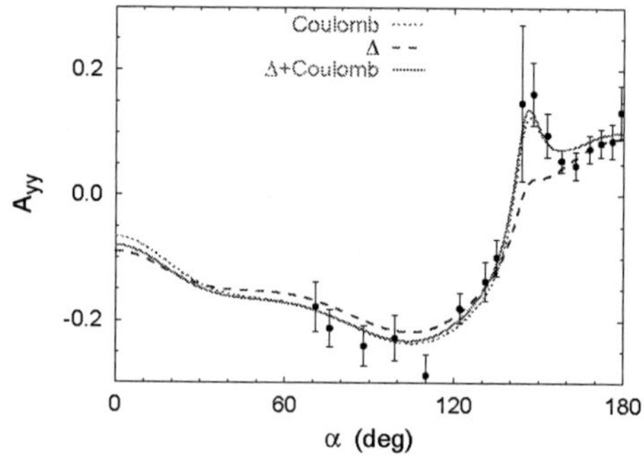

FIGURE 2. Same as in Fig. 1 for deuteron analyzing power A_{yy} resulting from dp breakup at $E_d = 94.5$ MeV in SCRE geometry. The dots are data from Ref. [16].

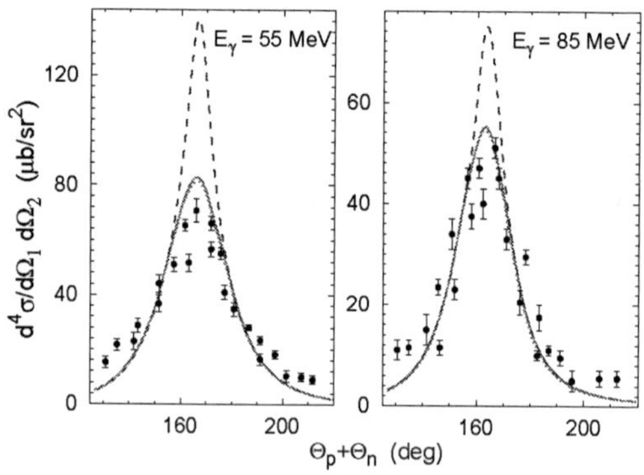

FIGURE 3. The semi-inclusive fourfold differential cross section for ${}^3\mathrm{He}(\gamma, pn)p$ reaction at 55- and 85-MeV photon lab energy as a function of the np opening angle $\theta_p + \theta_n$ with $\theta_p = 81°$. Curves as in Fig. 1. The experimental data are from Ref. [17].

Coulomb interaction included are, therefore, misleading vis-a-vis the validation of $2N+3N$ force models relative to the data.

FOUR-NUCLEON SCATTERING

The $4N$ scattering problem gives rise to the simplest set of nuclear reactions that shows the complexity of heavier systems. In $4N$ scattering the Coulomb interaction is paramount not only to treat p-^{3}He, p-^{3}H and d-d but also to separate the n-^{3}He threshold from p-^{3}H and at the same time avoid an excited state of the α particle a few keV bellow the lowest scattering threshold. All these complex features make the $4N$ scattering problem a natural theoretical laboratory to test different force models of the nuclear interaction, after the $3N$ system [3, 10].

The equations we solve are the four-body AGS equations of Ref. [18] whose symmetrized form was used in Ref. [19] to study n-^{3}H elastic scattering and in Ref. [20] to calculate p-^{3}He with the Coulomb force included between all three-protons. As in (2a) the four-body AGS transition operators in the presence of screened Coulomb become R dependent. The transition operators $\mathscr{U}^{\beta\alpha}_{(R)}$ where $\alpha(\beta) = 1$ and 2 corresponds to initial/final $1+3$ and $2+2$ two-cluster states, respectively, satisfy the symmetrized AGS equations

$$\mathscr{U}^{11}_{(R)} = -(G_0 T^{(R)} G_0)^{-1} P_{34} - P_{34} U^1_{(R)} G_0 T^{(R)} G_0 \mathscr{U}^{11}_{(R)} + U^2_{(R)} G_0 T^{(R)} G_0 \mathscr{U}^{21}_{(R)}, \tag{8a}$$

$$\mathscr{U}^{21}_{(R)} = (G_0 T^{(R)} G_0)^{-1} (1 - P_{34}) + (1 - P_{34}) U^1_{(R)} G_0 T^{(R)} G_0 \mathscr{U}^{11}_{(R)}. \tag{8b}$$

Here G_0 is the free four-particle Green's function and $T^{(R)}$ the two-nucleon t-matrix derived from nuclear potential plus screened Coulomb between pp pairs. The operators $U^{\alpha}_{(R)}$ obtained from

$$U^{\alpha}_{(R)} = P_\alpha G_0^{-1} + P_\alpha T^{(R)} G_0 U^{\alpha}_{(R)}, \tag{9a}$$

$$P_1 = P_{12} P_{23} + P_{13} P_{23}, \tag{9b}$$

$$P_2 = P_{13} P_{24}, \tag{9c}$$

are the symmetrized AGS operators for the $1+3$ and $2+2$ subsystems and P_{ij} is the permutation operator of particles i and j. Defining the initial/final $1+3$ and $2+2$ states with relative two-body momentum $\mathbf{p}$

$$|\phi^{(R)}_\alpha(\mathbf{p})\rangle = G_0 T^{(R)} P_\alpha |\phi^{(R)}_\alpha(\mathbf{p})\rangle, \tag{10}$$

the amplitudes for $1+3 \to 1+3$ and $1+3 \to 2+2$ are obtained as $\langle \mathbf{p}_f | T^{\beta\alpha}_{(R)} | \mathbf{p}_i \rangle = S_{\beta\alpha} \langle \phi^{(R)}_\beta(\mathbf{p}_f) | \mathscr{U}^{\beta\alpha}_{(R)} | \phi^{(R)}_\alpha(\mathbf{p}_i) \rangle$ with $S_{11} = 3$ and $S_{21} = \sqrt{3}$.

In close analogy with pd elastic scattering, the full scattering amplitude, when calculated between initial and final p-^{3}He states, may be decomposed as follows

$$T^{11}_{(R)} = T^{\text{c.m.}}_R + [T^{11}_{(R)} - T^{\text{c.m.}}_R], \tag{11}$$

with the long-range part $T^{\text{c.m.}}_R$ being the two-body t-matrix derived from the screened Coulomb potential of the form (1) between the proton and the c.m. of ^{3}He, and the remaining Coulomb distorted short-range part $[T^{11}_{(R)} - T^{\text{c.m.}}_R]$ as demonstrated before for pd. Applying the renormalization procedure, i.e., multiplying both sides of Eq. (11) by the renormalization factor Z_R^{-1}, in the $R \to \infty$ limit, yields the full $1+3 \to 1+3$ transition amplitude in the presence of Coulomb

$$\langle \mathbf{p}_f | T^{11} | \mathbf{p}_i \rangle = \langle \mathbf{p}_f | T^{\text{c.m.}}_C | \mathbf{p}_i \rangle + \lim_{R\to\infty} \left\{ Z_R^{-\frac{1}{2}}(p_f) \langle \mathbf{p}_f | [T^{11}_{(R)} - T^{\text{c.m.}}_R] | \mathbf{p}_i \rangle Z_R^{-\frac{1}{2}}(p_i) \right\}. \tag{12}$$

Although the limit in the second term of Eq. (12) is carried out numerically, it is reached for finite R as shown in Fig. 4, where one sees that Coulomb effects are large. Once

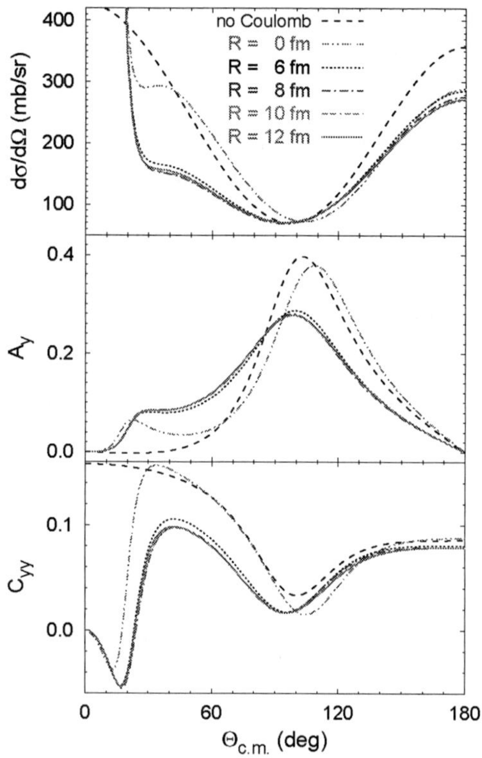

FIGURE 4. (Color online) Convergence of the p-^{3}He scattering observables with screening radius R. Results for the differential cross section, proton analyzing power A_y, and p-^{3}He spin correlation coefficient C_{yy} at 4 MeV proton lab energy obtained with screening radius $R = 0$ fm (dashed-double-dotted curves), 6 fm (dotted curves), 8 fm (dashed-dotted curves), 10 fm (double-dashed-dotted curves), and 12 fm (solid curves) are compared. Results without Coulomb (dashed curves) are given as reference for the size of the Coulomb effect. The $R = 0$ curve corresponds to adding the nuclear amplitude in the absence of Coulomb to the pure Coulomb amplitude [20] .

again we find that Coulomb effects are large enough to mask $3N$ and $4N$ force effects and for this reason the inclusion of Coulomb is paramount to meaningful calculations.

In Fig. 5-6 we show results for p-^{3}He and n-^{3}He elastic scattering respectively using AV18 [21], CD-Bonn [13], N3LO [22] and INOY04 [23] potentials between $2N$ pairs plus Coulomb. No $3N$ force is included at this stage though its scaling effects are simulated by INOY04 given that the binding energy of ^{3}H and ^{3}He are well reproduced by this potential. Results indicate that in p-^{3}He there is a stronger A_y deficiency than in n-^{3}He. Other examples may be found in Refs. [19, 20] and [24] that demonstrate that $2N$ forces alone fail to describe existing $4N$ data such as the total n-^{3}H cross section σ_t in the resonance region around $E_n = 3.5$ MeV, or the small tensor analysing powers in dd elastic scattering. These findings may reveal the need for stronger isospin dependent forces but further studies are needed such as including a $3N$ force.

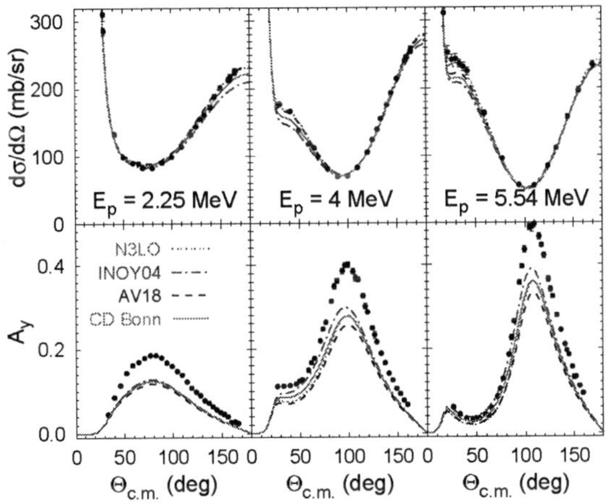

FIGURE 5. The differential cross section and proton analyzing power A_y of elastic p-^{3}He scattering at 2.25, 4.0, and 5.54 MeV proton lab energy. Results including the Coulomb interaction obtained with potentials CD Bonn (solid curves), AV18 (dashed curves), INOY04 (dashed-dotted curves), and N3LO (dotted curves) are compared. The data are from Refs. [25, 26, 27].

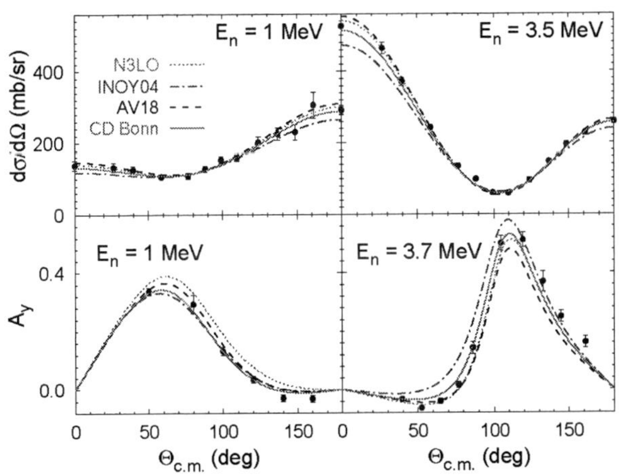

FIGURE 6. Differential cross section and neutron analyzing power of elastic n-^{3}He scattering at 1, 3.5, and 3.7 MeV neutron lab energy. The curves as in Fig. 5. The cross section data are from Ref. [28], A_y data are from Ref. [29] at 1 MeV and from Ref. [30] at 3.7 MeV.

There are three distinct ways for doing this: a) Add a static two-pion-exchange irreducible $3N$ force [31, 32, 33] to the underlying $2N$ forces; however, in this approach these two forces are not really consistent with each other. b) Use $2N+3N$ force models based on chiral effective field theory [34, 35] to guaranty consistency between the $2N$ and $3N$ forces. c) Extend the purely nucleonic model to allow the explicit excitation of a nucleon to a Δ isobar as was shown in the previous Section for the $3N$ systems using

CD-Bonn + Δ [14].

The studies of the $3N$ system reveal that all these different approaches lead to qualitatively similar results. In the $4N$ system the first one a) was already applied to n-^{3}H [5, 6] and p-^{3}He [7, 25] scattering. Therefore, following the work on the $3N$ system performed in Refs.[14, 10, 11], we use the last approach c) to study the ^{4}He binding energy and all $4N$ reactions below three-body breakup threshold. In the $3N$ system the excitation of a single nucleon to a Δ isobar yields an effective $3N$ force with components of Fujita-Miyazawa type [36] and much richer structures in a reducible form; besides pion (π) exchange, the $3N$ force has contributions of shorter range due to the exchange of heavier mesons. In the $4N$ system an effective $4N$ force arises that also has parts of shorter range than π exchange.

The equations we solve are extended to the ΔNNN sector; only states that are fully antisymmetric in all four baryons, nucleons and the Δ isobar arise. In Table 1 we collect the results for $3N$ and $4N$ binding energies. The inclusion of the Δ isobar increases the corresponding binding energies but is unable to reproduce the experimental values. Obviously, many-nucleon forces, not accounted for by the Δ isobar, make a rather significant contribution to $3N$ and $4N$ binding energies. Table 1 also splits up the total Δ-isobar effect into separate contributions obtained through incomplete calculations as discussed in Ref. [37]. (1) The $2N$ dispersion turns out to be massive in the $4N$ bound state with $\Delta E_2 = -2.80$ MeV; it arises mainly from the dispersion in the $^1S_0$ $2N$ partial wave. (2) The $3N$ force contribution of the Fujita-Miyazawa type $\Delta E_3^{\mathrm{FM}} = 2.25$ MeV is also quite large. The increase by the factor of 4.5 compared to the $3N$ bound state is understandable in terms of the different multiplicity with which the $3N$ force contributes: one in the $3N$ bound state and four in the $4N$. The observed factor of ≈ 4.5 comes from the fact that ^{4}He, being a denser system than ^{3}He or ^{3}H, squeezes out more binding from the underlying force. (3) The contribution of the higher order $3N$ force terms mediated by the diagonal $N\Delta$ potential $v_{\Delta\Delta}$ is $\Delta E_3^{\mathrm{h.o.}} = 1.30$ MeV comparable to the one of the Fujita-Miyazawa type. (4) Finally, in contrast to the complete $3N$ force contribution $\Delta E_3^{\mathrm{FM}} + \Delta E_3^{\mathrm{h.o.}} = 3.55$ MeV, the contribution arising from the effective $4N$ force $\Delta E_4 = 0.17$ MeV is indeed rather small.

In Fig. 7 we study the energy dependence of the total n-^{3}H cross section. The Δ-isobar excitation increases the $3N$ binding energy and through scaling improves the description of the data around threshold. However, there is quite a significant non-beneficial Δ effect in the region of the resonance which is strongly driven by n-^{3}H relative P waves. In Ref. [37] we split this effect into $2N$ dispersion, $3N$ and $4N$ force contributions, the last of which is negligible. Furthermore, when we split the total n-^{3}H cross section σ_t into the S- and P-wave contributions σ_t^S and σ_t^P we find that the Δ effects on the ^{3}H binding energy ε_t and the S-wave cross section are correlated by scaling in the same way as it has been observed in Ref. [19], i.e., σ_t^S decreases when ε_t increases. In contrast, there is no such a correlation in P waves where both $2N$ dispersion and effective $3N$ force decrease the cross section while having opposite effects on ε_t. Similar findings are observed for the p-^{3}He differential cross section [37] where disagreement with data increases at forward and backward angles when the Δ is included. This behavior has also been observed by the Pisa group [25] when the Urbana IX $3N$ force is added to AV18, but not in the work of Lazauskas & Carbonnel [5]. These differences need to be sorted out in the near future.

TABLE 1. Binding energies for ^{3}H, ^{3}He, and ^{4}He derived from the potentials CD Bonn and CD Bonn + Δ and the corresponding experimental values are given in the first three rows. The last four rows split the complete Δ effect up into $2N$ dispersion ΔE_2, Fujita-Miyazawa type $3N$ force effect ΔE_3^{FM}, higher order $3N$ force effect $\Delta E_3^{\mathrm{h.o.}}$, and $4N$ force effect ΔE_4 for ^{4}He. All results are given in MeV.

	^{3}H	^{3}He	^{4}He
CD Bonn	8.00	7.26	26.18
CD Bonn + Δ	8.28	7.53	27.10
exp	8.48	7.72	28.30
ΔE_2	-0.51	-0.48	-2.80
ΔE_3^{FM}	0.50	0.48	2.25
$\Delta E_3^{\mathrm{h.o.}}$	0.29	0.27	1.30
ΔE_4			0.17

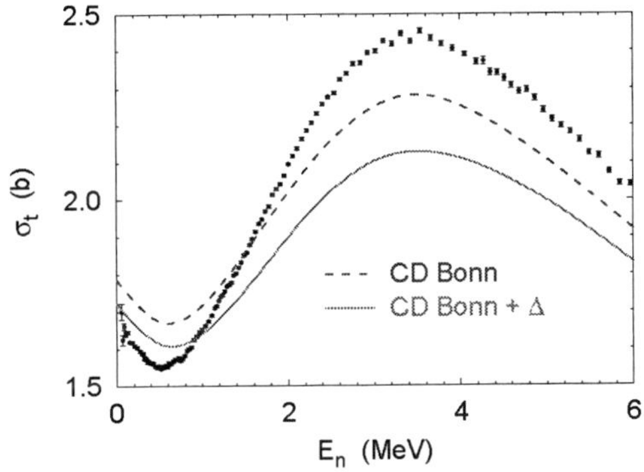

FIGURE 7. (Color online) Total cross section for n-^{3}H scattering as function of neutron lab energy calculated with the CD Bonn (dashed curve) and CD Bonn + Δ (solid curve) potentials. Experimental data are from Ref. [38].

CONCLUSIONS

We have reviewed the most recent work on *ab initio* scattering calculations for three- and four-nucleon reactions using realistic pairwise interactions and the Coulomb force between protons. For the first time $4N$ force effects are determined for the ^{4}He binding energy and $4N$ scattering observables using the coupled channel approach involving a single Δ excitation of a nucleon. The calculations with the inclusion of the Coulomb potential are now as accurate as those without, and provide an opportunity for a comprehensive test of hadronic interactions in the form of $2N$, $3N$ and $4N$ force models by comparing theory with experimental results obtained with charge particle reactions where data is abundant and error bars smaller. Since Coulomb effects are in many kinematic situations larger than $3N$ and $4N$ force effects, their inclusion is necessary to avoid the masking of many-nucleon forces.

REFERENCES

1. S. C. Pieper, V. R. Pandharipande, R.B. Wiringa and J. Carlson, *Phys. Rev.* **C64**, 014001 (2001).
2. K. Varga, S. C. Pieper, Y. Suzuki and R. B. Wiringa, *Phys. Rev.* **C66**, 044310 (2002).
3. W. Glöckle *et al.*, *Phys. Rep.* **274**, 107 (1996).
4. A. Kievsky, M. Viviani and S. Rosati, *Phys. Rev.* **C64**, 024002 (2001); M. Viviani *et al.*, *Phys. Rev.* **C61**, 064001 (2000).
5. R. Lazauskas and J. Carbonell, *Phys. Rev.* **C70**, 044002 (2004).
6. R. Lazauskas *et al.*, *Phys. Rev.* **C71**, 034004 (2005).
7. M. Viviani *et al.*, *Phys. Rev. Lett.* **86**, 3739 (2001).
8. J. R. Taylor, *Nuovo Cimento* **B23**, 313 (1974); M. D. Semon and J. R. Taylor, *ibid.* **A26**, 48 (1975).
9. E. O. Alt, W. Sandhas and H. Ziegelmann, *Phys. Rev.* **C17**, 1981 (1978).
10. A. Deltuva, A. C. Fonseca and P. U. Sauer, *Phys. Rev.* **C71**, 054005 (2005); **C72**, 054004 (2005); **C73**, 057001 (2006).
11. A. Deltuva, A. C. Fonseca and P. U. Sauer, *Phys. Rev. Lett.* **95**, 092301 (2005).
12. E. O. Alt, P. Grassberger and W. Sandhas, *Nucl. Phys.* **B2**, 167 (1967).
13. R. Machleidt, *Phys. Rev.* **C63**, 024001 (2001).
14. A. Deltuva, R. Machleidt and P. U. Sauer, *Phys. Rev.* **C68**, 024005 (2003).
15. St. Kistryn *et al.*, *Phys. Lett.* **B641**, 23 (2006).
16. D. A. Low *et al.*, *Phys. Rev.* **C44**, 2276 (1991).
17. N. R. Kolb, P. N. Dezendorf, M. K. Brussel, B. B. Ritchie, and J. H. Smith, *Phys. Rev.* **C44**, 37 (1991).
18. P. Grassberger and W. Sandhas, *Nucl. Phys.* **B2**, 181 (1967); E. O. Alt, P. Grassberger, and W. Sandhas, JINR report No. E4-6688 (1972).
19. A. Deltuva and A. C. Fonseca, *Phys. Rev.* **C75**, 014005 (2007).
20. A. Deltuva and A. C. Fonseca, *Phys. Rev. Lett.* **98**, 162502 (2007).
21. R. B. Wiringa *et al.*, *Phys. Rev.* **C51**, 38 (1995).
22. D. R. Entem and R. Machleidt, *Phys. Rev.* **C68**, 041001 (2003).
23. P. Doleschall, *Phys. Rev.* **C69**, 054001 (2004).
24. A. Deltuva and A. C. Fonseca, *Phys. Rev.* **C76**, 021001(R) (2007).
25. B. M. Fisher *et al.*, *Phys. Rev.* **C74**, 034001 (2006).
26. D. G. McDonald, W. Haeberli, and L. W. Morrow, *Phys. Rev.* **133**, B1178 (1964).
27. M. T. Alley and L. D. Knutson, *Phys. Rev.* **C48**, 1890 (1993).
28. J. D. Seagrave, L. Cranberg, and J. E. Simmons, *Phys. Rev.* **119**, 1981 (1960).
29. P. Jany *et al.*, *Nucl. Phys.* **A483**, 269 (1988).
30. H. O. Klages *et al.*, *Nucl. Phys.* **A443**, 237 (1985).
31. H. T. Coelho, T. K. Das , M. R. Robilotta, *Phys. Rev.* **C28**, 1812 (1983).
32. S. A. Coon, M. D. Scadron, P. C. McNamee, B. R. Barrett, D. W. E. Blatt, and B. H. J. McKellar, *Nucl. Phys.* **A317**, 242 (1979).
33. B. S. Pudliner, V. R. Pandharipande, J. Carlson, S. C. Pieper, and R. B. Wiringa, *Phys. Rev.* **C56**, 1720 (1997).
34. U. van Kolck, *Prog. Part. Nucl. Phys.* **43**, 337 (1999).
35. E. Epelbaum, W. Glöckle, and U.-G. Meissner, *Nucl. Phys.* **A671**, 295 (2000).
36. J. Fujita and H. Miyazawa, *Prog. Theor. Phys.* **17**, 360 (1957).
37. A. Deltuva, A. C. Fonseca and P. U. Sauer, submitted to Phys. Lett. B.
38. T. W. Phillips, B. L. Berman, and J. D. Seagrave, *Phys. Rev.* **C22**, 384 (1980).

Analyzing powers in the dd → ^{3}Hen(^{3}Hp) reactions at intermediate energies

V.P.Ladygin*, T.Uesaka†, T.Saito**, M.Janek*,‡, A.S.Kiselev*, A.K.Kurilkin*, T.A.Vasiliev*, M.Hatano**, A.Yu.Isupov*, H.Kato**, N.B.Ladygina*, Y.Maeda§, A.I.Malakhov*, J.Nishikawa¶, T.Ohnishi||, H.Okamura††, S.G.Reznikov*, H.Sakai‡‡, S.Sakoda**, N.Sakamoto||, Y.Satou§§, K.Sekiguchi||, K.Suda††, A.Tamii††, N.Uchigashima** and K.Yako**

*Joint Institute for Nuclear Research, Dubna, Moscow region, Russia
†Center for Nuclear Study, University of Tokyo, Bunkyo, Japan
**Department of Physics, University of Tokyo, Bunkyo, Japan
‡University of P.J. Šafárik, Košice, Slovakia
§Kyushi University, Hakozaki, Japan
¶Department of Physics, Saitama University, Urawa, Japan
||RIKEN (the Institute for Physical and Chemical Research), Wako, Japan
††Research Center for Nuclear Physics, Osaka University, Japan
‡‡University of Tokyo, Bunkyo, Japan
§§Tokyo Institute of Technology, Tokyo, Japan

Abstract. Data for the deuteron analyzing powers in the $dd \to {}^3$Hen(^{3}Hp) reactions obtained at 140-270 MeV are discussed. The observed negative sign of the tensor analyzing powers A_{yy}, A_{xx} and A_{xz} at small angles clearly demonstrate the sensitivity to the ratio of the D and S state components of the ^{3}He wave function. The behavior of the tensor analyzing powers at backward angles is sensitive to the short-range spin structure of the deuteron. However, the one-nucleon exchange calculations using standard ^{3}He and deuteron wave functions fail to reproduce the strong variation of the tensor analyzing powers as a function of angle in the cms. Sensitivity to relativistic effects is also discussed.

Keywords: Few-body systems, analyzing powers, deuteron, triton, ^{3}He
PACS: 24.70.+s, 25.10.+s, 21.45.+v

INTRODUCTION

During last several years a new generation of NN potentials were constructed (Nijmegen, CD-Bonn, AV-18 etc.). These potentials reproduce the NN scattering data up to 350 MeV with very good accuracy. But these potentials cannot reproduce the triton binding energy (underbinding is 0.8 MeV for CD-Bonn), deuteron-proton scattering, and breakup data. Incorporation of a three nucleon force (3NF), where the interaction depends on the quantum numbers of the all three nucleons, allows one to reproduce the triton binding energy and unpolarized deuteron-proton scattering and breakup data (see [1] and references therein).

However, the use of different 3NF models in Faddeev calculations cannot reproduce the polarization data accumulated during the last decade at different facilities [2]–[8]. The reasons for the discrepancy can be due to the inadequate description of the spin-

CP1011, *New Facet of Three Nucleon Force—50 Years of Fujita Miyazawa Three Nucleon Force (FM50), The International Symposium,* edited by H. Sakai, K. Sekiguchi, and B. F. Gibson

dependent part of the 3NF, as well as to the importance of relativistic and off-shell effects in the 2NF forces. In this respect, new data sensitive to the spin structure of ^{3}He(^{3}H) and 3NF are necessary to construct an adequate model of the 3NF.

The spin structure of light nuclei has been extensively investigated during the last decades. However, there are only few data sensitive to the spin structure of the three-nucleon bound state at short distances. Most precise data were obtained with hadronic probes [9, 10]. The calculations performed within a one-nucleon exchange (ONE) model [11] using standard three-nucleon bound state and deuteron wave functions predict that the tensor analyzing powers for the $dd \rightarrow {}^3\mathrm{He}n$ and $dd \rightarrow {}^3\mathrm{H}p$ reactions at forward scattering are sensitive to the D/S state ratio in ^{3}He and ^{3}H, respectively. Therefore, a new experiment has been proposed [12] to measure the energy and angular dependence of the tensor analyzing powers in the $dd \rightarrow {}^3\mathrm{He}n({}^3\mathrm{H}p)$ process at RIKEN under the condition that the contribution from the D- state in ^{3}He(^{3}H) becomes large.

EXPERIMENTAL PROCEDURE

The experiment was performed at the RIKEN Accelerator Research Facility using the SMART spectrograph. The details of the experiment are described elsewhere [13, 14].

The high-intensity polarized deuteron beam was produced by the polarized ion source (PIS) and accelerated by the AVF and Ring Cyclotrons up to the required energy (140, 200 or 270 MeV). The direction of the symmetric axis of the beam polarization was controlled with a Wien-filter located at the exit of the PIS. The polarization of the deuteron beam was measured with two beam-line polarimeters. using dp elastic scattering with known large values of the tensor and vector analyzing powers A_y, A_{yy}, A_{xx} and A_{xz} [2, 3].

The SMART (Swinger and Magnetic Analyzer with a Rotator and a Twister) spectrograph was used for the measurements. In this system, the incident beam can be swung by rotating the swinger magnet, while the magnetic analyzer is fixed. Measurement of the particles' momentum and separation from the primary beam was achieved by the magnetic system of the SMART spectrograph consisting of two dipole and three quadrupole magnets (Q-Q-D-Q-D). The deuteron beam intensity monitored by the Faradey cup located in the first dipole magnet was between 1 and 120 nA. The detection solid angle was 10^{-2} sr. Two deuterated polyethylene (CD_2) sheets 54 mg/cm^2 and 32 mg/cm^2 thick placed in the scattering chamber of the SMART were used as the deuterium targets. A carbon foil 34 mg/cm^2 thick was used for measurement of background spectra. The measurements on CD_2 and carbon targets were made for each setting to obtain the effect on deuterium via the CD_2–C subtraction procedure.

Multiwire drift chamber information was used to reconstruct the particles' trajectories in the focal plane. Three plastic scintillation counters were used in coincidence for the trigger and to provide the information about time-of-flight and energy losses of the particles. The time difference between the trigger signal and the radio-frequency signal of the cyclotron was used as the time-of-flight information (TOF). The distance between the target and the detection point was about 17 m, which is enough to separate tritons, deuterons and protons with the same momentum from the TOF. Pulse heights of the plastic scintillation counters were used to select the particle of interest at the trigger

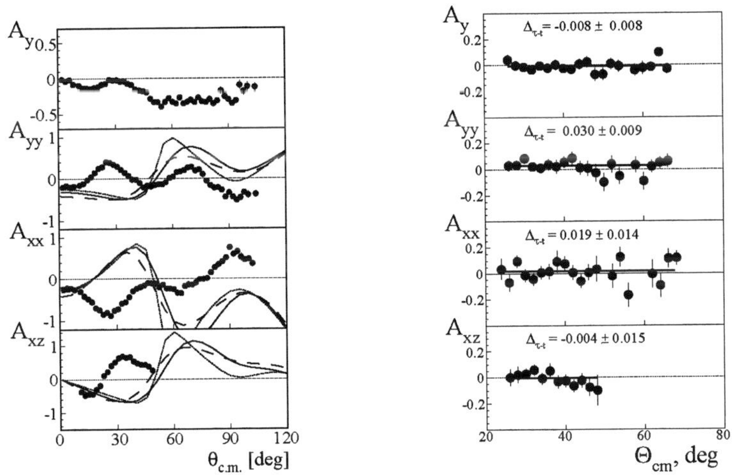

FIGURE 1. Results for the vector A_y and tensor A_{yy}, A_{xx} and A_{xz} analyzing powers of the $dd \to {}^3$Hen reaction [14] at energy T_d=270 MeV versus angle in the cms are presented in the left panel. The solid, dashed and dotted curves are the results of the non-relativistic ONE calculations [11, 12] using Urbana [16], Paris [17] and RSC [18] ^{3}He wave functions, respectively. The differences in the analyzing powers $\Delta_{\tau-t}$ for the charge symmetric $dd \to {}^3$Hen and $dd \to {}^3$Hp reactions at 270 MeV are presented in the right panel.

level. Particle identification was made using the information on the energy loss in the plastic scintillators and time-of-flight between the target and the detection point. The event was considered as a specific type of particle (^{3}He, ^{3}H or proton) only in the case when information on the pulse height in all tree scintillation counters was correlated.

RESULTS AND DISCUSSION

The results for the tensor analyzing power T_{20} of $dd \to {}^3$He(0°)n and $dd \to {}^3$H(0°)p were obtained for the energies 140–270 MeV [13]. The T_{20} values for both charge symmetric ^{3}H-p and ^{3}He-n channels at 140 and 200 MeV in this experiment were in good agreement with one another within the experimental accuracy. No evidence of charge symmetry breaking (CSB) was found in these processes. The positive sign of T_{20} in the energy domain explored reflects the positive sign of the D/S state ratio in ^{3}He(^{3}H) in momentum space. In this respect one can conclude that the T_{20} data are sensitive to the D-state in ^{3}He(^{3}H). The data obtained are in good agreement with the T_{20} data for d^3He-backward elastic scattering [10] when plotted versus relativistic internal momentum k [15].

The data on the tensor A_{yy}, A_{xx} and A_{xz} and vector A_y analyzing powers of the $dd \to {}^3$Hen reaction [14] obtained at 270 MeV are shown in the left panel of Fig. 1. The solid, dashed, and dotted curves are the results of the non-relativistic ONE calculations [11, 12] using Urbana [16], Paris [17] and RSC [18] ^{3}He wave functions, respectively.

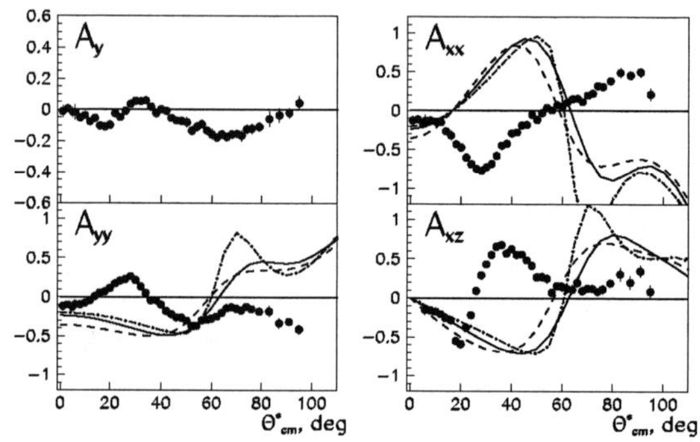

FIGURE 2. Results for the vector A_y and tensor A_{yy}, A_{xx}, A_{xz} analyzing powers of the $dd \to {}^3\mathrm{H}p$ reaction at energy T_d=200 MeV versus angle in the cms. The solid, long-dashed, and dotted curves are the results of ONE calculations using Urbana [16], Paris [17] and RSC [18] three-nucleon bound system wave functions.

The Paris deuteron wave function (DWF) [19] was used for the deuteron structure description. The experimental data on the tensor analyzing powers at small scattering angles are in qualitative agreement with the ONE calculations. The negative sign of the tensor analyzing powers A_{yy} and A_{xx} in this angular domain reflects the positive sign of the D/S state ratio in ^{3}He in tmomentum space. In this respect one can conclude that the obtained data are sensitive to the D-state in the ^{3}He in this region. However, the tensor analyzing powers demonstrate strong disagreement with the predictions of the ONE calculations at angles larger than 15° in the cms. The discrepancy between the data and the calculations found in Fig. 1 can be due both to the contribution from the reaction mechanism other than ONE and to the inadequate description of the short-range ^{3}He spin structure.

The results for the analyzing power differences in the ${}^3\mathrm{He}n$–${}^3\mathrm{H}p$ channels, $\Delta_{\tau-t}$, are shown in the right panel of Fig. 1 as a function of the scattering angle in the cms. The $\Delta_{\tau-t}$ values demonstrate no dependence on the scattering angle for all the analyzing powers. Averaged values of $\Delta_{\tau-t}$ are comparable with zero within the experimental accuracy. Therefore, one can conclude that no effect of CSB at the nuclear level in the analyzing powers of the $dd \to {}^3\mathrm{H}p$ and $dd \to {}^3\mathrm{He}n$ reactions at 270 MeV is observed.

The difference between the data obtained at 270 MeV and the calculations may indicate that knowledge of the short-range structure and the reaction mechanism used are not adequate. To understand what is the reason of the deviation, data at different energies are necessary. For this purpose, the analyzing powers for the $dd \to {}^3\mathrm{H}p$ reaction at 200 MeV have been measured. The experimental results on the vector A_y and tensor A_{yy}, A_{xx}, A_{xz} analyzing powers of the $dd \to {}^3\mathrm{H}p$ reaction at an incident deuteron energy of 200 MeV are presented in Fig. 2 as a function of the ^{3}H scattering angle in the cms. One can see that the experimental data for the analyzing powers A_{yy}, A_{xx}, A_{xz} strongly disagree with the predictions of ONE model calculations, which reproduce qualitatively the behavior of the tensor analyzing powers at small angles. The ONE calculation predicts a zero value of the vector analyzing power, but some structure in the angular distribution of the vector analyzing power is observed. Therefore, this result can

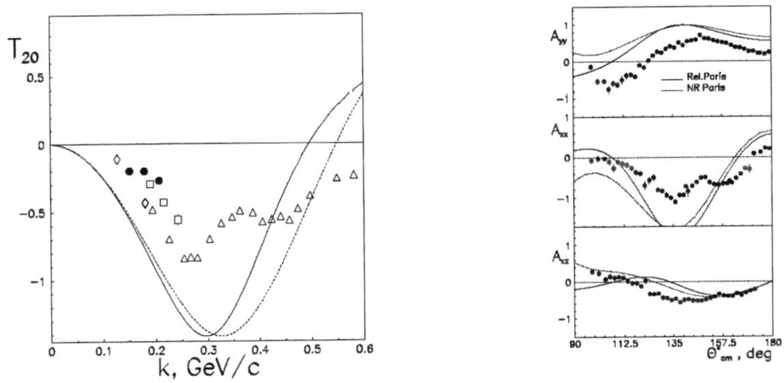

FIGURE 3. The left panel: the data for the tensor analyzing power T_{20} for $dd \rightarrow p^3$H (closed circles), $dp \rightarrow pd$ [3, 20] (open diamonds and triangles) and d^3He $\rightarrow p^4$He [21] (open squares) reactions plotted versus internal momentum k [15]. The right panel represents the results for the tensor analyzing powers A_{yy}, A_{xx} and A_{xz} in the $dd \rightarrow p^3$H reaction at 270 MeV versus the ^{3}H scattering angle in the cms. The curves are explained in the text.

be a clue that one should investigate reaction mechanisms in addition to ONE.

The left panel of Fig. 3 represents the results for the tensor analyzing power T_{20} in the $dd \rightarrow p^3$H reaction from the present experiment (closed circles) plotted versus internal momentum k [15] and compared with the data obtained in other (d,p) processes: dp-backward elastic scattering [3, 20] (open diamonds and triangles) and d^3He $\rightarrow p^4$He [21] (open squares) reaction. The solid and dashed curves are the results of ONE calculations using Paris [19] and Bonn [22] DWFs, respectively. The negative sign of T_{20} in the energy domain explored reflects the sensitivity to the deuteron internal structure; however, the strong disagreement of the data with the ONE predictions requires further theoretical investigation.

The results for the tensor analyzing powers A_{yy}, A_{xx} and A_{xz} in the $dd \rightarrow {}^3$Hp reaction at 270 MeV at backward ^{3}H emission angles are shown in the right panel of Fig. 3. The solid and dashed curves are the results of relativistic (using the minimal scheme [15]) and non-relativistic ONE calculations using the Paris DWF [19]. The relativistic corrections are smaller than at the forward angles [14], but should be taken into account for the description of the data at large p_T. The deviation of the tensor analyzing powers from ONE calculations, as well as the non-zero value of the vector analyzing power A_y requires that one consider mechanisms beyond ONE.

CONCLUSIONS

High precision data for the analyzing powers in the $dd \rightarrow {}^3$Hen and $dd \rightarrow {}^3$Hp reactions are obtained at intermediate energies using the SMART spectrograph.

The analyzing power data obtained for both $dd \rightarrow {}^3$Hen and $dd \rightarrow {}^3$Hp channels are in good agreement. The data demonstrate sensitivity to the ^{3}He(^{3}H) and deuteron spin structures, when the 3N system is emitted in the forward and backward directions in the cms, respectively, being in qualitative agreement with naive ONE expectations [11]. However, further theoretical studies are required for a quantitative description of the data.

Continuation of the analyzing powers measurements at energies above 500 MeV should be very interesting to study possible manifestations of non-nucleonic degrees of freedom [23].

ACKNOWLEDGMENTS

The authors express their thanks to the staff of the RARF for providing of excellent conditions for the R308n experiment. They are grateful to H.Kumasaka, R.Suzuki and R.Taki for their help during the experiment. The Russian part of collaboration thanks the RIKEN Directorate for the kind hospitality during the experiment. The investigation has been supported in part by the Russian Foundation for Basic Research (grant No.07-02-00102-a) and by the Grant Agency for Science at the Ministry of Education of the Slovak Republic (grant No.1/4010/07). V.P.L. thanks the Organizing Committee of the FM50 Conference for financial support.

REFERENCES

1. W. Glöckle, H. Witala, D. Hüber, H. Kamada, and J. Golak, *Phys.Rep.* **274**, 107–285 (1996).
2. N. Sakamoto, et al., *Phys.Lett. B* **367**, 60–64 (1996).
3. K. Sekiguchi, et al., *Phys.Rev. C* **65**, 034003 (2002).
4. E.J. Stephenson, et al., *Phys.Rev. C* **60**, 061001 (1999); R. Bieber, et al., *Phys.Rev.Lett.***84**, 606–609 (2000); H. Sakai, et al., *Phys.Rev.Lett.* **84**, 5288–5291 (2000).
5. R.V. Cadman, et al., *Phys.Rev.Lett.* **86**, 967–970 (2001).
6. K. Sekiguchi, et al., *Phys.Rev. C* **70**, 014001 (2004).
7. K. Hatanaka, et al., *Phys.Rev. C* **66**, 044002 (2002).
8. K. Ermish, et al., *Phys.Rev. C* **71**, 064004 (2005).
9. M.A. Miller, et al., *Phys. Rev. Lett.* **74**, 502–505 (1995).
10. M. Tanifuji, et al., *Phys.Rev. C* **61**, 024602 (2000).
11. V.P. Ladygin, and N.B. Ladygina, *Phys.Atom.Nucl.* **59**, 789–794 (1996); *ibid.* **65**, 1609–1615 (2002).
12. V.P. Ladygin, N.B. Ladygina, H. Sakai, and T. Uesaka, *Part.Nucl.Lett.* **3[100]-2000**, 74–85 (2000).
13. V.P. Ladygin, et al., *Phys.Lett.B***598**, 47–54 (2004); *Phys.Atom.Nucl.* **69**, 1271–1278 (2006).
14. M. Janek, et al., *Eur.Phys.J. A* **33**, 39–46 (2007).
15. P.A.M. Dirac, *Rev.Mod.Phys.* **21**, 392–399 (1949); S. Weinberg, *Phys.Rev.***150**, 1313–1318 (1966); L.L. Frankfurt, and M.I. Strikman, *Phys.Rep.* **76**, 215–347 (1981).
16. R. Schiavilla, V.R. Pandharipande, and R.B. Wiringa, *Nucl.Phys. A* **449**, 219–242 (1986).
17. J.-M. Laget, J.F. Lecolley, and F. Lefebvres, *Nucl.Phys. A* **370**, 479–495 (1981).
18. F.D. Santos, A.M. Eiro, and A. Barosso, *Phys.Rev. C* **19**, 238–243 (1979).
19. M. Lacombe, B. Loiseau, R. Vinh Mau, J. Cote, P. Pires, and R. de Tourreil, *Phys.Lett. B* **101**, 139–140 (1981).
20. V. Punjabi, et al., *Phys.Lett. B* **350**, 178–183 (1995).
21. T. Uesaka, et al., *Phys.Lett. B***467**, 199–204 (1999); *ibid.* **533**, 1–7 (2002).
22. R. Machleidt, K. Holinde, and C. Elster, *Phys.Rept.* **149**, 1–89 (1987).
23. Yu.N. Uzikov, *JETP. Lett.* **81**, 303–306 (2005).

Measurement of analyzing powers for d-p elastic scattering at the Internal Target Station of the Nuclotron

K. Suda*, T. Uesaka†, V.P. Ladygin**, P.K. Kurilkin**, Yu.V. Gurchin**, A.Yu. Isupov**, K. Itoh‡, M. Janek**,§, J.-T. Karachuk**,¶, T. Kawabata†, A.N. Khrenov**, A.S. Kiselev**, V.A. Kizka**, J. Kliman**, V.A. Krasnov**, N.B. Ladygina**, A.N. Livanov**, Y. Maeda‖, A.I. Malakhov**, V. Matoucek††, M. Morhac††, S.M. Piyadin**, S.G. Reznikov**, S. Sakaguchi†, H. Sakai‡‡,†, Y. Sasamoto†, K. Sekiguchi§§, M.A. Shikhalev**, I. Turzo††, T.A. Vasiliev** and H. Witała¶¶

**Research Center for Nuclear Physics, Osaka University, Ibaraki, Osaka 567-0047, Japan*
†Center for Nuclear Study, University of Tokyo, Bunkyo, Tokyo 113-0033, Japan
***Joint Institute for Nuclear Research, 141980 Dubna, Moscow region, Russia*
‡Department of Physics, Saitama University, Saitama 338-8570, Japan
§P.J.Safarik University, 041 80 Kosice, Slovakia
¶Advanced Research Institute for Electrical Engineering, 313 Splaiul Unirii, Bucharest, Romania
‖Department of Physics, Kyushu University, Higashi, Fukuoka 812-8581, Japan
††Institute of Physics Slovak Academy of Sciences, 84511 Bratislava, Slovakia
‡‡Department of Physics, University of Tokyo, Bunkyo, Tokyo 113-0033, Japan
§§RIKEN (The Institute for Physical and Chemical Research), Wako, Saitama 351-0198 Japan
¶¶Institute of Physics, Jagiellonian University, PL-30059 Cracow, Poland

Abstract. Measurements of analyzing powers A_y, A_{yy}, and A_{xx} for d-p elastic scattering were performed at E_d=0.88 and 2.0 GeV using a high-energy deuteron beam polarimeter at the Internal Target Station of the Nuclotron at JINR. The analyzing power data at 0.88 GeV are compared with several theoretical calculations.

Keywords: polarized deuteron beam, deuteron-proton elastic scattering
PACS: 29.27.Hj, 25.10.+s, 25.60.Bx

INTRODUCTION

Intensive studies of three-nucleon force (3NF) effects were performed using Nucleon-deuteron elastic scattering. The data for the cross section and vector analyzing powers in $\vec{d}$-p elastic scattering are well reproduced up to 135 MeV/u if one includes a 3NF [1]. However, the tensor analyzing powers are not always reproduced. At higher energies, the Faddeev calculations fail to reproduce the cross section data. For example at 250 MeV in $\vec{p}$-d elastic scattering, discrepancies remain at angles larger than 120° when compared with theoretical predictions including a 3NF [2]. Also, for the vector analyzing power, a large discrepancy remains at backward angles. These discrepancies become larger as the energy increases to 392 MeV [3]. It is pointed out that relativistic effects play a role. Therefore, measurements of the energy dependence of the $\vec{d}$-p elastic scattering

CP1011, *New Facet of Three Nucleon Force—50 Years of Fujita Miyazawa Three Nucleon Force (FM50), The International Symposium,* edited by H. Sakai, K. Sekiguchi, and B. F. Gibson

observables are desired.

For this purpose, and for the polarization measurements at high energies up to 2 GeV, we have constructed a high-energy deuteron beam polarimeter based on *d*-*p* elastic scattering at the internal target station (ITS) of the Nuclotron at JINR [4, 5]. The polarization measurements at high energies are essential for the spin physics program at JINR. In this program, an established measurement of deuteron polarization is required to deduce values of polarization observables reliably. In particular, it is necessary to measure both vector and tensor polarizations. We proposed to use *d*-*p* elastic scattering at backward angles ($\theta_{\text{c.m.}} > 60°$) as a polarimeter at energies of 0.88 – 2 GeV. The reaction has advantages as a beam-line polarimeter. Firstly, both vector and tensor analyzing powers for the reaction can attain large values [6, 7, 8, 9]. Secondly, a kinematic coincidence measurement of deuteron and proton with simple plastic scintillation detectors suffices for event identification. This is mainly due to a small background event rate, compared with the forward angles. The use of this reaction as a deuteron polarimeter at 140 – 270 MeV at large angles was established at RIKEN [1, 10, 11].

The analyzing power measurements were performed at E_d=880 MeV and 2.0 GeV in June, 2005. The analyzing power data at 880 MeV, which corresponds to the maximum deuteron beam energy at RIBF in Japan, is useful for development of the deuteron beam polarimeter at RIBF. In this paper, we report on the preliminary results at 880 MeV.

EXPERIMENT

Measurement of the analyzing powers for the *d*-*p* elastic scattering by using the ITS polarimeter was performed at 880 MeV and 2 GeV at JINR. The beam polarizations were measured using the polarimeter at 270 MeV, where the absolute calibration of the analyzing powers was performed at RIKEN [11]. The vector and tensor polarized deuteron beams were provided by the cryogenic atomic beam source POLARIS [12]. Two modes of vector and tensor polarizations were used as well as an unpolarized mode, $(P_Z,P_{ZZ}) = (+1/3,1),(+1/3,-1)$, and $(0,0)$, where the values are theoretical maxima. Typical beam intensity was 3×10^7 particles/spill, and the spill duration was 8 seconds. Figure 1 shows the schematic view of the ITS polarimeter. The ITS is composed of a spherical hull, a beam duct and the target stage where the target holder carrying up 6 different targets is located [13]. The ITS is well suited for the measurements of the *d*-*p* elastic scattering at wide angular range in the center of mass system. A polyethylene sheet of 10 μm thickness was used as a target as well as carbon wires for the background measurement. At each spill, the target was swept from the high- to low momentum sides of the Nuclotron. In order to estimate the effect of false asymmetry, target position was monitored by counting the number of pulses transmitted to the step motor attached to the target holder.

The scattered deuterons and recoil protons were detected in coincidence. Detectors were placed on the detector stage at the downstream of the target. Each detector consisted of a plastic scintillation counter coupled to a photomultiplier tube through a light guide. Deuteron and proton detectors were placed in the directions of left, right, and up around the beam axis at scattering angles ranging from $\theta_{\text{c.m.}} = 60°–140°$. The quasi-elastic *p*-*p* scattering was used to monitor relative luminosity. One pair of detectors were

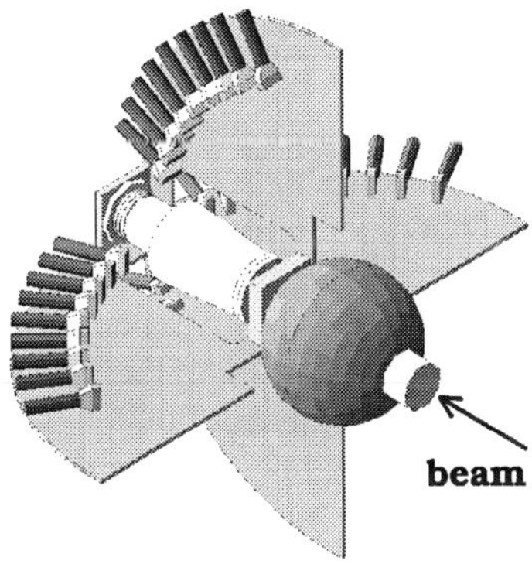

FIGURE 1. Schematic view of the detector setup.

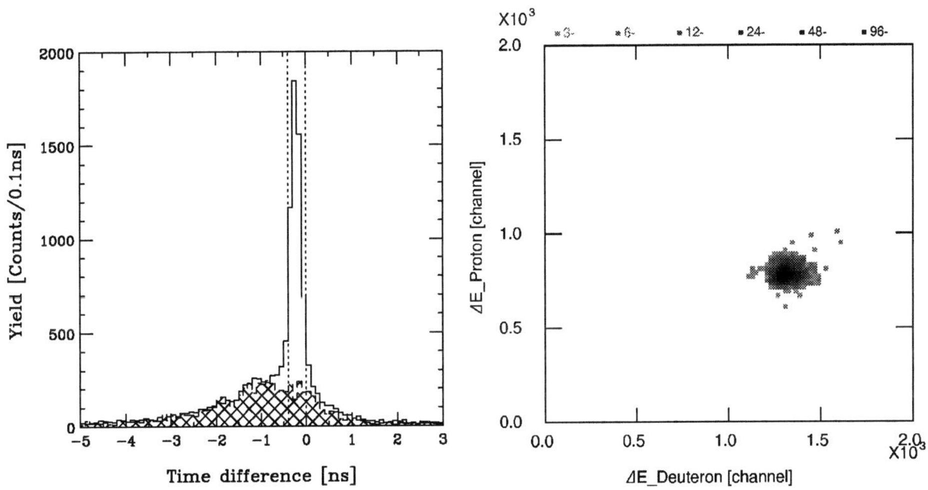

FIGURE 2. Difference of time-of-flight between the deuterons and protons (left panel) and energy loss correlations of the deuterons and protons gated by the time difference (right panel). In the left panel, a solid line indicates the spectrum for the polyethylene target, and a dashed line for carbon (hatched). The *d-p* elastic scattering events are indicated by dotted lines. In the right panel, a locus corresponding to *d-p* elastic scattering can be seen clearly.

placed in the directions of left and right at $\theta_{c.m.} = 90°$, where the vector analyzing power vanishes.

ANALYSIS

The data at 880 MeV were analyzed. The events of the *d-p* elastic scattering were identified by energy losses and a difference of time-of-flight for the deuterons and protons.

Figure 2 shows the spectra of the time-of-flight differences between the deuterons and protons. A peak corresponding to *d-p* elastic scattering was clearly observed. The carbon contribution was found to be not negligible at this energy. Therefore, the yield for the carbon was subtracted from that for the polyethylene. The events were selected by a narrow gate indicated in Fig. 2 in order to increase the S/N ratio. The energy loss correlation between deuteron and proton gated by the time difference is also shown in Fig. 2. A clear locus corresponding to the *d-p* elastic scattering was observed. No cuts of the target position were applied for the present analysis. The analyzing powers at 880 MeV were obtained from the asymmetries of yields and the beam polarizations measured at 270 MeV.

RESULTS AND DISCUSSION

Preliminary results for the analyzing powers of the *d-p* elastic scattering at 880 MeV are shown in Fig. 3. Only statistical errors are shown. The data are compared with several

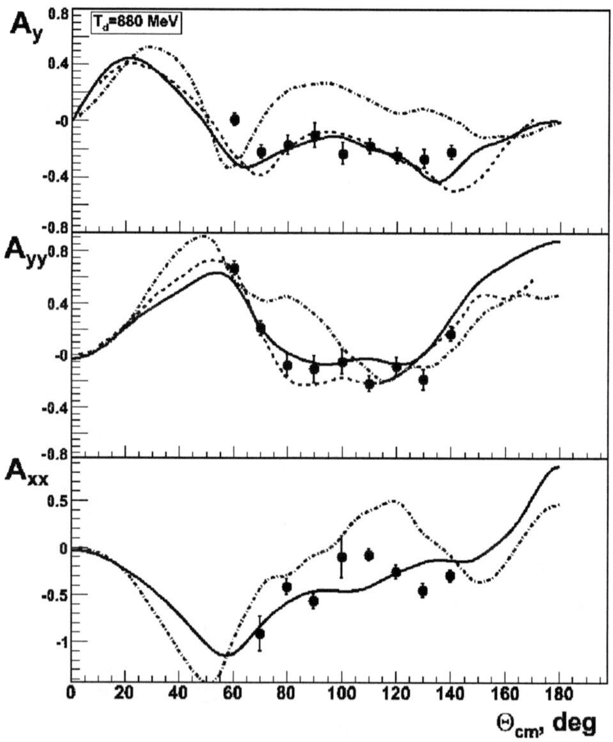

FIGURE 3. Analyzing powers A_y, A_{yy}, and A_{xx} for *d-p* elastic scattering at $E_d = 880$ MeV. Only statistical errors are shown. Solid lines show the Faddeev calculations based on the CD-Bonn NN potential. The dashed lines and dot-dashed lines represent the multiple scattering model calculations using CD-Bonn deuteron wave functions (DWF) [14] and optical-potential calculations using a Dibaryon DWF [15], respectively.

theoretical calculations. The solid lines show the Faddeev calculations based on the CD-Bonn NN potential. The calculation reproduces the analyzing power data well even though the energy is well above the pion production threshold and the NN potential can be used only up to 350 MeV. The dashed lines and dot-dashed lines represent the multiple scattering model calculations using CD-Bonn deuteron wave functions (DWF) [14] and optical-potential calculations using Dibaryon DWF [15], respectively. The multiple scattering model calculations well reproduce the experimental data.

In Fig. 4, the calculated differential cross sections are compared with the experimental data for the *p-d* elastic scattering at 425 and 470 MeV [16, 17]. The Faddeev calculations

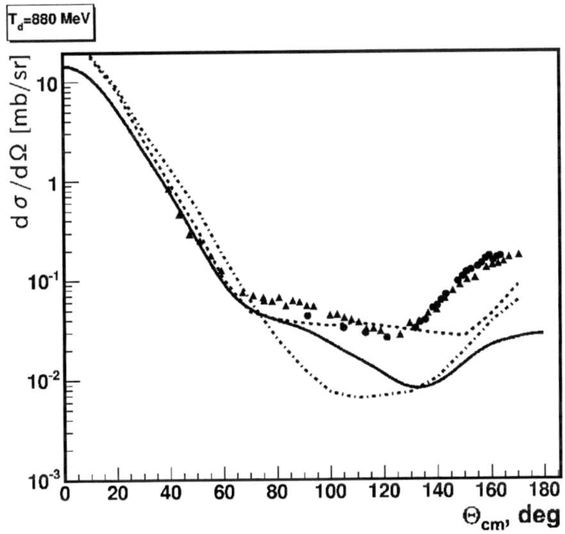

FIGURE 4. Differential cross sections for *d-p* elastic scattering. The theoretical curves are the same as in Fig. 3. The experimental data for the *p-d* elastic scattering at 425 and 470 MeV [16, 17] are shown by triangles and circles, respectively.

fail to reproduce the data at the angles larger than 90°. In order to describe the data in the region of the cross section minimum, one needs to take into account NN scattering amplitude up to 1 GeV. On the other hand, the multiple scattering model calculations are in good agreement in the angular range from 30° – 130°. The inclusion of 3NF may improve the description at backward angles.

ACKNOWLEDGMENTS

The authors would like to express their gratitude to the Nuclotron accelerator staff. They thank Yu.S. Anisimov, A.F. Elishev, Z.P. Kuznezova, A.P. Laricheva, A.G. Litvinenko, V.G. Perevozchikov, and V.N. Zhmyrov for the help during the preparation and performance of the experiment. This work has been supported partly by the Grant-in-Aid for Scientific Research N°14740151 of the Ministry of Education, Culture, Sports, Science, and Technology of Japan, by the Russian Foundation for Basic Research (grant N°07-

02-00102a), and by the Grant Agency for Science at the Ministry of Education of the Slovak Republic (Grant N°1/4010/07).

REFERENCES

1. K. Sekiguchi *et al.*, Phys. Rev. C **65**, 034003 (2002).
2. K. Hatanaka *et al.*, Phys. Rev. C **66**, 044002 (2002).
3. A. Tamii *et al.*, AIP Conf. Proc. **915**, 765 (2007).
4. T. Uesaka, V. P. Ladygin *et al.*, Phys. Part. Nucl. Lett. **3**, 305 (2006).
5. T. Uesaka *et al.*, CNS Annual Report 2004, 81 (2005).
6. M. Bleszynski *et al.*, Phys. Lett. B **87**, 178 (1979).
7. M. Bleszynski *et al.*, Phys. Lett. B **106**, 42 (1981).
8. M. Haji-Saied *et al.*, Phys. Rev. C **36**, 2010 (1987).
9. V. Ghazikhanian *et al.*, Phys. Rev. C **43**, 1532 (1991).
10. N. Sakamoto *et al.*, Phys. Lett. B **367**, 60 (1996).
11. K. Suda *et al.*, Nucl. Instrum. Methods A **572**, 745 (2007).
12. N.G. Anishenko *et al.*, AIP Conf. Proc. **95**, 445 (1983).
13. A.I. Malakhov *et al.*, Nucl. Instrum. Methods A **440**, 320 (2000); Yu.S. Anisimov *et al.*, Proc. of the 7th Int. Workshop on Relativistic Nuclear Physics, Stara Lesna, Slovak Republic, 117 (2003).
14. N.B. Ladygina, nucl-th/0705.3149.
15. M.A. Shikhalev, nucl-th/0710.4040.
16. N.E. Booth *et al.*, Phys. Rev. D4, 1261 (1971).
17. J.C. Alder *et al.*, Phys. Rev. C6, 2010 (1972).

Triton and hypertriton binding energies with SU_6 quark-model baryon-baryon interactions

Y. Fujiwara*, Y. Suzuki†, M. Kohno** and K. Miyagawa‡

*Department of Physics, Kyoto University, Kyoto 606-8502, Japan
†Department of Physics, and Graduate School of Science and Technology, Niigata University, Niigata 950-2181, Japan
**Physics Division, Kyushu Dental College, Kitakyushu 803-8580, Japan
‡Department of Applied Physics, Okayama Science University, Okayama 700-0005, Japan

Abstract. Previously we calculated the binding energies of the triton and hypertriton, using an SU_6 quark-model interaction which is obtained by a resonating-group method for two baryon clusters. In contrast to the previous calculations employing the energy-dependent interaction kernel, we present new results using a renormalized interaction which is energy-independent and still preserves all the two-baryon data. The new binding energies are slightly smaller than the previous values. In particular the triton binding energy turns out to be 8.14 MeV with a charge-dependence correction of the two-nucleon force, 190 keV, being included. This indicates that the energy to be accounted for by three-body forces is about 350 keV.

Keywords: quark model, baryon-baryon interaction, triton, hypertriton, Faddeev calculation
PACS: 21.45.+v, 13.75.Cs, 13.75.Ev

INTRODUCTION

The QCD-inspired spin-flavor SU_6 quark model (QM) for the baryon-baryon interaction, developed by the Kyoto-Niigata group, has achieved accurate descriptions of available NN and YN experimental data [1]. In particular, the most recent model fss2 produces an accuracy comparable to modern realistic meson-exchange NN potentials. Since the QM description of the short-range part is quite different from that of meson-exchange potentials, it is interesting to use the QM interaction to calculate some properties of three-baryon systems, *e.g.*, the binding energies of the triton (^{3}H) and hypertriton ($^3_\Lambda$H). For this purpose, we developed in Refs. [2, 3] a three-cluster equation, called an εK prescription, which employs energy-*dependent* two-cluster exchange kernels of the resonating-group method (RGM). Applying the εK formulation to the case of the quark-exchange kernels, we obtained the following results for fss2; the triton binding energy is $B_t = 8.518$ MeV [4, 5], and the Λ separation energy of $^3_\Lambda$H is $B_\Lambda = 289$ keV [6]. The QM triton binding energy is much larger than that of the standard meson-exchange NN models, which yield at most about 8.0 MeV when the well-known charge-dependence effect of the order of 190 keV is taken into account. For the $^3_\Lambda$H result, it should be noted that most of the Nijmegen models fail to reproduce enough Λ-separation energy, except for the NSC89 and the more recent soft-core models NSC97 (e) and (f) [7, 8].

Recently, important progress has been made by using an energy-*independent* RGM kernel to the 3α system [9] and other three-cluster systems [10], through a standard procedure [11] for eliminating the energy dependence of the RGM kernel. This renor-

CP1011, *New Facet of Three Nucleon Force—50 Years of Fujita Miyazawa Three Nucleon Force (FM50), The International Symposium,* edited by H. Sakai, K. Sekiguchi, and B. F. Gibson

malized kernel will naturally give results different from the εK prescription when it is applied to few-body systems interacting via the QM baryon-baryon interactions. We will report these new results in this paper.

FORMULATION

The energy-independent renormalized RGM kernel $V^{\rm RGM}$ for a two-cluster system reads

$$V^{\rm RGM} = V_{\rm D} + G + W\ , \tag{1}$$

where $V_{\rm D}$ is the direct potential, and G is the sum of the exchange kinetic-energy and interaction kernels. It is the kernel W that appears through the elimination of the energy-dependence, and it is given by

$$W = \Lambda \frac{1}{\sqrt{1-K}} h \frac{1}{\sqrt{1-K}} \Lambda - h\ . \tag{2}$$

Here K is the exchange normalization kernel, h denotes $h_0 + V_{\rm D} + G$ with h_0 being the kinetic energy for the two-cluster relative motion, and $\Lambda = 1 - |u\rangle\langle u|$ is a two-cluster Pauli projection operator, where $|u\rangle$ is a Pauli-forbidden state satisfying $K|u\rangle = |u\rangle$. An advantage of using $V^{\rm RGM}$ is that the two-cluster RGM equation takes the form of the usual Schrödinger equation in the Pauli-allowed model space, and the relative wave function is properly normalized [11]. This Schrödinger-type equation for the relative wave function gives the same asymptotic behavior as the original RGM equation, thus preserving the phase shifts and physical observables for the two-cluster system. The difference between the previous energy-dependent RGM kernel, $V^{\rm RGM}(\varepsilon) = V_{\rm D} + G + \varepsilon K$, and $V^{\rm RGM}$ in Eq. (1) is essentially a replacement of $\Lambda(\varepsilon K)\Lambda$ with W. Here ε is the two-cluster relative energy measured from its threshold. The value is, however. not defined in the three-cluster system and it was determined in a self-consistent procedure in the εK treatment.

The three-cluster equation to be solved reads

$$P\left[E - H_0 - V^{\rm RGM}_\alpha - V^{\rm RGM}_\beta - V^{\rm RGM}_\gamma\right] P\Psi = 0\ , \tag{3}$$

where α, β, and γ denote three independent pairs of two-cluster subsystems, and where H_0 is the free three-body kinetic-energy operator, and $V^{\rm RGM}_\alpha$ stands for the RGM kernel in Eq. (1) for the α-pair, *etc.* The three-body operator P projects on the Pauli-allowed space with a proper symmetry of clusters, and it is constructed from the orthogonality constraint that each pair of two-cluster subsystems is free from any Pauli-forbidden states [12, 13, 14]. This definition of the three-cluster Pauli-allowed space may not be exactly equivalent to the standard definition given by the three-cluster normalization kernel. We, however, employed this orthogonality condition in the εK prescription. See [2, 3] for detail. We use the same definition of P in this paper as well.

RESULTS AND DISCUSSION

Table 1 lists three-nucleon bound state properties of fss2, predicted by the Faddeev calculations in the truncated model spaces. The *np* interaction, up to the total angular-momentum J, is employed in the isospin basis. The charge root mean square (rms) radii of ^{3}H and ^{3}He are calculated with the finite size corrections of the proton and neutron through, $\langle r^2\rangle_{^3\mathrm{H}} = \left[R_C(^3\mathrm{H})\right]^2 + (0.8750)^2 + 2(-0.1161)$ and $\langle r^2\rangle_{^3\mathrm{He}} = \left[R_C(^3\mathrm{He})\right]^2 + (0.8750)^2 + \frac{1}{2}(-0.1161)$, where ${R_C}^2$ (in fm^2) stands for the squared charge radius for each isospin state of the point nucleons. The Coulomb force and the relativistic correction terms [15] of the charge current operator are entirely neglected.

The final fss2 prediction for the triton energy is -8.326 MeV, which is higher by 192 keV than the $E^{\varepsilon K}$ value -8.518 MeV. Compared to the experimental value $E^{\exp}(^3\mathrm{H}) = -8.482$ MeV, the calculated value is high by 156 keV. In fact, we have to take into account the effect of the charge dependence of the two-nucleon force, which is estimated to result in an energy loss of about 190 keV [16]. Therefore, our calculation using the QM potential misses the triton binding energy by 346 keV, namely, about 350 keV. Comparing the energy convergence between the present and $E^{\varepsilon K}$ methods in Table 1, we note that, in the present treatment, the 5-channel calculation already gives an energy close to the converged value. Namely, the energy gain from the 5-channel to 50-channel calculation is only 65 keV, while it is 329 keV in the εK prescription.

For a realistic calculation of the ^{3}H binding energy, it is important to use an *NN* interaction which reproduces various deuteron properties and some typical *NN* scattering data, such as the ε_1 parameter, which are strongly correlated to the D-state probability, P_D, of the deuteron [17]. The effective range parameters of the $^1S_0$ scattering also play an essential role. In order to demonstrate the correlation between B_t and P_D, we display in Fig. 1 the updated energies of fss2 and FSS together with the values calculated by various realistic *NN* potentials. We find that fss2 gives a larger binding energy than the modern realistic meson-exchange potentials like Bonn-C and AV18, while the result of FSS is not very far from that of Bonn-C. It is interesting to note that our QM points are apparently off the line on which the data points of the modern meson-exchange potentials fall. The five-channel calculation of the model QCM-A by Takeuchi *et al.* [21]

TABLE 1. The energy and charge rms radius of three-nucleon ground-state calculated using the energy-independent renormalized RGM kernel derived from the QM fss2 potential.

n	$E(^3\mathrm{H})$ (MeV)	$E^{\varepsilon K}(^3\mathrm{H})$ (MeV)	$\sqrt{\langle r^2\rangle_{^3\mathrm{H}}}$ (fm)	$\sqrt{\langle r^2\rangle_{^3\mathrm{He}}}$ (fm)
2 (S)	−7.952	−7.807	1.80	1.95
5 (S,D)	−8.261	−8.189	1.76	1.92
10 $(J \leq 1)$	−7.962	−8.017	1.77	1.95
18 $(J \leq 2)$	−8.228	−8.439	1.75	1.93
26 $(J \leq 3)$	−8.313	−8.503	1.75	1.92
34 $(J \leq 4)$	−8.322	−8.514	1.75	1.92
42 $(J \leq 5)$	−8.326	−8.517	1.75	1.92
50 $(J \leq 6)$	−8.326	−8.518	1.75	1.92

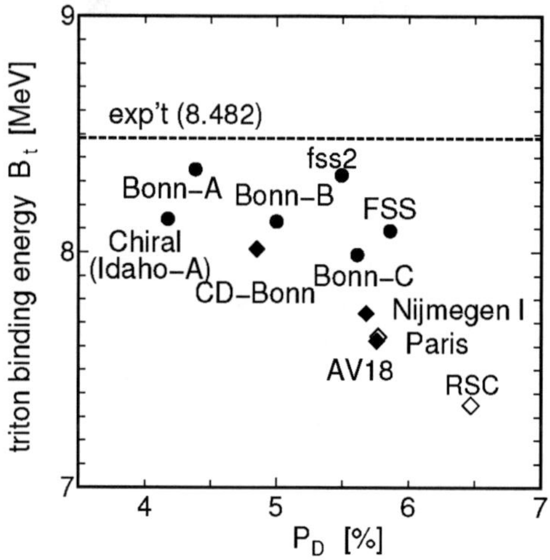

FIGURE 1. Calculated ^3H binding energies B_t as a function of the deuteron D-state probability P_D. Calculations are made in the isospin basis, using the np interaction, for fss2, FSS, Bonn-A, B, C and Chiral (denoted by black circles). The CD-Bonn, Nijmegen I and AV18 (denoted by black diamonds) take into account the effect of charge-dependence of the interaction. In the Paris and RSC results (denoted by the open diamonds), the $^1S_0$ interaction is determined from the pp scattering data. The calculated values are taken from [16], [17] (RSC, Bonn-A, B, C), from [18] (AV18, CD-Bonn) from [19] (Nijmegen I) and from [20] (Chiral).

gives almost the same result as Bonn-C.

The results of the hypertriton Faddeev calculations are listed in Table 2 for the model fss2. The Λ separation energy of the hypertriton is $B_\Lambda = 262$ keV, which is 27 keV less than the εK value, 289 keV. So far all the Faddeev calculations, using the energy-independent renormalized RGM kernels, yield less binding than the εK prescription, as long as the full model space with enough angular-momenta is taken into account. Compared with the experimental value, ${B_\Lambda}^{\rm exp} = 130 \pm 50$ keV, the model fss2 produces overbinding by 82 - 182 keV. In spite of the large error bar of the experiment, we may conclude that the ΛN interaction of fss2 is probably slightly too attractive. The accuracy of the present-day YN interaction makes it impossible for us to discuss the effect of the ΛNN three-body force and the charge-symmetry breaking effect of the Λp and Λn interactions.

We find in Table 2 that the 15-channel calculation with S- and D-states only is a good approximation to the full calculation. The value of B_Λ is 226 keV in the 15-channel calculation, and the energy gain extended to the full calculations is only 36 keV. The corresponding energy gain in the previous εK prescription is as large as 91 keV. The admixture of the ΣNN component, P_Σ, is 0.83% for fss2, which was previously 0.80% in the εK prescription.

As to the overbinding of the model fss2, we have discussed in Ref. [6] that a slight increase of the κ meson mass will improve the fit to the experimental value, without

TABLE 2. Hypertriton ($^3_\Lambda$H) properties calculated using the energy-independent renormalized RGM kernels derived from the QM fss2 potential.

n	E (MeV)	B_Λ (keV)	$B_\Lambda^{\varepsilon K}$ (keV)	P_Σ (%)
6 (S)	−2.392	167	137	0.566
15 (SD)	−2.451	226	198	0.775
30 ($J \leq 1$)	−2.404	179	178	0.679
54 ($J \leq 2$)	−2.467	243	273	0.792
78 ($J \leq 3$)	−2.483	259	285	0.824
102 ($J \leq 4$)	−2.486	261	288	0.828
126 ($J \leq 5$)	−2.487	262	289	0.830
150 ($J \leq 6$)	−2.487	262	289	0.830

changing the good reproduction of the low-energy ΛN cross section data. If we modify the κ-meson mass from the value used in fss2, $m_\kappa = 936$ MeV, to 995 MeV, we would obtain $B_\Lambda = 134$ keV with $P_\Sigma = 0.56\,\%$, which is very close to the NSC89 prediction $B_\Lambda = 143$ keV with $P_\Sigma = 0.5\,\%$ [7, 8].

Rather moderate modification of the present results from the previous εK prescription is related to a simple structure of the quark-exchange normalization kernel $\Lambda K \Lambda$ in the Pauli-allowed space. For the NN interaction, $\Lambda = 1$ since there is no Pauli-forbidden state. For the positive-parity states, the largest eigenvalue of K in absolute magnitude is 1/9 for the $(0s)$ harmonic-oscillator state. Although most Pauli-forbidden states appear in the P-states, such partial waves give minor contributions to the binding energy of the triton. For the ΛN–ΣN interaction, we have a Pauli-forbidden state classified by the SU_3 label $(11)_s$. Once this component is properly eliminated, the eigenvalues of $\Lambda K \Lambda$ also become very small. These are the main reasons why the present treatment by the energy-independent renormalized RGM kernel gives the results that differ little from the previous energy-dependent εK prescription. On the other hand, the difference between $\Lambda(\varepsilon K)\Lambda$ and W becomes sometimes large in nuclear cluster systems, because the number of Pauli-forbidden states in general increases and they play more complicated roles in determining the structure of the nuclear systems. For example, the energy difference of the ^{9}Be ground state, calculated in $n\alpha\alpha$ three-cluster model, is more than 1 MeV [22, 10] depending on whether it is calculated in the renormalized RGM approach or in the εK prescription.

SUMMARY

We have recalculated the triton and hypertriton binding energies in a new semi-microscopic three-cluster equation, using the energy-independent renormalized RGM kernels of the quark-model baryon-baryon interactions. This formulation produces slightly less attraction in the three-baryon systems, compared to the previous energy-dependent treatment of the two-cluster RGM kernels. For the triton binding energy, the fss2 quark model gives 8.326 MeV, which is smaller by 156 keV than the experimental value of 8.482 MeV. By taking into account the charge-dependence correction of 190 keV, we conclude that the quark-model potential underbinds the triton by approximately

350 keV. Thus the energy to be accounted for by three-nucleon forces may not be as large as 0.5 - 1 MeV, which most of standard meson-exchange potentials [18] predict. The above different predictions for the contributions of the three-body force prompt us to examine what results the present quark-model *NN* interaction produces in other three-nucleon observables, especially in the *nd* and *pd* scattering. This will be examined soon and reported elsewhere. For the Λ separation energy of the hypertriton, the overbinding of the model fss2 is slightly reduced. We still have large ambiguities in the ^1S and ^3S ΛN interactions, before further details, such as the charge symmetry breaking in the Λp, Λn interactions and the ΛNN three-body force, come into play. The comparison of the fss2 value, 262 keV, with the experimental one, $B_\Lambda^{\rm exp} = 130 \pm 50$ keV, indicates that the ^1S interaction of fss2 is still too attractive, which could be corrected by choosing a slightly heavier κ-meson mass.

ACKNOWLEDGMENTS

This work was supported by Grants-in-Aid for Scientific Research (C) (Grant Nos. 18540261 and 17540263) and Bilateral Joint Research Projects (2006-2008) from the Japan Society for the Promotion of Science (JSPS).

REFERENCES

1. Y. Fujiwara, Y. Suzuki, and C. Nakamoto, *Prog. Part. Nucl. Phys* **58**, 439 (2007).
2. Y. Fujiwara, H. Nemura, Y. Suzuki, K. Miyagawa, and M. Kohno, *Prog. Theor. Phys.* **107**, 745 (2002).
3. Y. Fujiwara, Y. Suzuki, K. Miyagawa, M. Kohno, and H. Nemura, *Prog. Theor. Phys.* **107**, 993 (2002).
4. Y. Fujiwara, K. Miyagawa, M. Kohno, Y. Suzuki, and H. Nemura, *Phys. Rev. C* **66**, 021001(R) (2002).
5. Y. Fujiwara, K. Miyagawa, Y. Suzuki, M. Kohno, and H. Nemura, *Nucl. Phys.* **A721**, 983c (2003).
6. Y. Fujiwara, K. Miyagawa, M. Kohno, and Y. Suzuki, *Phys. Rev. C* **70**, 024001 (2004).
7. K. Miyagawa, H. Kamada, W. Glöckle, and V. Stoks, *Phys. Rev. C* **51**, 2905 (1995).
8. A. Nogga, H. Kamada, and W. Glöckle, *Phys. Rev. Lett.* **88**, 172501 (2002).
9. Y. Suzuki, H. Matsumura, M. Orabi, Y. Fujiwara, P. Descouvemont, M. Theeten, and D. Baye, *Phys. Lett. B* **659**, 160 (2008).
10. M. Theeten, H. Matsumura, M. Orabi, D. Baye, P. Descouvemont, Y. Fujiwara, and Y. Suzuki, *Phys. Rev. C* **76**, 054003 (2007).
11. W. Timm, H. R. Fiebig, and H. Friedrich, *Phys. Rev. C* **25**, 79 (1982).
12. H. Horiuchi, *Prog. Theor. Phys.* **51**, 1266 (1974).
13. H. Horiuchi, *Prog. Theor. Phys.* **53**, 447 (1975).
14. Y. F. Smirnov, I. T. Obukhovsky, Y. M. Tchuvil'sky, and V. G. Neudatchin, *Nucl. Phys.* **A235**, 289 (1974).
15. K. T. Kim, Y. E. Kim, D. J. Klepacki, R. A. Brandenburg, E. P. Harper, and R. Machleidt, *Phys. Rev. C* **38**, 2366 (1988).
16. R. Machleidt, Adv. Nucl. Phys. **19**, 189 (1989).
17. R. A. Brandenburg, G. S. Chulick, R. Machleidt, A. Picklesimer, and R. M. Thaler, *Phys. Rev. C* **37**, 1245 (1988).
18. A. Nogga, H. Kamada, and W. Glöckle, *Phys. Rev. Lett.* **85**, 944 (2000).
19. A. Nogga (2001), Ph. D. thesis, Ruhr-Universität Bochum (2001).
20. D. R. Entem, and R. Machleit, *Phys. Lett.* **B524**, 93 (2002).
21. S. Takeuchi, T. Cheon, and E. F. Redish, *Phys. Rev. Lett.* **B280**, 175 (1992).
22. M. Theeten, D. Baye, and P. Descouvemont, *Phys. Rev. C* **74**, 044304 (2006).

Miyazawa Supersymmetry

Sultan Catto

University Center and The Graduate School City University of New York
365 Fifth Avenue, New York, NY 10016

Abstract. Quark model with potentials derived from QCD, including the quark-diquark model for excited hadrons leading to Miyazawa symmetry between mesons and baryons gives mass formulae in very good agreement with experiment and goes a long way in explaining the approximate symmetries and supersymmetries of the hadronic spectrum, including the symmetry breaking mechanism.

Keywords: Supersymmetry, Superalgebra
PACS: 11.30.Pb 12.60.Jv 21.60.fv

Introduction: Miyazawa, in a series of papers written in 1966 and 1968 extended the $SU(6)$ group of Gürsey, Radicati [1] and Sakita [2] to the supergroup $SU(6/21)$. Elements of this supergroup transform constituent quarks and diquarks to each other. In particular, in his seminal articles [3,4], he found: *(i)* A general definition of $SU(m/n)$ superalgebras, expressing the symmetry between m bosons and n fermions, with Grassmann valued parameters, *(ii)* a derivation of the super-Jacobi identity, and *(iii)* the relation of the baryon mass splitting to the meson mass splitting through the new mass formula. This work contained the first classification of superalgebras (later rediscovered by mathematicians in the seventies). Because of the field theoretic prejudice against $SU(6)$, Miyazawa's work was generally ignored. Supersymmetry was, of course, rediscovered in the seventies within the context of the dual resonance models by Ramond; Neveu and Schwarz (1971), with the important later contributions of Gliozzi, Scherk and Olive. Golfand and Likthman (1971) and independently Volkov and Akulov (1973) proposed the extension of the Poincaré group to the super-Poincaré group. Examples of supersymmetric field theories were given and the general method based on the super-Poincaré group was discovered by Wess and Zumino (1974). The super-Poincaré group contains transformations between fields associated with different spins 0, $\frac{1}{2}$ and 1. The Coleman-Mandula theorem was amended in 1975 by Haag, Lopuszanski and Sohnius who showed that the [super-Poincaré group $\times G_{int}$] is the maximal symmetry of the S-matrix. Unfortunately, $SU(6)$ symmetry was still forbidden. There is a good phenomenological evidence that in a rotationally excited baryon a quark-diquark $(q-D)$ structure is favored over a three quark (qqq) structure. Regge trajectories for mesons and baryons are closely parallel; both have a slope of about $0.9GeV^{-2}$. At large spin two of the quarks form a diquark ($D = qq$), a bilocal object at one end of a bag, the remaining quark being at the other. For the light quarks we had shown ([5],[6],[7]) that the underlying quark-diquark symmetry leads to supersymmetric $SU(6/21)$ symmetry between mesons and baryons. This new scheme uses split octonion algebra that produces the algebraic description of color degrees of freedom, supresses color-symmetric and space-symmetric quark configurations, and leads to existence of exotic meson (diquark-

CP1011, *New Facet of Three Nucleon Force—50 Years of Fujita Miyazawa Three Nucleon Force (FM50), The International Symposium,* edited by H. Sakai, K. Sekiguchi, and B. F. Gibson

antidiquark) states for which there is now some experimental evidence. In this note we shall deal with derivation of hadronic mass formulae only. A new extended relativistic mass formulae based on octonionic supersymmetry of hadrons will be discussed in another publication ([8]).

Mass Formula: The ground state energy eigenvalue E of the Hamiltonian can be estimated by using the Heisenberg uncertainty principle. This leads to the replacement of r by Δr and p_r by

$$\Delta p_r = \frac{1}{2}(\Delta r)^{-1}, \qquad (h=1). \tag{1}$$

Then E as a function of Δr is minimized for the value of r_0 of Δr. The r_0 corresponds to the Bohr radius for the bound state. The confining energy associated with this Bohr radius is obtained from the linear confining potential $S(r) = br$, so that the effective masses of the constituents become

$$M_1 = m_1 + \frac{1}{2}S_0, \quad M_2 = m_2 + \frac{1}{2}S_0, \quad (S_0 = br_0) \tag{2}$$

For a meson m_1 and m_2 are the current quark masses while M_1 and M_2 can be interpreted as the constituent quark masses. Note that even in the case of vanishing quark masses associated with perfect chiral symmetry, confinement results in non zero constituent masses that spontaneously break the $SU(2)\times SU(2)$ symmetry of the u, d quarks.

Let us illustrate this method on the simplified spin free Hamiltonian involving only the scalar potential. In the center of mass system, $p^{(1)} + p^{(2)} = 0$, or $p^{(1)} = -p^{(2)} = p$. The semi-relativistic hamiltonian of the system is then given by

$$E_{12}\ \Phi = \sum_{i=1}^{2}\sqrt{(m_i + \frac{1}{2}br)^2 + p^2}\ \Phi. \tag{3}$$

Taking $m_1 = m_2 = m$ for the quark-antiquark system, we have

$$E_{12}\ \Phi = 2\sqrt{(m + \frac{1}{2}br)^2 + p_r^2 + \frac{\ell(\ell+1)}{r^2}}\ \Phi, \tag{4}$$

where we have written the momentum part in spherical coordinates.

Putting $b = \mu^2,\ \ \rho = \mu\, r$, for the $q-\bar{q}$ system we find E_{12} by minimizing the function

$$E_{q\bar{q}} = 2\sqrt{(m + \frac{1}{2}\mu\rho)^2 + \frac{\mu^2}{\rho^2}(\ell + \frac{1}{2})^2}. \tag{5}$$

For u and d quarks, m is small and can be neglected so that

$$E^2 = \mu^2[\rho^2 + \rho^{-2}(2\ell+1)^2] \tag{6}$$

which has a minimum for

$$\rho^2 = \rho_0^2 = 2\ell + 1, \tag{7}$$

giving

$$E^2_{min} = E^2(\rho_0) = 4\mu^2(\ell + \frac{1}{2}). \tag{8}$$

Thus, we obtain a linear Regge trajectory with

$$\alpha' = \frac{1}{4}\mu^{-2} = \frac{b}{4}. \tag{9}$$

$\alpha' = 0.88(GeV)^{-2}$ for mesons giving the value 0.54 GeV for μ. A more accurate calculation (see [4]) gives

$$\alpha' = (2\pi\mu^2)^{-1}, \quad \mu \sim 0.43 GeV. \tag{10}$$

The constituent quark mass can be defined in two ways

$$M_c(\ell) = \frac{1}{2}E_{min} = \mu\sqrt{\ell+\frac{1}{2}}, \tag{11}$$

or

$$m_c'(\ell) = S_0 = \frac{1}{2}\mu\rho_0 = \frac{\mu}{\sqrt{2}}\sqrt{\ell+\frac{1}{2}}. \tag{12}$$

The first definition gives for $\ell = 0$,

$$M_c = 0.31 GeV \quad for \quad \mu = 0.43 \tag{13}$$

in the case of u and d quarks.

When the Coulomb like terms are introduced in the simplified Hamiltonian (4) with negligible quark masses one obtains

$$E = \frac{\mu}{\rho}[-\bar{\alpha} + \sqrt{\rho^4 + (2\ell+1)^2}] \tag{14}$$

with

$$\bar{\alpha} = \frac{4}{3}\alpha_s \quad for(q\bar{q}), \quad \bar{\alpha} = \frac{2}{3}\alpha_s \quad for(qq). \tag{15}$$

In the energy range around 1 GeV, α_s is of order of unity. Estimates range from 0.3 to 3. Minimization of E gives

$$E_0 = \mu u_0^{\frac{-1}{4}}(-\bar{\alpha} + \sqrt{u_0 + (2\ell+1)^2}) \tag{16}$$

where

$$u_0(\varepsilon) = \rho_0^4 = (2\ell+1)^2(1+\frac{1}{2}\beta^2 + \varepsilon\sqrt{2}\beta\sqrt{\ell+\frac{1}{8}\beta^2}), \tag{17}$$

$$\varepsilon = \pm 1, \quad \beta = \frac{\bar{\alpha}}{(2\ell+1)}. \tag{18}$$

The minimum E_0 is obtained for $\varepsilon = -1$, giving to second order in β:

$$E_0 = \mu\sqrt{2(2\ell+1)}(1 - \frac{\beta}{\sqrt{2}} - 3\frac{\beta^2}{8}). \tag{19}$$

Linear Regge trajectories are obtained if β^2 is negligible. Then for mesons

$$E_0^2 = 4\mu^2\ell + 2\mu^2(1-\sqrt{2}\bar{\alpha}). \tag{20}$$

The β^2 is negligible for small ℓ only if we take the lowest estimate for α_s, giving 0.4 for $\bar{\alpha}$ in the $q\bar{q}$ case. For mesons with u, d constituents, incorporating their spins through the Breit term we obtain approximately

$$m_\rho \simeq m_\omega = E_0 + \frac{c}{4}, \quad m_\pi = E_0 - \frac{3c}{4}, \quad c = K\,\frac{\Delta V}{M_q^2} \tag{21}$$

where M_q is the constituent quark mass. This gives

$$E_0 = \frac{(3m_\rho + m_\pi)}{4} = 0.61 GeV. \tag{22}$$

The Regge slope being of the order of $1GeV$ an average meson mass of the same order is obtained from Eq.(19) in the linear trajectory approximation. To this approximation $\bar{\alpha}$ should be treated like a parameter rather than be placed by its value derived from QCD under varying assumptions. Using Eq.(10) for μ one gets a better fit to the meson masses by taking $\alpha_s \sim 0.2$. Turning now to baryon masses, we must first estimate the diquark mass. We have for the qq system

$$M_D = \mu(\sqrt{2} - \frac{2}{3}\alpha_s), \tag{23}$$

that is slightly higher than the average meson mass

$$\tilde{m} = \mu(\sqrt{2} - \frac{4}{3}\alpha_s). \tag{24}$$

Here we note that E is not very sensitive to the precise value of the QCD running coupling constant in the GeV range. Taking $\alpha_s \sim 0.3$ changes E^{qq} from 0.55 to $0.56GeV$.

Note that Eq.(23) gives $m_D = 0.55GeV$. For excited $q-\bar{q}$ and $q-D$ systems if the rotational excitation energy is large compared with μ, then both the m_D and the Coulomb term $-\frac{4}{3}\frac{\alpha_s}{r}$ (same for $q-D$ and $q-\bar{q}$ systems) can be neglected. Thus, for both $(q-D)$ [excited baryon] and $q-\bar{q}$ [excited meson] systems we have Eq.(8), namely

$$(E^{q-D})^2 \sim (E^{q-\bar{q}})^2 \sim 4\mu^2\ell + 2\mu^2 \tag{25}$$

giving again Eq.(9), i.e.

$$(\alpha')_{q-D} = (\alpha')_{q-\bar{q}} \cong \frac{1}{4\mu^2} \quad or \quad (\frac{1}{2\pi\mu^2}) \tag{26}$$

as an explanation of hadronic supersymmetry in the nucleon and meson Regge spectra. We also have, extrapolating to small ℓ:

$$\Delta(M^2)^{q-D} = \Delta(m^2)^{q-\bar{q}} = 4\mu^2\Delta\ell = \frac{1}{\alpha'}\Delta\ell. \tag{27}$$

For $\Delta\ell = 1$ we find

$$m_{\Delta}^2 - m_N^2 = m_{\rho}^2 - m_{\pi}^2. \tag{28}$$

This relationship is same as the one proved in our earlier paper ([1]) through the assumption that $U(6/21)$ symmetry is broken by an operator that behaves like $s = 0$, $I = 0$ member of 35×35 representations of $SU(6)$, which is true to 5%. It corresponds to a confined quark approximation with $\alpha_s = 0$.

The potential model gives a more accurate symmetry breaking ($\alpha_s \sim 0.2$):

$$\frac{9}{8}(m_{\rho}^2 - m_{\pi}^2) = m_{\Delta}^2 - m_N^2 \tag{29}$$

with an accuracy of 1%.

This mass squared formula arises from the second order iteration of the $q - D$, $q - \bar{q}$ Dirac equation. The factor $\frac{9}{8}$ comes from

$$\frac{1}{2}(\frac{4}{3}\alpha_s)^2 = \frac{8}{9}\alpha_s^2. \tag{30}$$

At this point it is more instructive to derive a first order mass formula. Since the constituent quark mass M_q is given by Eq.(11) ($\ell = 0$), we have

$$M_q = \frac{\mu}{\sqrt{2}}, \tag{31}$$

so that

$$\bar{m} = 2M_q(1 - \frac{\sqrt{2}}{3}\alpha_s) \simeq 1.9M_q. \tag{32}$$

When the baryon is regarded as a $q - D$ system, each constituent gains an effective mass $\frac{1}{2}\mu\rho_0$ which was approximately the effective mass of the quark in the meson. Hence, the effective masses of q and D in the baryon are

$$m_q' \simeq M_q, \quad m_D' = M_D + M_q \simeq 3M_q. \tag{33}$$

The spin splittings for the nucleon N and the Δ are given by the Breit term

$$\Delta M = K\Delta V \frac{S_q \cdot S_D}{m_q' m_D'}. \tag{34}$$

For the nucleon with spin $\frac{1}{2}$ the term $\mathbf{S}_q \cdot \mathbf{S}_D$ gives -1 while it has the value $\frac{1}{2}$ for Δ with spin $\frac{3}{2}$. Using the same K for mesons and baryons which are both considered to be a bound state of a color triplet with a color antitriplet we can relate the baryon splitting ΔM to the meson splitting Δm for which $\mathbf{S}_q \cdot \mathbf{S}_{\bar{q}}$ takes the values $\frac{1}{4}$ and $\frac{-3}{4}$. Hence we find

$$\Delta M = M_{\Delta} - M_N = \frac{3}{2} \cdot \frac{K\Delta V}{m_q' m_D'} = \frac{1}{2} \cdot \frac{K\Delta V}{M_q^2}, \quad \text{and} \quad \Delta m = \frac{K\Delta V}{M_q^2} \tag{35}$$

which leads to a linear mass formula $\Delta M = \frac{1}{2}\Delta m$ which is well satisfied, and has been verified before using the three quark constituents for the baryon([3]).

The formation of diquarks which behave like antiquarks as far as QCD is concerned is crucial to hadronic supersymmetry and to quark dynamics for excited hadrons. The splittings in the mass spectrum are well understood on the basis of spin-dependent terms derived from QCD. This approach to hadronic physics has led to many in depth investigations recently. For extensive references we refer to recent papers by Lichtenberg and collaborators ([9]) and by Klempt ([10]). To see the symmetry breaking effect, note that the mass of a hadron will take the approximate form

$$m_{12} = m_1 + m_2 + K\frac{\mathbf{S}_1 \cdot \mathbf{S}_2}{m_1 m_2} \tag{36}$$

where m_i and $\mathbf{S}_i$ $(i = 1,2)$ are respectively the constituent mass and the spin of a quark or a diquark. The spin-dependent Breit term will split the masses of hadrons of different spin values. If we assume $m_q = m_{\bar{q}} = m$, where m is the constituent mass of u or d quarks, and denote the mass of a diquark as m_D, then this approximation gives

$$m_\pi = (m_{q\bar{q}})_{s=0} = 2m - K\frac{3}{4m^2}, \quad m_\rho = (m_{q\bar{q}})_{s=1} = 2m + K\frac{1}{4m^2} \tag{37}$$

$$m_\Delta = (m_{qD})_{s=3/2} = m + m_D + K\frac{1}{2mm_D} \tag{38}$$

$$m_N = (m_{qD})_{s=1/2} = m + m_D - K\frac{1}{mm_D}, \tag{39}$$

Eliminating m, m_D and K, we obtain a mass relation

$$\frac{8}{3}\cdot\frac{2m_\Delta + m_N}{3m_\rho + m_\pi} = 1 + \frac{3}{2}\cdot\frac{m_\rho - m_\pi}{m_\Delta - m_N} \tag{40}$$

which agrees with experiment to 13%.

Acknowledgments: I thank Professor Miyazawa-sensei for wonderful discussions, and Professor Sakai for inviting me to this very exciting conference and his hospitality at University of Tokyo. This work is supported by DOE grants DE-AC-0276-ER3074 and 3075, and PSC-CUNY Research Awards.

REFERENCES

1. F. Gürsey and L.A. Radicati, Phys. Rev. Lett. **13** (1964) 299.
2. B. Sakita, Phys. Rev. **136** (1964) B1756.
3. H. Miyazawa, Prog. Theo. Phys. **36** (1966) 1266.
4. H. Miyazawa, Phys. Rev. **170** (1968) 1586.
5. Catto, S and Gürsey, F 1985 Nuovo Cim. 86 A, 201-218
6. Catto, S and Gürsey, F 1988 Nuovo Cim. 99A, 685-699
7. Catto, S 1994 Colored Supersymmetry of Mesons and Baryons Based on Octonionic Algebras. Symmetries in Science *VI*, 129-148. Ed. B. Gruber (Plenum Publishers)
8. Catto, S 2008 Octonionic Supersymmetry. Proceedings of the Group Theoretical Methods in Physics. Ed. by J. Birman, S. Catto and B. Nicolescu (Springer Publishers)
9. M. Anselmino et.al., Rev. Mod Phys.**65** (1993) 1199.
10. O.Klempt Baryon Resonances and strong QCD. ArXiv:nucl-ex/0203002

Study of the tensor correlation in a neutron-rich *sd*-shell region with mean-field and beyond-mean-field methods

Satoru Sugimoto*, Hiroshi Toki† and Kiyomi Ikeda**

*Kyoto University, Kyoto 606-8502, Japan
†Research Center for Nuclear Physics (RCNP), Osaka University, Osaka 567-0047, Japan
**The Institute of Physical and Chemical Research (RIKEN), Saitama 351-0198, Japan

Abstract. We study the effect of the tensor force on nuclear structure with mean-field and beyond-mean-field methods. An important correlation induced by the tensor force is a two-particle–two-hole ($2p2h$) correlation, which cannot be treated with a standard mean-field method. To treat the $2p2h$ tensor correlation, we develop a new framework [charge- and parity-projected Hartree-Fock (CPPHF) method], which is a beyond-mean-field method. In the CPPHF method, we introduce single-particle states with parity and charge mixing. The parity and charge projections are performed on a total wave function before variation. We apply the CPPHF method to oxygen isotopes including neutron-rich ones. The potential energy from the tensor force has the same order of magnitude as that from the LS force and becomes smaller with neutron number, which indicates that excess neutrons do not contribute to the $2p2h$ tensor correlation significantly. We also study the effect of the tensor force on spin-orbit-splitting (ls-splitting) in a neutron-rich fluorine isotope ^{23}F. The tensor force reduces the ls-splitting for the proton d-orbits by about 3MeV. This effect is important to reproduce the experimental value.

Keywords: tensor force, Hartree-Fock method, spin-orbit-splitting
PACS: 21.10.Dr,21.10.Pc,21.60.Jz

INTRODUCTION

The strong tensor force is characteristic of nuclear many-body systems and is related to the Fujita-Miyazawa three nucleon force through the pion. The study of nuclear matter using the Brueckner G-matrix theory demonstrated that the tensor force gives large attractive energy and plays an important role for the saturation mechanism of nuclear binding energy and density [1]. Sophisticated calculations for light nuclei with realistic nuclear forces have actually revealed that the tensor force produces a large attractive energy in finite nuclei [2]. Recently, the relation between shell evolution with neutron or proton number and the tensor force has been widely discussed [3]. The effect of the tensor force on a halo structure in ^{11}Li has also been discussed [4].

The tensor force induces two-particle–two-hole ($2p2h$) correlations, which are responsible for a large attractive energy in nuclei. The $2p2h$ tensor correlation is a kind of over-shell correlation and cannot be treated with a simple mean-field model. To treat the $2p2h$ tensor correlation, we need a framework beyond mean-field. The tensor force is mediated by the pion, which is a pseudoscalar and isovector particle. Considering these characters, we introduce single-particle states with parity and charge mixing [5, 6]. Because a total wave function consisting of the single-particle states with parity and charge

CP1011, *New Facet of Three Nucleon Force—50 Years of Fujita Miyazawa Three Nucleon Force (FM50), The International Symposium*, edited by H. Sakai, K. Sekiguchi, and B. F. Gibson

mixing, does not have good parity and a definite charge number, we perform the parity and charge projections on it before variation. By taking a variation with respect to a single-particle state, we obtain a Hartree-Fock-like equation (charge- and parity-projected Hartree-Fock equation) [7, 8]. Therefore, we call our method the charge- and parity-projected Hartree-Fock (CPPHF) method.

The CPPHF method is applied to both non-relativistic [7, 9] and relativistic calculations [8, 10]. In the present contribution, we discuss our result for oxygen isotopes including neutron-rich ones [9]. We also discuss the effect of the tensor force on spin-orbit-splitting (*ls*-splitting) in ^{23}F, which has a proton coupled to a neutron-rich nucleus ^{22}O [11].

STUDY OF OXYGEN ISOTOPES WITH THE CPPHF METHOD

Before discussing the result for the oxygen isotopes, we discuss the result for ^{16}O. Spherical symmetry is assumed in the present calculation. We use the modified Volkov force No. 1 (MV1) [12] for the central force and the G3RS force [13] for the LS and tensor forces. The Majorana parameter in the MV1 force is fixed to 0.6. We multiply the ^{3}E central force by 1.025 and the three-body force by 1.25 in the MV1 force, which are determined to reproduce the binding energy and the radius of ^{16}O in the presence of the tensor force. The LS force is multiplied by 2 to make its strength comparable to that used in standard mean-field calculations.

In Table 1, we show the results for ^{16}O with the simple Hartree-Fock (HF) and CPPHF methods. In the HF case, the potential energy from the tensor force is negligibly small, because ^{16}O has LS-closed-shell structure. In the CPPHF case, the potential energy from the tensor force becomes finite. It indicates that the CPPHF method can treat the $2p2h$ tensor correlation, which cannot be treated with the simple HF method. Even if parity mixing is allowed for single-particle states in the HF calculation, we only obtain a result with no parity-mixing. It means that the projection before variation is essential to treat the $2p2h$ tensor correlation with the CPPHF method.

The $2p2h$ tensor correlation not only affects the binding energy but also induces high-momentum components in the single-particle wave functions [7, 9]. To see the effect of the high-momentum components, the absolute value of the charge form factor for ^{16}O as a function of momentum squared is shown in Fig. 1. By comparing with the CPPHF case and the HF case, it is seen that the $2p2h$ tensor correlation enhances the charge form factor in the high momentum region. Although the charge form factor approaches the

TABLE 1. Results for ^{16}O with the HF and CPPHF methods. E and T are the total energy and the total kinetic energy respectively. V_{C}, V_{3B}, V_{T}, V_{LS} and V_{Coul} are the potential energies for the central, three-body, tensor, LS, and Coulomb forces, respectively. These are given in MeV. R_{c} is the root-mean-square charge radius in fm.

	E	T	$V_{\mathrm{C}}+V_{\mathrm{3B}}+V_{\mathrm{Coul}}$	V_{T}	V_{LS}	R_{c}
HF	-124.1	230.0	-353.2	0.0	-0.9	2.73
CPPHF	-127.1	237.1	-351.6	-11.7	-1.0	2.73

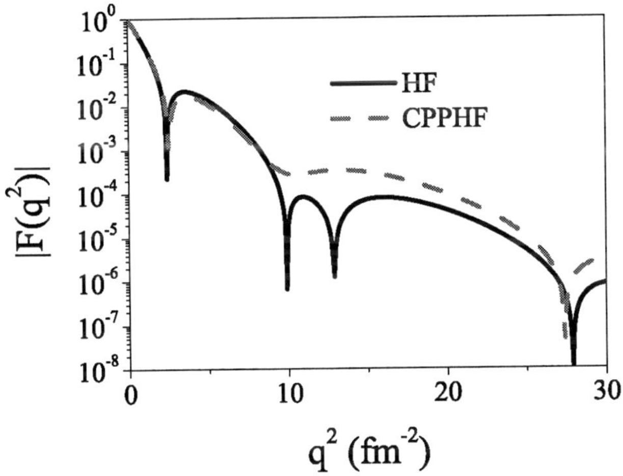

FIGURE 1. Absolute values of the charge form factor for ^{16}O with the Hartree-Fock (HF) and charge- and parity-projected Hartree-Fock (CPPHF) methods as a function of momentum squared (q^2). The solid and dashed curves correspond to the HF and CPPHF methods, respectively.

experimental data because the $2p2h$ tensor correlation in included in the CPPHF method, we need further work before comparing the present result with the experimental data, taking into account other correlations which cannot be treated with the present method such as short-range correlations and meson exchange current effects.

We apply the CPPHF method to closed-shell and closed-subshell oxygen isotopes, ^{14}O, ^{16}O, ^{22}O, ^{24}O, and ^{28}O, which are assumed to have neutron closed-shell or closed-subshell structures up to the $0p_{3/2}$, $0p_{1/2}$, $0d_{5/2}$, $1s_{1/2}$, and $0d_{3/2}$ orbits, respectively. The proton shell structure is fixed to have a closed-shell structure up to the $0p_{1/2}$ orbit. Because both parities are mixed in a single-particle state in the CPPHF method, we cannot simply label the single-particle orbits as above. But parity mixing is not so large (see Fig. 3), so that we can approximately label single-particle orbits as usual.

In Fig. 2, we show the potential energies per particle from the tensor and LS fores as a function of mass number. The potential energy from the LS force is plotted only for the CPPHF case because that for the HF case is almost the same as the CPPHF case. While the potential energy from the tensor force in the HF case is very small and repulsive, that in the CPPHF case is relatively large and attractive. In fact, the potential energy from the tensor force is comparable to that from the LS force. The potential energy from the LS force becomes negligibly small for LS-closed-shell nuclei and large for jj-closed-shell nuclei. The potential energy from the tensor force behaves differently. It becomes smaller with neutron number. Because the $2p2h$ tensor correlation is mainly induced by the ^3E tensor force and there are no protons in the *sd* shell, excess neutrons around the ^{16}O core hardly contribute to the $2p2h$ tensor correlation. It is the reason why the tensor

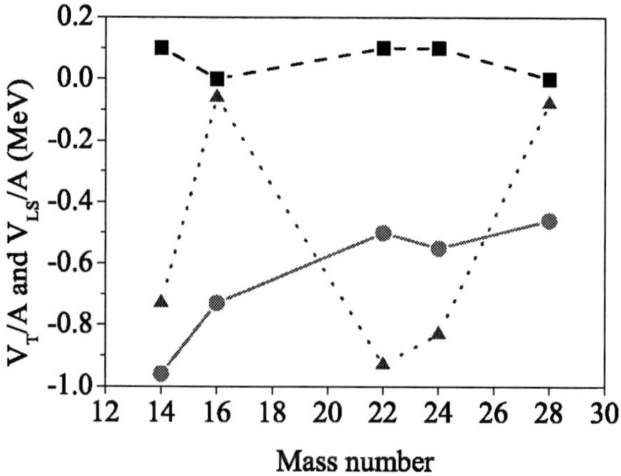

FIGURE 2. Potential energies from the tensor and LS forces in the oxygen isotopes divided by mass number. The horizontal axis indicates mass number. The circle and triangle symbols correspond to the potential energies from the tensor and LS forces calculated with the CPPHF method. The square symbols correspond to the potential energy from the tensor force calculated with the HF method.

potential energy decreases with neutron number.

In Fig. 3, the mixing probabilities of the opposite-parity components for single-particle states which have neutron components as dominant ones are shown. They are extracted from the intrinsic wave functions obtained with the CPPHF method. The mixing probability of the opposite parity component is one of the measures for the contribution of each single-particle orbit to the tensor correlation, because the tensor correlation causes the parity mixing in the CPPHF method. In ^{16}O, the mixing probabilities for the $0s_{1/2}$ and $0p_{1/2}$ orbits are around 15% and that for the $0p_{3/2}$ orbit is very small. It indicates that the tensor correlation in ^{16}O is mainly produced by the correlation involving the $0s_{1/2}$ and $0p_{1/2}$ orbits. When the neutron $0d_{5/2}$ orbit is fully occupied (^{22}O), the mixing probabilities for the orbits which are already occupied in ^{16}O do not change significantly. The mixing probability for the $0d_{5/2}$ orbit is very small. This fact means that the neutrons in the $0d_{5/2}$ orbit do not affect the tensor correlation in the ^{16}O core and do not contribute to the tensor correlation as discussed in the previous paragraph. If two neutrons are added in the neutron $1s_{1/2}$ orbit, the mixing probabilities in the ^{16}O core drastically change. Because the $0s_{1/2}$, $0p_{1/2}$, and $1s_{1/2}$ orbits have the same total angular momentum 1/2, a kind of a blocking effect works among them. It is interesting that only the two neutrons largely change the correlation in the ^{16}O core. This result indicates that the tensor correlation will be affected by excess neutrons in a neutron-rich nucleus.

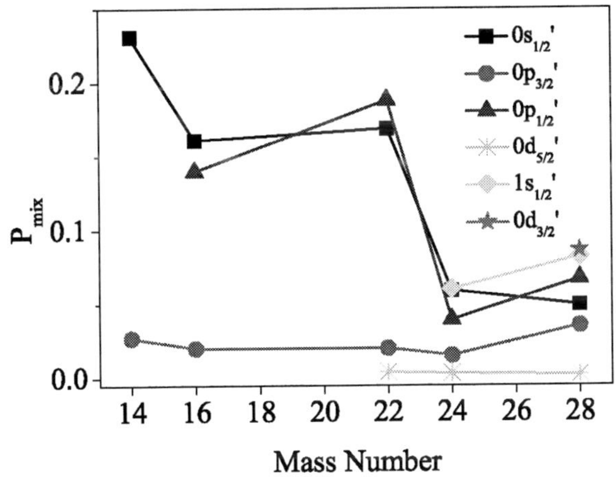

FIGURE 3. Mixing probabilities of the opposite-parity components for single-particle states which have neutron components as the dominant ones. The horizontal axis represents mass number. The square, circle, triangle, asterisk, diamond, and star symbols correspond to the first j=1/2 ($0s'_{1/2}$; $s_{1/2}$ dominant), the first j=3/2 ($0p'_{3/2}$; $p_{3/2}$ dominant), the second j=1/2 ($0p'_{1/2}$; $p_{1/2}$ dominant), the first j=5/2 ($0d'_{5/2}$; $d_{5/2}$ dominant), the third j=1/2 ($1s'_{1/2}$; $s_{1/2}$ dominant), and the second j=3/2 ($0d'_{3/2}$; $d_{3/2}$ dominant), respectively. Here, j is the total angular momentum $l+s$.

EFFECT OF THE TENSOR FORCE ON ls-SPLITTING IN ^{23}F

It is known that the tensor force or the pion field affects single-particle ls-splitting in jj-closed-shell nuclei at the mean field level [3, 14, 15]. At the mean field level, the tensor force does not influence the ls splitting in ^{16}O but has an effect on ls-splitting in ^{22}O, where the neutron $0d_{5/2}$ orbit is fully occupied. Recently, an experiment to study single-particle states in ^{23}F was performed [16]. The experiment shows that the ls-splitting for the proton $0d$ orbits are reduced by about 1MeV from that in ^{17}F. To see the effect of the tensor force on the ls-splitting in ^{23}F, we perform the Hartree-Fock calculation with the tensor force and the results are shown in Table 2. Our calculation reveals that the tensor

TABLE 2. ls splitting for the proton $0d$ orbits in ^{23}F. $\Delta(V_{\rm LS})$, $\Delta(V_{\rm T})$, and Δ(others) are the contributions to the ls splitting from the LS force, the tensor force, and the other forces including the kinetic term, respectively. Those are given in MeV. The experimental value for $\Delta(0d_{3/2}-0d_{5/2})$ = 4.06MeV [16].

	$\Delta(0d_{3/2}-0d_{5/2})$	$\Delta(V_{\rm LS})$	$\Delta(V_{\rm T})$	Δ(others)
MV1 with the tensor force	4.2	7.8	−3.1	−0.5
MV1 without the tensor force	7.2	8.3	0.0	−1.1

force reduces the ls-splitting by about 3MeV. With this reduction, the ls splitting for the proton $0d$ orbits becomes close to the experimental value. This effect is caused by the interaction between a valence proton and the excess neutrons around ^{16}O through the tensor force. Our result indicates that the tensor force will actually change shell structure in neutron-rich nuclei.

SUMMARY

We studied the effect of the tensor force with mean-field (HF) and beyond mean field (CPPHF) methods. The $2p2h$ tensor correlation induces high-momentum components in single-particle states and the potential energy from the tensor force has the same order of magnitude as that from the LS force in oxygen isotopes. The $2p2h$ tensor correlation energy becomes smaller with neutron number, because excess neutrons around the ^{16}O core do not contribute to the $2p2h$ tensor correlation. We found a kind of blocking effect on the tensor correlation, which works among single-particle orbits with the same total angular momentum.

We also studied the effect of the tensor force on ls-splitting in ^{23}F. The tensor force reduces the ls-splitting for the proton $0d$ orbits by about 3MeV. With this reduction, the ls-splitting for the proton $0d$ orbits becomes close to the experimental value.

ACKNOWLEDGMENTS

This work is supported by Grant-in-Aid for the 21st Century COE "Center for Diversity and Universality in Physics" and Grant-in-Aid for Scientific Research 19740140 from the Ministry of Education, Culture, Sports, Science and Technology (MEXT), Japan.

REFERENCES

1. H.A. Bethe, *Annu. Rev. Nucl. Sci.* **21**, 93 (1971).
2. S. C. Pieper and R.B. Wiringa, *Annu. Rev. Nucl. Part. Sci.*, **51** (2001) 53.
3. T. Otsuka, T. Suzuki, R. Fujimoto, H. Grawe, and Y. Akaishi, *Phys. Rev. Lett.* **95**, 232502 (2005).
4. T. Myo, K. Katō, H. Toki, and K. Ikeda, *Phys. Rev. C*, **76**, 024305 (2007).
5. H. Toki, S. Sugimoto, and K. Ikeda, *Prog. Theor. Phys.* **108** 903 (2002).
6. Y. Ogawa, H. Toki, S. Tamenaga, H. Shen, A. Hosaka, S. Sugimoto, and K. Ikeda, *Prog. Theor. Phys.* **111**, 75 (2004).
7. S. Sugimoto, K. Ikeda, and H. Toki, *Nucl. Phys. A* **740**, 77 (2004).
8. Y. Ogawa, H. Toki, S. Tamenaga, S. Sugimoto, and K. Ikeda, *Phys. Rev C* **73**, 034301 (2006).
9. S. Sugimoto, K. Ikeda, and H. Toki, *Phys. Rev. C* **75**, 014317 (2007).
10. Y. Ogawa, H. Toki, and S. Tamenaga, *Phys. Rev. C* **76**, 014305 (2007).
11. S. Sugimoto, H. Toki, and K. Ikeda, *Phys. Rev. C* **76** 054310.
12. T. Ando, K. Ikeda, and A. Tohsaki-Suzuki, *Prog. Theor. Phys.* **64**, 1608 (1980).
13. R. Tamagaki, *Prog. Theor. Phys.* **39**, 91 (1968).
14. R. M. Tarbutton and K. T. R. Davies, Nucl Phys. A **120**, 1 (1968).
15. M. Löpez-Quelle, N. Van Giai, S. Marcos, and L. N. Savushkin, Phys. Rev. C **61**, 064321 (2000).
16. S. Michimasa *et al.*, Phys. Lett. B **638**, 146 (2006); Nucl. Phys. A **787**, 569c (2007).

Three-body force and Coulomb force in three-body systems

S. Oryu, Y. Hiratsuka, S. Nishinohara*, S. Gojuki† and S. Chiba**

*Department of Physics, Tokyo University of Science, Noda, Chiba 278-8510, Japan
†SGI, Yebis, Shibuya-ku Tokyo, Japan
**Advanced Science Research Center, Japan Atomic Energy Research Institute, Tokai, Ibaraki, 319-1195, Japan

Abstract. We review a novel, rigorous three-body Coulomb treatment that incorporates the mesonic three-body force and the excited-nucleon three-body force such as the Fujita-Miyazawa type. Our method is mainly based on many-potential theory and the coupled-channels method. We clarify the difference between our theory and the approximate Coulomb treatment in the renormalization technique of Alt-Sandhas-Ziegelmann (ASZ), by examining the two-body and three-body equations term by term. Furthermore, we demonstrate why the ASZ method can numerically converge to a finite value regardless of its validity.

Keywords: new coupled channel three-body Faddeev-type equation, three-body force, a new rigorous Coulomb treatment
PACS: 21.45.+v,24.10.-i,25.10.+s,34.50.-s

INTRODUCTION

The standard Faddeev equations have been solved for two-body forces. If the two-body forces are reliable regarding the on-shell and off-shell behavior, any discrepancy between the numerical results and experiment is attributed to a three-body force. In the three-nucleon system, several-meson exchanges among nucleons occur in the low energy region, while at higher energies it contains the effects of the excited-nucleon states, and relativistic effects. Therefore, we have multiple origins of the three-body force (3BF). Regarding the mesonic 3BF, it comes from many-body effects which contain three nucleons plus mesons where the time-independent approximation of meson-exchanges causes a contact-type 3BF with ladder-diagrams. This is one of the mesonic 3BFs.

In the higher energy region, there exist nucleon excitations such as the Δ-isobar resonances which have masses several hundred MeV more than the nucleon mass. It was pointed out fifty years ago by Fujita and Miyazawa that the resonance and/or the resonance tail create a three-nucleon force. We would like to treat the excited-nucleon 3BF separately from the mesonic 3BF mentioned above, because the mesonic 3BF may be rather isotropic, while the massive Δ-isobar transfer could affect the kinematic angular dependence. This means that the 3BF is energy dependent. One can reproduce the n-d doublet scattering length by including a 3BF, while the problem of describing backward p-d scattering at several hundred MeV has not been resolved. We take into account the excited baryons by means of a coupled-channel Faddeev formalism.

We review our recent progress in the Coulomb problem for two- and three-body systems with the type-two 3BF. We stress the importance of the half off-shell Coulomb

CP1011, *New Facet of Three Nucleon Force—50 Years of Fujita Miyazawa Three Nucleon Force (FM50), The International Symposium,* edited by H. Sakai, K. Sekiguchi, and B. F. Gibson

amplitude in the three-nucleon system for the breakup process and also the interference with the 3BF. We include a special subsection to highlight the approximations of the ASZ-renormalization method compared with our rigorous many-potential theory. Finally, we extract and show the nucleon-excitation 3BF effect from our coupled-channel equations in the last section.

REVIEW OF THE COULOMB PROBLEM

Novel Coulomb Treatment

The long range Coulomb difficulty was rediscovered by L. D. Faddeev [1] around 50 years ago. We would like to clarify the difficulty of non-L^2 kernels and present a novel theory to overcome the difficulty, comparing with historical works. A promising approach was given in ASZ [2], however, they could not overcome the non-L^2 (non compact) problem but adopted a drastic approximation in taking the free Green's function $G_0(E) \equiv [E - H_0]^{-1}$ instead of the resolvent: $G^{\phi}(E) \equiv [E - H_0 - (V^C - V^R)]^{-1}$. It seems that $\lim_{R\to\infty} G^{\phi}(E)$ has an appropriate limit, as well as $\lim_{R\to\infty} V^R \to V^C$ without any obstacle. Unfortunately, their treatment misleads people to a false paradise where they might receive satisfactory converged numbers. One should be convinced that one cannot take those "limits" separately in the integral for the screened Coulomb potential V^R, the resolvent function $G^{\phi}(E)$, and the transition matrix $T^R(E)$. Those limits should be treated as a "trinity", not term by term, when the Lippmann-Schwinger equation is to be solved. As a consequence, the numerical method used on a computer would lead to a divergence. In computer science, we must pre-process the equations in a mathematically rigorous way to obtain an L^2-kernel, because the computer is a tool which can calculate many although finite numbers. Recently, we obtained a "theorem" by which one can solve the Coulomb problem rigorously. Here, we will review our new theory [3],[4]. Let us we start from two-potential theory, where the potential is given by $V^{(C)} = V^s + V^C = (V^s + V^R) + (V^C - V^R) = V^{(R)} + V^{\phi}$ with the short range nuclear potential V^s, and where V^C, V^R are the Coulomb and the screened Coulomb potentials, respectively. Therefore, $V^{(R)} = (V^s + V^R)$ is a kind of short range potential, and $V^{\phi} = (V^C - V^R)$ is an auxiliary potential. For these three different potentials, we would like to apply the two- (many-) potential theory; however, the two-body Lippmann-Schwinger equation is not valid for the non-L^2 kernel. Therefore, we adopt a certain finite small parameter λ for the potential V^C which means that we use $V^C(\lambda)$ instead of the genuine Coulomb potential. Therefore, $V^C(\lambda)$ is a kind of screened Coulomb potential with the range $1/\lambda > R$ for the present.

$$
\begin{aligned}
T^{(C)}(\lambda) &= V^{(C)}(\lambda) + V^{(C)}(\lambda) G_0 T^{(C)}(\lambda) &(1)\\
&= (V^{(R)} + V^{\phi}(\lambda)) + (V^{(R)} + V^{\phi}(\lambda)) G_0 T^{(C)}(\lambda) &(2)\\
&= \overline{\omega}^{\phi}(\lambda) T^{(R)\phi}(\lambda) \omega^{\phi}(\lambda) + T^{\phi}(\lambda) &(3)\\
&= \overline{\omega}^{\phi}(\lambda) \left[\overline{\omega}^{R\phi}(\lambda) t^{sR\phi}(\lambda) \omega^{R\phi}(\lambda) + T^{R\phi}(\lambda) \right] \omega^{\phi}(\lambda) + T^{\phi}(\lambda) &(4)\\
&= \overline{\omega}^{C}(\lambda) t^{sC}(\lambda) \omega^{C}(\lambda) + T^{C}(\lambda). &(5)
\end{aligned}
$$

Therefore, the two- (many-) potential theory gives us an onion-like structure in (4) where $\overline{\omega}^{\phi}(\lambda)\overline{\omega}^{R\phi}(\lambda)=\overline{\omega}^{C}(\lambda)$, $\omega^{R\phi}(\lambda)\omega^{\phi}(\lambda)=\omega^{C}(\lambda)$, and $t^{sR\phi}(\lambda)=t^{sC}(\lambda)$, with

$$\begin{aligned}
T^{C}(\lambda) &= (V^{R}+V^{\phi}(\lambda))+(V^{R}+V^{\phi}(\lambda))G_{0}T^{C}(\lambda) && (6)\\
&= \overline{\omega}^{\phi}(\lambda)T^{R\phi}(\lambda)\omega^{\phi}(\lambda)+T^{\phi}(\lambda) && (7)\\
T^{\phi}(\lambda) &= V^{\phi}(\lambda)+V^{\phi}(\lambda)G_{0}T^{\phi}(\lambda) && (8)\\
&= V^{\phi}(\lambda)(1+G_{0}T^{\phi}(\lambda))\equiv V^{\phi}(\lambda)\omega^{\phi}(\lambda) && (9)\\
&= (1+T^{\phi}(\lambda)G_{0})V^{\phi}(\lambda)\equiv\overline{\omega}^{\phi}(\lambda)V^{\phi}(\lambda) && (10)\\
T^{R\phi}(\lambda) &= V^{R}+V^{R}G^{\phi}(\lambda)T^{R\phi}(\lambda) && (11)\\
&= V^{R}(1+G^{\phi}(\lambda)T^{R\phi}(\lambda))\equiv\overline{\omega}^{R\phi}(\lambda) && (12)\\
G^{\phi}(\lambda) &\equiv \frac{1}{E-H_{0}-V^{\phi}(\lambda)} && (13)\\
T^{(R)\phi}(\lambda) &= (V^{s}+V^{R})+(V^{s}+V^{R})G^{\phi}(\lambda)T^{(R)\phi}(\lambda) && (14)\\
&= \overline{\omega}^{R\phi}(\lambda)t^{sR\phi}(\lambda)\omega^{R\phi}(\lambda)+T^{R\phi}(\lambda) && (15)\\
t^{sR\phi}(\lambda) &= V^{s}+V^{s}G^{R\phi}(\lambda)t^{sR\phi}(\lambda) && (16)\\
G^{R\phi}(\lambda) &= \frac{1}{E-H_{0}-V^{R}-V^{\phi}(\lambda)}=\frac{1}{E-H_{0}-V^{C}(\lambda)} && (17)\\
t^{sC}(\lambda) &= V^{s}+V^{s}G^{C}(\lambda)t^{sC}(\lambda). && (18)
\end{aligned}$$

From (15), (4), and (7), we can prove that the difference between two amplitudes leads to a finite value by eliminating the auxiliary amplitude $T^{\phi}(\lambda)$

$$\begin{aligned}
T^{(R)\phi}(\lambda)-T^{R\phi}(\lambda) &= \overline{\omega}^{R\phi}(\lambda)t^{sR\phi}(\lambda)\omega^{R\phi}(\lambda) && (19)\\
T^{(C)}(\lambda)-T^{C}(\lambda) &= \overline{\omega}^{\phi}(\lambda)\left[\overline{\omega}^{R\phi}(\lambda)t^{sR\phi}(\lambda)\omega^{R\phi}(\lambda)\right]\omega^{\phi}(\lambda) && (20)\\
&= \overline{\omega}^{\phi}(\lambda)\left[T^{(R)\phi}(\lambda)-T^{R\phi}(\lambda)\right]\omega^{\phi}(\lambda). && (21)
\end{aligned}$$

By adopting the theorem which was proposed in Ref. 3, one obtains for the on-shell momentum k, and a critical range R_c, $\lim_{\lambda\to 0}\lim_{R\to R_c}T^{\phi}(p,p';\lambda)=0$ for $p=k$ and/or $p'=k$, and it becomes $\lim_{\lambda\to 0}\lim_{R\to R_c}K^{\phi}(p,p';\lambda)$ for $p\neq k, p'\neq k$, where $K^{\phi}(\lambda)$ is the K-matrix for $V^{\phi}(\lambda)$. Furthermore, we take $V^{C}(p,p';\lambda)=V^{R}(p,p')$ for $(p=p')$, and $V^{C}(p,p')$ for $(p\neq p)$. By this definition, we can redefine $\tilde{V}^{\phi}(\lambda)=V^{C}(\lambda=1/R)-V^{R}=0$ for $(p=p')$, and $\tilde{V}^{\phi}(\lambda)=V^{C}(\lambda=0)-V^{R}(p,p')$ for $(p\neq p')$, respectively. Finally, we can find R_c when $\tilde{K}^{\phi}(k,k;\lambda=1/R_c)=0$. This defines a boundary on which one can solve the Coulomb-like *LS* equation with the screened Coulomb potential. It should be stressed that any screened Coulomb potential "could not be solved" for an "arbitrary finite range". Here $\tilde{K}^{\phi}$ should be used only inside of the integral. Fortunately, the two potential theory puts $\tilde{\omega}^{\phi}$, and $\overline{\tilde{\omega}}^{\phi}$ as well as $\tilde{K}^{\phi}$ inside of the integral, i.e.,

$$T^{(C)}(p,p') = \int_0^{\infty}\int_0^{\infty}\overline{\tilde{\omega}}^{C}(p,p'')\tilde{t}^{sC}(p'',p''')\tilde{\omega}^{C}(p''',p')\frac{p''^2p'''^2}{(2\pi^2)^2}dp''dp'''$$

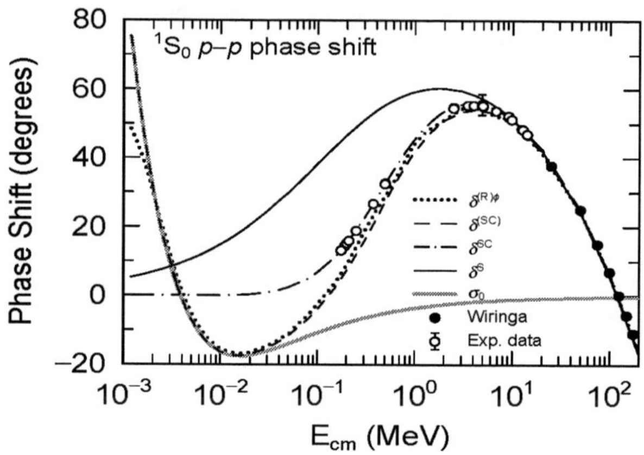

FIGURE 1. The $^1S_0$-wave proton-proton phase shift $\delta^{(R)\phi}$ by our theory (dots) for the Reid Soft Core (RSC) NN potential + the Coulomb potential is plotted along with the r-space calculational result ($\delta^{(SC)}$: dashed-line) in the energy range from the keV region to the several hundred MeV region. A simple approximation for the boundary range parameter $R_c = e^{\gamma}/2k$ is adopted. σ_0, shown by the grey curve, denotes the Coulomb phase shift. The dash-dotted line, $\delta^{SC} = \delta^{(SC)} - \sigma_0$, is obtained as the Coulomb-modified phase shift for the *p-p* $^1S_0$ state, which is to be compared with experimental data [5][6], and theoretical value[6]. The solid line denotes the phase shift for the RSC potential alone, which was obtained with a boundary condition for the short-range force. The difference between the solid line and the dash-dotted line designates the effect of Coulomb distortion for the nucleon-nucleon force.

$$+ \quad T^C(p,p') \tag{22}$$

$$\begin{aligned} T^C(p,p') &= \int_0^\infty \int_0^\infty \overline{\tilde{\omega}}^\phi(p,p'')\tilde{T}^{R\phi}(p'',p''')\tilde{\omega}^\phi(p''',p')\frac{p''^2 p'''^2}{(2\pi^2)^2}dp''dp''' \\ &+ \Big(\tilde{K}^\phi(p,p') - \tilde{V}^\phi(p,p') + V^\phi(p,p')\Big)(1-\delta_{pk})(1-\delta_{p'k}). \end{aligned} \tag{23}$$

where $\delta_{pk} = 1$ for $p = k$ and zero for $p \neq k$.

Approximations in the renormalization method of the Alt-Sandhas-Ziegelmann equation

The renormalization method which is used in the Alt-Sandhas-Ziegelmann equation utilized an approximation for (21) for the on-shell case [2][7]:

$$T^{(C)}(\lambda) - T^C(\lambda) = \overline{\omega}^\phi(\lambda)\Big[T^{(R)\phi}(\lambda) - T^{R\phi}(\lambda)\Big]\omega^\phi(\lambda) \tag{24}$$

$$\approx Z^{-1/2}\Big[T^{(R)} - T^R\Big]Z^{-1/2}, \tag{25}$$

where $Z^{-1/2} = \exp(-i\phi)$, and ϕ is the renormalization phase shift. A drastic approximation of (14), (11) was made by using G_0 instead of $G^{\phi}(\lambda)$

$$T^{(R)} = V^{(R)} + V^{(R)} G_0 T^{(R)}, \qquad T^R = V^R + V^R G_0 T^R. \tag{26}$$

This process avoids the non-L^2 kernel by an approximation. The second approximation exists in the transition from (24) to (25), where the term could be calculated by double integrals, i.e.,for the on-shell case,

$$\left\{\overline{\omega}^{\phi}(\lambda)\left[T^{(R)\phi}(\lambda) - T^{R\phi}(\lambda)\right]\omega^{\phi}(\lambda)\right\}_{kk} \equiv \int_0^{\infty}\int_0^{\infty} \frac{p^2 p'^2}{(2\pi^2)^2} dp dp'$$

$$\times \overline{\omega}^{\phi}(k,p;\lambda)\left[T^{(R)\phi}(p,p';\lambda) - T^{R\phi}(p,p';\lambda)\right]\omega^{\phi}(p',k;\lambda) \tag{27}$$

$$\approx \exp[-i\phi(k)]\left[T^{(R)}(k,k) - T^R(k,k)\right]\exp[-i\phi(k)], \tag{28}$$

where, again, a drastic approximation is made separately term by term. These are typical approximations of Alt-Sandhas-Ziegelmann theory. If one would apply eq.(28) to the off-shell and half- off-shell cases, further approximations would be indispensable. For example, the three-body breakup theory needs the half off-shell three-body amplitude $X^{(C)}_{n,m}(q_0,q';E)$, where the three-body rearrangement amplitudes given by $X^{(C)}_{n,m}(q,q';E)$, and $X^{(C)}_{n,m}(q_0,q_0;E)$ is the on-shell amplitude. The breakup amplitude is given by using the propagator τ^{sC}_m and the breakup form-factor $g^{(C)}_m$,

$$\sum_m X^{(C)}_{n,m}(q_0,q';E)\tau^{sC}_m(q';E)g^{(C)}_m(q',p') \tag{29}$$

$$\approx \sum_m \Big\{ \exp[-i\Phi(q_0)][X^{(R)}_{n,m} - X^R]_{q_0,q'}\exp[-i\Phi(q')]$$

$$+\exp[-i\Phi(q_0)]X^R_{q_0,q'}\exp[-i\Phi(q')]\Big\}\tau^{sC}_m(q';E)g^{(C)}_m(q',p') \tag{30}$$

$$\approx \sum_m \Big\{ \exp[-i\Phi(q_0)][X^{(R)}_{n,m} - X^R_{n,m}]_{q_0,q'}\exp[-i\Phi(q')]$$

$$+[X^C_{n,m}]_{q_0,q'}\Big\}\tau^{sC}_m(q';E)g^{(C)}_m(q',p'), \tag{31}$$

where the final momentum set (q',p') moves on the ellipse defined by three-body energy conservation: $q'^2/2\mu + p'^2/2\nu = E$. The rearrangement amplitude $X^{(C)}_{n,m}$ is half-off-shell.

MULTI-CHANNEL FADDEEV-LIKE EQUATION AND THE THREE-BODY FORCE EFFECT

We proposed a novel three-body equation with charged particles plus the mesonic three-body force amplitude T^0 in the Faddeev-type formulation [3]. The three-body amplitude $x^{(C)}(q,q';E)$ with the nuclear potential V^s, the Coulomb $V^C(\lambda)$-type and T^0

is given for the three-body Jacobi Channels: $\alpha : p_1(p_2n), \beta : p_2(p_1n), \gamma : n(p_1p_2)$,

$$x^{(C)}_{\alpha,\beta} = Z^{(C)}_{\alpha,\beta} + \sum_{\gamma} Z^{(C)}_{\alpha,\gamma}\tau^{sC}_{\gamma}x^{(C)}_{\gamma,\beta} \tag{32}$$

where the Born term is given by $Z^{(C)}_{\alpha,\beta} = \langle\psi^s_\alpha|[G_0^{-1}\overline{\delta}_{\alpha\beta} + (\Sigma_{\gamma\delta}\,\tilde{\delta}_{\alpha\gamma}T^C_{\gamma\delta}\tilde{\delta}_{\delta\beta} - T^C_\alpha\delta_{\alpha\beta}) + \Sigma_{\eta\gamma\delta\xi}\,\overline{\omega}^C_{\alpha\eta}\tilde{\delta}_{\eta\gamma}T^0_{\gamma\delta}\tilde{\delta}_{\delta\xi}\omega^C_{\xi\beta}]|\psi^s_\beta\rangle$, and with $1 = \delta_{\alpha\beta} + \overline{\delta}_{\alpha\beta} \equiv \tilde{\delta}_{\alpha\beta}$ [3]. We can extend the Jacobi-channels to the cases with the isobar resonances: $\alpha1 :\ p_1(p_2n)$, $\alpha2 :\ p_1(p_2\Delta^0)$, $\alpha3 :\ p_1(p_2\Delta^+)$, $\beta1 :\ p_2(p_1n)$, $\beta2 :\ p_2(p_1\Delta^0)$, $\beta3 :\ p_2(p_1\Delta^+)$, $\gamma1 :\ n(p_1p_2)$, $\gamma2 :\ \Delta^0(p_1p_2)$, $\gamma3 :\ \Delta^+(p_1p_2)$. Therefore, the two-body propagators which could survive are $\tau^{sC}_{\alpha1,\alpha1} = \tau^s_{\alpha1,\alpha1}$, $\tau^{sC}_{\beta2,\beta2} = \tau^s_{\beta2,\beta2}$, $\tau^{sC}_{\gamma3,\gamma3}$, $\tau^{sC}_{\alpha1,\alpha2} = \tau^{sC}_{\alpha2,\alpha1} = \tau^s_{\alpha2,\alpha1}$, $\tau^{sC}_{\alpha1,\alpha3} = \tau^{sC}_{\alpha3,\alpha1} = \tau^s_{\alpha3,\alpha1}$, $\tau^{sC}_{\alpha2,\alpha3} = \tau^{sC}_{\alpha3,\alpha2} = \tau^s_{\alpha3,\alpha2}$, $\tau^{sC}_{\beta1,\beta2} = \tau^{sC}_{\beta2,\beta1} = \tau^s_{\beta2,\beta1}$, $\tau^{sC}_{\gamma1,\gamma3} = \tau^{sC}_{\gamma3,\gamma1}$. The multi-channel equations are given by taking a, b, c = 1, 2, 3

$$X^{(C)}_{\alpha a,\beta b} = Z^{(C)}_{\alpha a,\beta b} + \sum_{\gamma c,d} Z^{(C)}_{\alpha a,\gamma c}\left(\tau^{sC}_{\gamma c}\delta_{cd} + \overline{\tau}^{sC}{}_{\gamma c,\gamma d}\right)2X^{(C)}_{\gamma d,\beta b} \tag{33}$$

$$X^{(C)}_{\alpha a,\beta b} = x^{(C)}_{\alpha a,\beta b} + \sum_{\gamma c,d} x^{(C)}_{\alpha a,\gamma c}\overline{\tau}^{sC}{}_{\gamma c,\gamma d}X^{(C)}_{\gamma d,\beta b} \tag{34}$$

$$X^{(C)}_{\alpha a,\beta b} = x^{(C)}_{\alpha a,\beta b} + \sum_{\gamma,c,d}\sum_{\xi,j} x^{(C)}_{\alpha a,\gamma c}\overline{\tau}^{sC}{}_{\gamma c,\gamma d}\left([1 - x^{(C)}\overline{\tau}^{sC}]^{-1}\right)_{\gamma d,\xi j}x^{(C)}_{\xi j,\beta b}, \tag{35}$$

with $\tau^{sC}_{\gamma c,\gamma c} \equiv \tau^{sC}_{\gamma c}\delta_{cd}$, $\overline{\tau}^{sC}_{\gamma c,\gamma d} \equiv \tau^{sC}_{\gamma c,\gamma d}(1 - \delta_{cd})$. The second term of (35) is the 3BF-effect. We obtained one to one correspondence between the multi-channel Faddeev formalism with the isobar system and the 3BF potential with isobar inclusion. The interference between the Coulomb interaction and the 3BF is apparent in our formalism.

REFERENCES

1. L. D. Faddeev, Sov. Phys. JETP **12**, 1014 (1961); L.D.Faddeev, S.P.Merkuriev, Quantum Scattering Theory for Several Particle Systems, Nauka pub.(1985), Kluwer Academic Pub.(1993).
2. E. O. Alt, W. Sandhas, and H. Ziegelmann, Phys. Rev. **C17**, 1981 (1978).
3. S. Oryu, Phys. Rev. **C73**, 054001 (2006).
4. S. Oryu, S. Nishinohara, N. Shiiki, and S. Chiba, **C75**, 021001 (2007).
5. J. R. Bergervoet, P. C. van Campen, W. A. van der Sanden, and J. J. de Swart, Phys. Rev. **C38**, 15 (1988).
6. R. B. Wiringa, V.G. J. Stoks and R. Schiavilla, Phys. Rev. **C51**, 38 (1995).
7. A. Fonseca, talk in this workshop (2007).

A proposal to observe a strong Van der Waals force in low energy proton-proton scattering

Tetsuo Sawada

Institute of Quantum Science, Nihon University, Tokyo, Japan

Abstract. We propose a means to test for the possible existence of a strong Van der Waals component in the nuclear force by observing the destructive interference pattern in the forward region of z in low energy proton-proton scattering.

INTRODUCTION

It is generally believed that interactions between hadrons are short range. However in the magnetic monopole model of the hadron (dyon model), hadrons are identified with the magnetic counterpart of ordinary atoms or positronium [1]. Exactly the same mechanism, as in the atomic or molecular physics, leads to the appearance of a strong Van der Waals interaction between hadrons. The starting point of the theory of the magnetic monopole is to symmetrize Maxwell's equations with respect to the interchange between the electric object and the magnetic object, and to treat electricity and magnetism on the same footing. Therefore the fundamental equations of electromagnetism become

$$div\vec{D} = 4\pi\rho \qquad rot\vec{H} - \frac{1}{c}\frac{\partial\vec{D}}{\partial t} = \frac{4\pi}{c}\vec{i}, \tag{1}$$

$$div\vec{B} = 4\pi\,{}^{\star}\rho \qquad rot\vec{E} + \frac{1}{c}\frac{\partial\vec{B}}{\partial t} = -\frac{4\pi}{c}\,{}^{\star}\vec{i}, \tag{2}$$

where ${}^{\star}\rho$ and ${}^{\star}\vec{i}$ are the magnetic charge density and its current density, respectively.

Let us consider a system in which the electric charge Q and the magnetic charge ${}^{\star}Q$ coexist. Since the angular momentum density of the electromagnetic field is $\vec{r'} \times (\vec{E}\times\vec{B})/c$, the angular momentum stored in the whole space in the form of the electromagnetic field is $(-{}^{\star}QQ/c)\hat{r}$ where $\hat{r}$ is the unit vector directed from ${}^{\star}Q$ to Q. If we remember that, in quantum mechanics, a component of the angular momentum can assume only integer multiples of $\hbar/2$, we can obtain the charge quantization condition of Dirac:

$$\frac{{}^{\star}QQ}{\hbar c} = \frac{1}{2}n \qquad \text{with} \qquad n = 0, \quad \pm 1, \quad \pm 2\cdots \tag{3}$$

From the value of the fine structure constant $e^2/\hbar c = 1/137$, we can determine the magnetic counterpart of the "fine structure constant", which is ${}^{\star}e^2/\hbar c = 137/4$. This indicates that the magnetic Coulomb force is super-strong. Therefore the magnetic monopole has a tendency to form a bound state of total magnetic charge zero, namely the "magnetic atom", and Schwinger identified such a composite state with the hadron.

CP1011, *New Facet of Three Nucleon Force—50 Years of Fujita Miyazawa Three Nucleon Force (FM50), The International Symposium,* edited by H. Sakai, K. Sekiguchi, and B. F. Gibson

It is instructive to remember the ordinary atom-atom interaction, which consists of two parts, namely the attractive Van der Waals potential and the inner repulsive core. The Lenard-Jones potential $V(r) = -c_1/r^6 + c_2/r^{12}$ is commonly used in describing such an interaction. Likewise in the dyon model of the hadron, the nuclear potential is the sum of the Van der Waals term and the inner repulsive core, which is modified by short range potentials arising from the exchange of mesons. In order to observe the strong Van der Waals force in the nuclear interaction, we shall explore the difference in the analytic structure of the scattering amplitudes.

EXTRA SINGULARITY ARISING FROM THE LONG RANGE FORCE

Because the basic magnetic Coulomb interaction is superstrong, the induced interaction such as the Van der Waals still stays order of the strong interaction. Contrary to the Yukawa potential $g^2 \exp[-\mu r]/r$, the asymptotic behavior of the Van der Waals potential is the power form, namely $V(R) \sim -C/R^{\alpha}$, where the power α is 6 for the London type and 7 for the Casimir-Polder type respectively. To see the singular behavior at $t = 0$ arising from the long range force, it is convenient to write a potential $V(r)$ as the superposition of the Yukawa potential of various force range:

$$V(r) = -\frac{1}{\pi m^2} \int_0^{\infty} dt' \rho(t') \frac{e^{-r\sqrt{t'}}}{r} \quad . \tag{4}$$

This form is generally valid because this is essentially the Laplace representation of $rV(r)$. What is important is that the weight function $\rho(t')$ is nothing but the spectral function $A_t(s,t')$ of the dispersion representation of the scattering amplitude $A(s,t)$

$$A(s,t) = \frac{1}{\pi} \int_0^{\infty} dt' \frac{A_t(s,t')}{t'-t} + \frac{1}{\pi} \int_0^{\infty} du' \frac{A_u(s,u')}{u'-u} \tag{5}$$

If we remember that the Fourier transformation of the potential $V(r)$ is the scattering amplitude in the Born approximation, and that the Fourier transformation of the Yukawa function $e^{-\sqrt{t'}r}/r$ is $4\pi/(t'-t)$, where the momentum transfer variable $\vec{q}$ relates to t by $t = -\vec{q}^{\,2}$, we can obtain the relation $A_t(4m^2, t') = \rho(t')$.

The behavior of the spectral function $A_t(4m^2, t')$ at $t' = 0$ determines the asymptotic behavior of the potential $V(r)$. In particular, if we choose $\rho(t') = \pi C' t'^{\gamma}$ in Eq.(4), the potential becomes

$$V(r) = -\frac{C'}{m^2 r} \int_0^{\infty} dt' t'^{\gamma} e^{-\sqrt{t'}r} = -\frac{2C'}{m^2 r^{2\gamma+3}} \Gamma(2\gamma+2) \quad . \tag{6}$$

When we write the asmpyotic form of the potential as $V(r) = -C/r^{\alpha}$, we get the important relations

$$\alpha = 2\gamma + 3 \qquad \textit{and} \quad C = \frac{2\Gamma(\alpha-1)}{m^2} C', \tag{7}$$

which connects the power α of the asymptotic behavior of the long range potential and the power γ of the threshold behavior at $t=0$ of the spectral function of the scattering amplitude $A_t(s,t)$. Therefore the singular behavior of the amplitude $A(s,t)$ at $t=0$ is

$$A(s,t) = -\frac{C'}{\sin \pi\gamma}(-t)^{\gamma} + \cdots, \tag{8}$$

where dots represent the back ground regular function. In particular for the Van der Waals interaction of the London type ($\alpha = 6$), the amplitude is $C'(-t)^{3/2} + \cdots$. When γ is an integer n, we must take the limit $\gamma \to n$ in estimating the singular behavior in Eq.(8). For example, when the Van der Waals potential is the Casimir-Polder type ($\alpha = 7$ namely $\gamma = 2$), the singularity becomes $A(s,t) = (C'/\pi)(-t)^2 \log(-t) + \cdots$.

There are two comments. The first one is that even we alter the spectrum $A_t(4m^2, t')$ at finite t', the asymptotic behavior of the potential does not change. Therefore the asymptotic behavior of $V(r)$ is controlled solely by the threshold behavior of the spectrum. The second one is that although the power γ of the singularities are determined by the first Born approximation, the second Born term has the singular term with the power $\gamma_2 = 2\gamma + 1$. So if $\gamma > -1$, the second Born term gives only the non-leading singularity.

EXTRA SINGULARITY OF THE PARTIAL WAVE AMPLITUDE AT $\nu = 0$

Since $-t = 2\nu(1-z)$, the leading singularity of $A(s,t)$ is $-(C'/\sin\pi\gamma)(2\nu)^{\gamma}(1-z)^{\gamma}$. The partial wave projection of this function is

$$h_\ell(\nu) = -\frac{C'}{\sin \pi\gamma}(2\nu)^{\gamma} w_\ell(\gamma) \cdots \quad with \quad w_\ell(\gamma) = \frac{1}{2}\int_{-1}^{1} dz P_\ell(z)(1-z)^{\gamma} \quad . \tag{9}$$

The explicit forms of the first few $w_\ell(\gamma)$ are $w_0(\gamma) = 2^{\gamma}/(\gamma+1)$, $w_1(\gamma) = -2^{\gamma}\gamma/(\gamma+1)(\gamma+2)$ and so on. When all the forces are short range, the behavior of the partial wave amplitude $h_\ell(\nu)$ at $\nu = 0$ is $c\nu^{\ell}$, therefore we can analyze $h_\ell(\nu)/\nu^{\ell}$ rather than $h_\ell(\nu)$, and which makes it easier to see the singularity at $\nu = 0$.

The singularity of the Van der Waals force of the London type is $C''_\ell \nu^{3/2}$ in h_ℓ. However for the P-wave we can consider $h_1(\nu)/\nu$, and its singularity behaves as $C''_1\sqrt{\nu}$. Since the coefficient C''_1 is negative, the singularity appears as a cusp which points upward at $\nu = 0$. In order to see the the extra singularity as clearly as possible, it is necessary to remove the near-by spectrum of the background function, and to prepare the wide domain of analyticity in the neighborhood of $\nu = 0$. The first thing to do is to remove the unitarity cut:

$$\frac{K_1(\nu)}{\nu} \equiv \frac{h_1(\nu)}{\nu} - \frac{1}{\pi}\int_0^L \frac{\operatorname{Im} h_1(\nu')}{\nu'(\nu'-\nu)} d\nu', \tag{10}$$

this function is called the Kantor amplitude and is regular at $\nu = 0$. The next near-by spectrum is that of the one-pion exchange, and which is removed by

$$\frac{\tilde{K}_1(\nu)}{\nu} = \frac{K_1(\nu)}{\nu} - \frac{1}{4}\frac{g^2}{4\pi}\frac{\mu^2}{2\nu^2} Q_1\left(1 + \frac{\mu^2}{2\nu}\right), \tag{11}$$

where μ is the neutral pion mass and $g^2/4\pi$ is the π-N coupling constant, which is 14.4. Since the phase shifts relate to the amplitudes by $h_\ell(\nu) = (\sqrt{m^2+\nu}/\sqrt{\nu})\exp[i\delta_\ell]\sin\delta_\ell$, the spectral function Im$h_1(\nu')$ in Eq.(10) is $(\sqrt{m^2+\nu}/\sqrt{\nu})\sin^2\delta_\ell$.

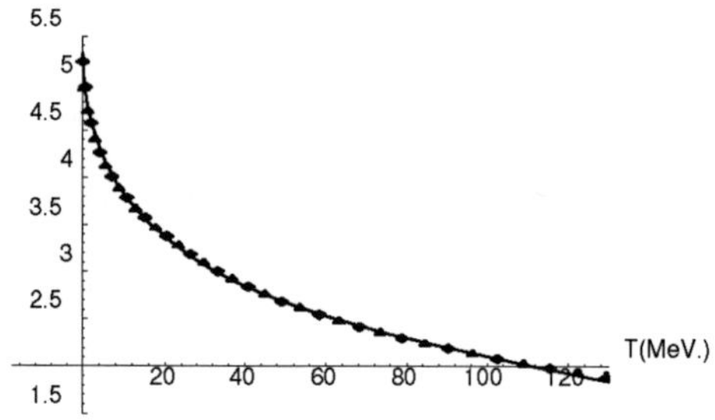

FIGURE 1. $-\tilde{K}_0^{once}(\nu)$, which is the once subtracted S-wave Kantor amplitude minus OPEC, is plotted against T_{lab}. The curve $(L)_3$ is the 3-parameter long range fit of Eq.(13). The values of the parameters of the best fit in 0.6MeV.$\leq T_{lab} \leq$125MeV. are given in Eq.(14). The long range potential is close to the Van der Waal force of the London type, because the power α of the asymtotic behavior $V(r) \sim -C/r^\alpha$ is 6.08

.

Although the Kantor amplitude of the P-wave is easy to compute, the P-wave phase shift data in the low energy region are not very precise. On the other hand, the low energy S-wave phase shifts of the proton-proton are most precise in the hadron physics. To do the same type of analysis as $h_1(\nu)/\nu$ we shall analyze the once subtracted S-wave amplitude introduced by

$$h_0^{once}(\nu) = \frac{h_0(\nu) - h_0(0)}{\nu}. \tag{12}$$

After removing the unitarity cut and the one-pion exchange term, we obtain $K_0^{once}(\nu)$. Since the S-wave phase shift is not small even in the low energy region, we need the smooth data of the energy dependent analysis to do the integration of Eq.(10). Details of such a calculation is given in a separate paper [2].

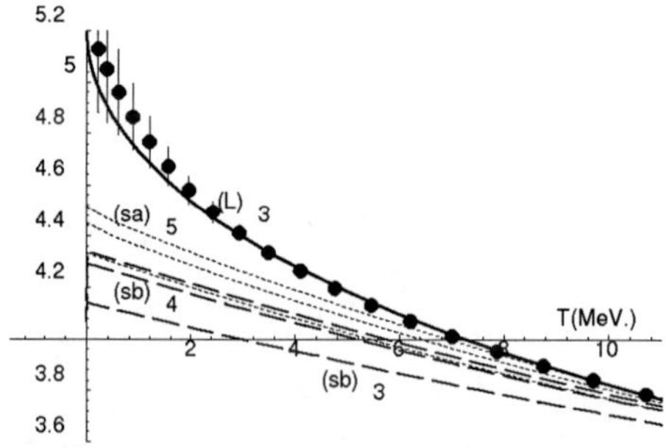

FIGURE 2. The same graph as fig.1 is plotted in $T_{lab} \leq$ 10MeV. . The curves $(sa)_n$ and $(sb)_n$ are the n-parameter short range fits, in which the spectrum $A_t(s,t)$ for two-pion exchange is represented as a sum of n delta functions. For $(sa)_n$, $A_t = \sum_{k=2}^{n+1} c_k\delta(t-k^2\mu^2)$, whereas for $(sb)_n$, $A_t = \sum_{k=3}^{n+2} c'_k\delta(t-k^2\mu^2)$. Data points in 0.6MeV.$\leq T_{lab} \leq$ 125MeV. are fitted to determine the coefficients c_k and c'_k.

In figure 1 and 2, $-K_0^{once}(\nu)$ is given with different scales. The curve $(L)_3$ is the three parameter fit of the long range force type:

$$A_t(s,t') = C' t'^{\gamma} e^{-\beta t'}, \tag{13}$$

and the parameters are determined by fitting to $-K_0^{once}(\nu)$ in the region 0.6MeV.$\leq T_{lab} \leq$125MeV.:

$$\gamma = 1.54, \quad \beta = 0.063 \quad \text{and} \quad C' = 0.35 \tag{14}$$

in the unit of the pion Compton wave length. In figure 2, n parameter short range fits $(sa)_n$ and $(sb)_n$ are also given. Contrarily to the long range fit, the short range fits cannot reproduce the data points in 0.6 MeV $< T_{lab} <$ 125 MeV well.

EXTRA SINGULARITY OF THE ANGULAR DISTRIBUTION AT $z = 1$

When we fix the incident energy T_{lab} properly, we must observe the singular behavior $(1-z)^{\gamma}$ at $z = 1$. Even if we fit it by a polynomial in $-z_1 \leq z \leq z_1$ by minimizing the maxmum deviation, the singular function must deviate abruptly from the polynomial in the small neighborhood of $z = 1$. On the other hand, since the two-pion exchange spectrum starts at $z = 1 + 4\mu^2/2\nu$ and the required extrapolation length is 0.137 on the z-plane when $\theta_1 = 30°$, first four or five terms of the polynomial are sufficient unless ν is very large. However we cannot choose ν too small, because the coefficient of $(1-z)^{\gamma}$ is proportional to $(2\nu)^{\gamma}$. Considering these conflicting requirements, I recommend to choose the incident energy in $\nu = \mu^2 \sim 3\mu^2$, which corresponds to $T_{lab} = 38.6$ MeV $\sim$ 115.8 MeV.

For the proton-proton scattering, we must include the term of the Coulomb repulsion $-e^2/t$. It is expected that the attractive Van der Waals force and the repulsive Coulomb force make the characteristic destructive interference pattern in the forward region because both forces are long range. It is convenient to write the cross section of the low energy p-p scattering, in which the one-pion exchange term is separated.

$$\nu\sigma(\theta) = \frac{1}{4}|f^s(\theta) + 2\exp[i\delta_0^E]\sin\delta_0^E + \alpha_e(z)|^2 + \tag{15}$$

$$+ \frac{3}{4}|f^t(\theta) + 6\delta_{1C} z + \alpha_o(z)|^2, \tag{16}$$

where $f^s(\theta)$ and $f^t(\theta)$ are the Coulomb plus the one-pion exchange terms of the spin-singlet and spin-triplet amplitude respectively [3].

$\alpha_e(z)$ and $\alpha_o(z)$ in Eqs.(15) and (16) are the even and odd polynomial of z respectively, which are to be determined from the cross section data in the middle region $-z_1 \leq z \leq z_1$, and we shall designate the cross section constructed in this way as $\sigma_{smooth}(\theta)$. When all the forces except that of Coulomb are short range, $\sigma_{smooth}(\theta)$ reproduces the cross section in all the physical region $-1 \leq z \leq 1$. On the other hand, when the long range force is acting whose spectral function given in Eq.(13), the amplitude must have the singularity of

$$A^{VdW}(s,t) = \frac{C'}{\pi}\int_0^\infty dt' \frac{t'^{\gamma} e^{-\beta t'}}{t'-t} \quad . \tag{17}$$

To make the cross section, we must choose $\alpha_e(z)$ and $\alpha_o(z)$ in Eqs.(15) and (16) as

$$\alpha_e(z) = A^{VdW}(s,t) + A^{VdW}(s,u) + \text{ even polynomial of z, and} \tag{18}$$
$$\alpha_o(z) = A^{VdW}(s,t) - A^{VdW}(s,u) + \text{ odd polynomial of z,} \tag{19}$$

where the coefficients of the polynomial are determined in $-z_1 \leq z \leq z_1$.

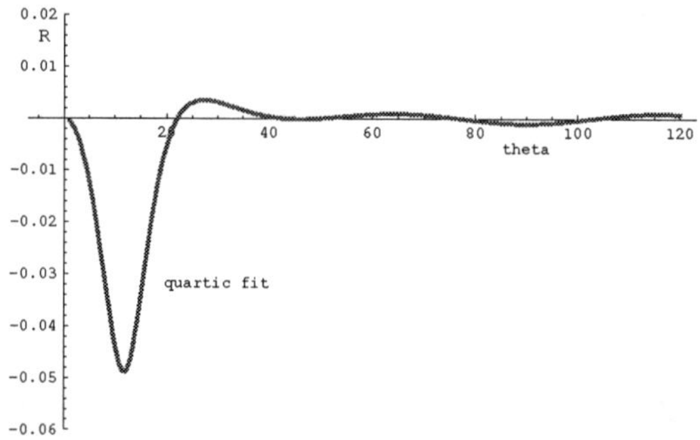

FIGURE 3. The relative deviation $R = \Delta\sigma/\sigma$ of the p-p scattering at T_{lab} =50MeV. is plotted against θ. The coefficients of the quartic polynomial of z are determined by the cross section data in the off-forward and the off-backward region $\theta = [30°, 150°]$

Since the p-p cross section is rapidly changing in the forward region, it is convenient to introduce the relative deviation $R(\theta)$ by

$$R(\theta) = \frac{\sigma^{VdW}(\theta) - \sigma_{smooth}(\theta)}{\sigma_{smooth}(\theta)} \quad . \tag{20}$$

Figure 3 is the plot of the relative deviation $R(\theta)$ against θ, the scattering angle in the center of mass system. The incident energy is fixed at $T_{lab} = 50$MeV., and the fitting region is $30° \leq \theta \leq 150°$. We can see a narrow dip at $\theta = 12°$ with the depth around 5%. This is the prediction on the angular distribution of the cross section, in which the long range parameters of Eq.(14) are used. At present there are no sufficient experimental data in the forward region [4]. Our proposal is to measure the p-p cross section with the precision $0.1 \sim 0.2\%$ in the middle region and 0.5% in the forward region in order to confirm the existence of the strong Van der Waals interaction in the nuclear force.

REFERENCES

1. J.Schwinger, Science **165**, 757, (1969); A.O.Barut, Phys.Rev. **D3**, 1747, (1971); T.Sawada, Phys.Lett. **B43**, 517, (1973).
2. T.Sawada, Int. Journ. Mod. Phys. **A11**, 5365, (1996); T.Sawada, "Present status of the long range component in the nuclear force", preprint NUP-A-2000-10, (2000).
3. T.Sawada and D.Kiang, Nuovo Cimento **77A**, 308, (1983).
4. A.Berdoz, F.Foroughi and C.Nussbaum, J. Phys. **G12**, L133, (1986).

ΛNN Three-Body Force due to Coherent Λ-Σ Coupling

Yoshinori Akaishi*,† and Khin Swe Myint**

*RIKEN Nishina Center, Wako 351-0198, Japan
†College of Science and Technology, Nihon University, Funabashi 274-8501, Japan
**Mandalay University, Mandalay, Union of Myanmar

Abstract. The overbinding problem of ${}^{5}_{\Lambda}$He is solved by introducing a concept of coherent Λ-Σ coupling which is equivalent to a ΛNN three-body force. This three-body force is coherently enhanced in the 0^+ states of ${}^{4}_{\Lambda}$H and ${}^{4}_{\Lambda}$He. The 0^+-1^+ splitting in these hypernuclei is mainly due to coherent Λ-Σ coupling and partly due to the ΛN spin-spin interaction. A ΛNN three-body potential is derived from the coupled-channel treatment. The origin of the repulsive and attractive nature of the three-body force is discussed. Coherent Λ-Σ coupling becomes more important in neutron-rich hypernuclei and especially in neutron-star matter at high densities. The possible existence of "hyperheavy hydrogen", ${}^{6}_{\Lambda}$H, is suggested.

Keywords: Three-body force, hypernuclei, neutron-rich nuclei

INTRODUCTION

There has been a long standing problem in reproducing systematically the experimental Λ-separation energies of ${}^{3}_{\Lambda}$H, ${}^{4}_{\Lambda}$H, ${}^{4}_{\Lambda}$He and ${}^{5}_{\Lambda}$He. It is the so-called overbinding problem of ${}^{5}_{\Lambda}$He. A ΛN interaction which reproduces the 3-body and 4-body hypernuclei overbinds ${}^{5}_{\Lambda}$He seriously. Dalitz *et al.* [1] carefully investigated this problem by using Nogami's ΛNN three-body force [2] of the Fujita-Miyazawa type [3]. They, however, could not solve the problem due to the lack of coherence of the three-body force.

In this paper we propose a solution to the problem by combining different treatments in two pictures [4]. One is a "single-channel" picture where the hyperon-nucleon (YN) channel is truncated to only the ΛN space interacting with a ΛN potential. The other is a "two-channel" picture where ΛN and ΣN spaces are connected by a ΛN-ΣN coupled-channel potential. Both the potentials are adjusted to reproduce the same ΛN phase shifts. In the two-channel picture the suppression of two-body ΛN-ΣN coupling solves the overbinding problem of ${}^{5}_{\Lambda}$He but it, in turn, causes a serious underbinding of the four-body hypernuclei. The shortfall in this binding is overcome by the introduction of Λ-Σ coupling which is equivalent to a ΛNN three-body force. This three-body force becomes sufficiently strong in the 0^+ states of ${}^{4}_{\Lambda}$H and ${}^{4}_{\Lambda}$He due to coherent enhancement. In Section 2 we solve the so-called overbinding problem of ${}^{5}_{\Lambda}$He (single-channel picture), or the underbinding problem of ${}^{4}_{\Lambda}$He (two-channel picture) by introducing a concept of coherent Λ-Σ coupling. In Section 3 a ΛNN three-body potential is constructed on the basis of the two-channel picture. The origin of the repulsive and attractive nature of the ΛNN three-body force is clearly explained.

The ΛNN three-body force due to coherent Λ-Σ coupling becomes more important

CP1011, *New Facet of Three Nucleon Force—50 Years of Fujita Miyazawa Three Nucleon Force (FM50), The International Symposium,* edited by H. Sakai, K. Sekiguchi, and B. F. Gibson

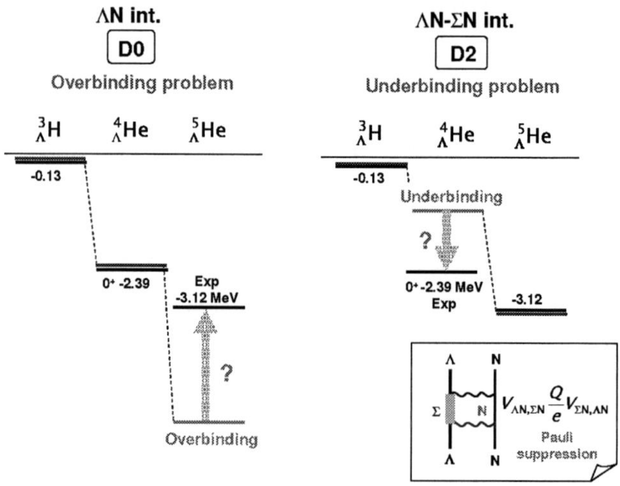

FIGURE 1. Two pictures for s-shell hypernuclei. The problem is "overbinding" of ${}^{5}_{\Lambda}$He in the D0 picture and "underbinding" of ${}^{4}_{\Lambda}$He in the D2 picture.

in neutron-star matter at high densities. The coherent coupling is nothing but a result of the mean field which causes isospin mixing of hyperons, $\Lambda(I=0)$ and $\Sigma(I=1)$, by the strong interaction. Some experimental information on isospin mixing with $(N-Z)^2$ collectivity could be obtained from neutron-rich hypernuclei. A large ΛNN three-body force effect is expected in ${}^{6}_{\Lambda}$H. The possible existence of ${}^{6}_{\Lambda}$H and ${}^{10}_{\Lambda}$Li is discussed in Section 4.

COHERENT Λ-Σ COUPLING IN *S*-SHELL HYPERNUCLEI

Let's consider the overbinding problem of ${}^{5}_{\Lambda}$He. We introduce two pictures as stated in the Introduction. One is the "single-channel" picture: only the ΛN space is considered with the ΛN potential, D0. The other is the "two-channel" picture, where ΛN and ΣN spaces are connected via the ΛN-ΣN coupled-channel potential, D2. Both the potentials are detailed in Ref. [4] and are adjusted so as to reproduce the ΛN phase shifts obtained from the Nijmegen D potential [5]. The Nijmegen SC potentials [6] are also used to check more realistic cases.

The Λ separation energies are calculated in the usual way by the Brueckner-Hartree-Fock (BHF) method. The D0 potential overbinds ${}^{5}_{\Lambda}$He, as shown in Fig. 1, just as Dalitz found 35 years ago. On the other hand, the D2 potential reproduces well the experimental Λ-separation energy in ${}^{5}_{\Lambda}$He, where Pauli suppression of the two-body ΛN-ΣN coupling solves the overbinding problem of ${}^{5}_{\Lambda}$He. However, the same suppression causes, in turn, serious underbinding of the Λ in the four-body hypernuclei. Thus, the problem is "overbinding" of ${}^{5}_{\Lambda}$He in the "single-channel" D0 picture, while it is "underbinding" of ${}^{4}_{\Lambda}$H and ${}^{4}_{\Lambda}$He in the "two-channel" D2 picture.

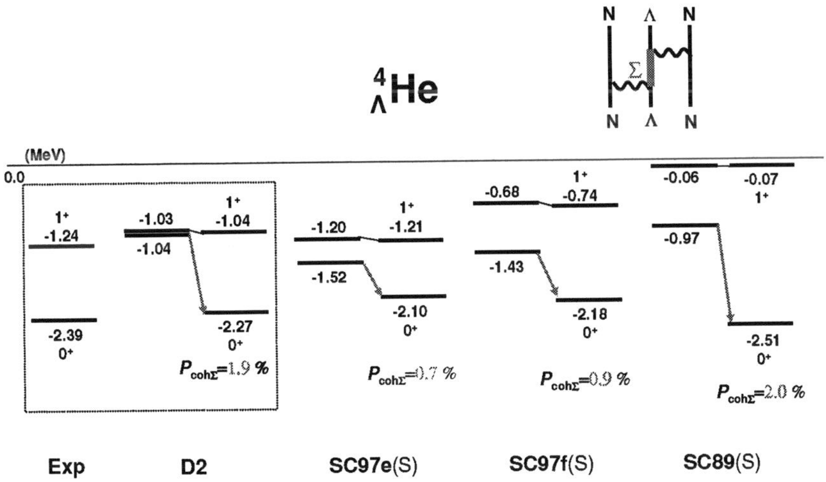

FIGURE 2. Λ energy levels calculated with the D2 and realistic Nijmegen potentials. The level shift shown by arrows is mainly due to the ΛNN three-body force coming from coherent Λ-Σ coupling.

We approach the problem from the two-channel D2 picture side. The shortfall in the binding is overcome by the ΛNN three-body force arising from the ΛN_1-ΣN_2 coupling which is not included in a usual BHF treatment. We divide the Λ-Σ coupling into incoherent and coherent parts. The process, $\Lambda N \to \Sigma N_{\text{excited}}$, is called "incoherent coupling" since the difference of N-hole states destroys the coherence, and it is incorporated into reaction matrix (g-matrix) of the BHF framework. The process, $\Lambda N \to \Sigma N_{\text{unexcited}}$, is referred to as *"coherent coupling"*, where a nucleon converts a Λ into a Σ and remains in its original level, giving all the other nucleons equal chance to interact with the converted Σ. Then, three nucleons in the ground-state configuration of $^4_\Lambda$H can coherently contribute to the Λ-Σ coupling.

This coherent effect is taken into account by introducing Σ-space components explicitly into the wave function of $^4_\Lambda$H [7].

$$|^4_\Lambda\text{H}> = \Phi_\Lambda(\vec{r})|^3\text{H}> - \sqrt{\frac{2}{3}}\Phi_{\Sigma^-}(\vec{r})|^3\text{He}> + \sqrt{\frac{1}{3}}\Phi_{\Sigma^0}(\vec{r})|^3\text{H}> . \tag{1}$$

Then the hyperon-nucleus mean-field potential in the four-body hypernucleus has a Λ-Σ coupling term, to which the two-body ΛN-ΣN effective interaction contributes as;

$$\begin{aligned} &\frac{3}{2}g_{\Sigma N,\Lambda N}(^3\text{E}) - \frac{1}{2}g_{\Sigma N,\Lambda N}(^1\text{E}) \quad \text{for} \quad 0^+, \\ &\frac{1}{2}g_{\Sigma N,\Lambda N}(^3\text{E}) + \frac{1}{2}g_{\Sigma N,\Lambda N}(^1\text{E}) \quad \text{for} \quad 1^+, \end{aligned} \tag{2}$$

where $g(^3\text{E})$ and $g(^1\text{E})$ are reaction matrices in the triplet-even and singlet-even states respectively. Since the ΛN-ΣN coupling interaction is much stronger in the ^{3}E state than in the ^{1}E state, the Λ-Σ coupling mean-field, $U_{\Lambda\Sigma}$, is approximately three times stronger in the 0^+ state than in the 1^+ state.

The results obtained for ${}^4_\Lambda$He are shown in Fig. 2. Without $U_{\Lambda\Sigma}$ the 0^+-1^+ level splitting is too small. Then, the coupling $U_{\Lambda\Sigma}$ term pushes down the 0^+ level close to the experimental one. Charge-base consideration gives us detailed information about the coherence mechanism. In the 0^+ state "proton-induced" transition, $\Lambda p \rightarrow \Sigma N$, and "neutron-induced" transition, $\Lambda n \rightarrow \Sigma N$, are coherently added to form a strong $U_{\Lambda\Sigma}$ mean field. On the other hand, in the 1^+ state both transitions are destructively summed. Since the energy gain is proportional to the square of $U_{\Lambda\Sigma}$, the attractive effect due to the coherent Λ-Σ coupling becomes one order of magnitude larger in the 0^+ state than in the 1^+ state. Thus, we find that the coherently enhanced Λ-Σ coupling solves the underbinding problem of ${}^4_\Lambda$He. This physical picture was confirmed by Nemura *et al.* [8] in full *ab initio* calculations of 3-5 body *s*-shell hypernuclei including Σ degrees-of-freedom with realistic *YN* interactions.

Evidence for coherent Λ-Σ coupling is obtained from the Σ hypernucleus, ${}^4_\Sigma$He [9, 10]. Harada carried out comprehensive analyses and succeeded in fitting various production spectra consistently over a wide range from the Λ region to the Σ region [11]. The coherent Λ-Σ coupling potential plays a crucial role in reproducing the spectra.

PROPERTIES OF THE Λ*NN* THREE-BODY FORCE

The significance of Λ-Σ coupling in hypernuclei has long been recognized by many authors [12, 2, 7, 4]. The Λ*NN* three-body force due to Λ-Σ coupling was introduced historically to be repulsive to remedy the overbinding of ${}^5_\Lambda$He. The same three-body force was found in the previous section to be strongly attractive in ${}^4_\Lambda$He(0^+). Now, the problem is how to reconcile the repulsive nature and the attractive nature of the Λ*NN* three-body force.

Figure 3 shows some diagrams in the two-channel and single-channel pictures, which contribute to the interaction energy of ${}^4_\Lambda$He. The single-channel diagram (a) includes the two-channel diagram (b) which gives a negative-definite contribution. However, diagram (b), which is a main cause of the overbinding appearing in the single-channel calculation, must be excluded from the calculation since it violates the Pauli exclusion principle between two spin-up protons in the Σ intermediate state. Diagram (c), which is the exchange of the two protons of the diagram (b) and is therefore positive-definite, is the origin of the repulsive Λ*NN* three-body force ("exclusive 3BF") as discussed by Nogami [2]. By the introduction of the "exclusive 3BF" in the single-channel calculation, the unphysical contribution of diagram (b) is cancelled.

If we antisymmetrize not only the Λ*NNN* initial state but also the Σ*NNN* intermediate states, the Pauli forbidden diagrams (b) and (c) cancel each other. But the Pauli allowed diagrams (d) and (e) survive; they are nothing but the coherent Λ-Σ coupling. Diagram (e) is the origin of the attractive Λ*NN* three-body force ("coherent 3BF") discussed in the previous section. This attractive Λ*NN* three-body force is the essential result of the Λ-Σ coupling.

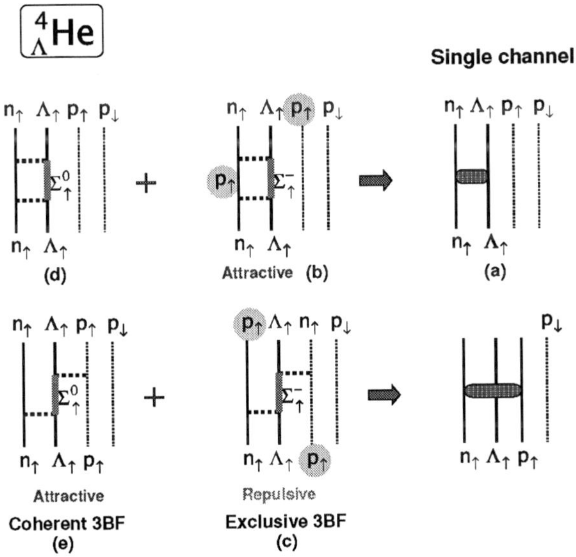

FIGURE 3. Diagrams representing two-body and three-body forces due to Λ-Σ coupling where subscript arrows show spin up or down. The origin of the exclusive and the coherent ΛNN three-body forces is depicted.

Spin dependence of the ΛNN three-body force

The Λ-Σ coupling effect is equivalently described by a ΛNN three-body potential. We derive the explicit form of the three-body potential. We consider the coherent Λ-Σ coupling where all nucleons always remain in the ground-state configuration. The following spin and isospin operators, $\vec{\sigma}_1\vec{\sigma}_2$, $\vec{\sigma}_\Lambda(\vec{\sigma}_1+\vec{\sigma}_2)$, $\vec{\tau}_1\vec{\tau}_2$, $(\vec{\tau}_1\vec{\tau}_2)(\vec{\sigma}_1\vec{\sigma}_2)$, and $(\vec{\tau}_1\vec{\tau}_2)$ $(\vec{\sigma}_\Lambda\ (\vec{\sigma}_1\ +\ \vec{\sigma}_2\ ))$ are used to represent the three-body ΛN_1N_2 potential. For s-shell hypernuclei the nucleon isospin operators are no longer independent of spin operators as seen from the relation between spin-exchange and isospin-exchange operators,

$$\frac{1+(\vec{\tau}_1\vec{\tau}_2)}{2}\frac{1+(\vec{\sigma}_1\vec{\sigma}_2)}{2}=-1. \tag{3}$$

Thus, the ΛNN three-body force can be completely represented by the remaining spin operators as

$$U_{\Lambda\mathrm{NN}}=\sum_{\alpha,\beta}W_3^{\alpha\beta}(r_{1\Lambda},r_{\Lambda 2})[a_{\alpha\beta}+b_{\alpha\beta}(\vec{\sigma}_1\vec{\sigma}_2)+c_{\alpha\beta}\frac{1}{2}\vec{\sigma}_\Lambda(\vec{\sigma}_1+\vec{\sigma}_2)], \tag{4}$$

$$W_3^{\alpha\beta}=V^{\alpha}_{\Lambda\mathrm{N},\Sigma\mathrm{N}}(r_{1\Lambda})\frac{1}{\bar{e}}V^{\beta}_{\Sigma\mathrm{N},\Lambda\mathrm{N}}(r_{\Lambda 2})+V^{\beta}_{\Lambda\mathrm{N},\Sigma\mathrm{N}}(r_{1\Lambda})\frac{1}{\bar{e}}V^{\alpha}_{\Sigma\mathrm{N},\Lambda\mathrm{N}}(r_{\Lambda 2}), \tag{5}$$

where $\alpha, \beta = t$(spin triplet) or s(spin singlet), and $\bar{e}$ is an average excitation energy. The spin coefficients are

$$\begin{pmatrix} a_{tt} & b_{tt} & c_{tt} \\ a_{ts} & b_{ts} & c_{ts} \\ a_{ss} & b_{ss} & c_{ss} \end{pmatrix} = \begin{pmatrix} \frac{7}{16} & \frac{3}{16} & \frac{3}{8} \\ \frac{1}{8} & \frac{1}{8} & -\frac{1}{4} \\ \frac{5}{48} & \frac{1}{48} & -\frac{1}{8} \end{pmatrix}. \tag{6}$$

This force structure is consistent with that of Dabrowski's three-body force [13] which was derived by perturbation treatment for a phenomenological YN interaction. Now, the feature of repulsion and attraction of the present ΛNN three-body force are shown quantitatively as follows: the three-body force contributes approximately 3.0 MeV repulsion to ${}^{5}_{\Lambda}$He, 1.0 MeV repulsion to ${}^{4}_{\Lambda}$He(1^+), -0.44 MeV attraction to ${}^{4}_{\Lambda}$He(0^+) and -0.05 MeV attraction to ${}^{3}_{\Lambda}$H, playing an essential role in solving the "overbinding / undebinding" problem of the s-shell hypernuclei.

Here, a comment should be made. The ΛNN three-body force having the coefficients of Eq.(6) must be used together with a ΛN two-body interaction in the single-channel picture. If one uses the g-matrix of a realistic YN interaction as a ΛN two-body effective interaction, one must remove the "exclusive 3BF" which is already taken into account in g as a Pauli suppression effect, and employ the following spin coefficients instead of Eq.(6),

$$\begin{pmatrix} a_{tt} & b_{tt} & c_{tt} \\ a_{ts} & b_{ts} & c_{ts} \\ a_{ss} & b_{ss} & c_{ss} \end{pmatrix} = \begin{pmatrix} \frac{(7-6\gamma)}{16} & \frac{3}{16} & \frac{(3-\gamma)}{8} \\ \frac{1}{8} & \frac{1}{8} & -\frac{1}{4} \\ \frac{(5-6\gamma)}{48} & \frac{1}{48} & -\frac{(1-\gamma)}{8} \end{pmatrix} \tag{7}$$

with γ = (number of ΛN pairs) / (number of ΛNN triplets). Then, the above-mentioned three-body force contribution changes as 3.0 MeV $\to$ 0 for ${}^{5}_{\Lambda}$He and -0.44 MeV $\to$ -1.2 MeV for ${}^{4}_{\Lambda}$He(0^+), which is the gain from the essential dynamics of the "coherent 3BF".

The most characteristic part of the ΛNN three-body force of Eq.(4) is the spin-spin term of $\vec{\sigma}_\Lambda(\vec{\sigma}_1 + \vec{\sigma}_2)$ coming just from the coherent Λ-Σ coupling. The importance of this term has been also stressed by Bodmer *et al.* [14], though the origin of Bodmer-Usmani's three-body force is the dispersive effect which is different from the coherent Λ-Σ coupling. Calculations with the ΛNN force of Eq.(4) show that the major part of the energy splitting between 0^+ and 1^+ states of ${}^{4}_{\Lambda}$He is attributed to the spin-spin term of the three-body force, which is 62% / 100% in the case of SC97f(S) / D2 shown in Fig. 2. This is in agreement with 70% obtained from Bodmer-Usmani's three-body force [14]. The present result justifies Gibson's statement [15] that the 0^+-1^+ difference is not a measure of ΛN two -body spin-spin interaction.

NEUTRON-RICH HYPERNUCLEI

If more neutrons are added to the 0^+ state of ${}^{4}_{\Lambda}$H, the coherent Λ-Σ coupling effect would be much more enhanced by the extra neutrons. In this respect we investigate neutron-rich hypernuclei and also neutron-star matter.

Figure 4 demonstrates the coherence mechanism of a Λ plus core-nucleus system. With regard to Λ-Σ coupling, the T(isospin)=0 nuclear core is inactive because of the

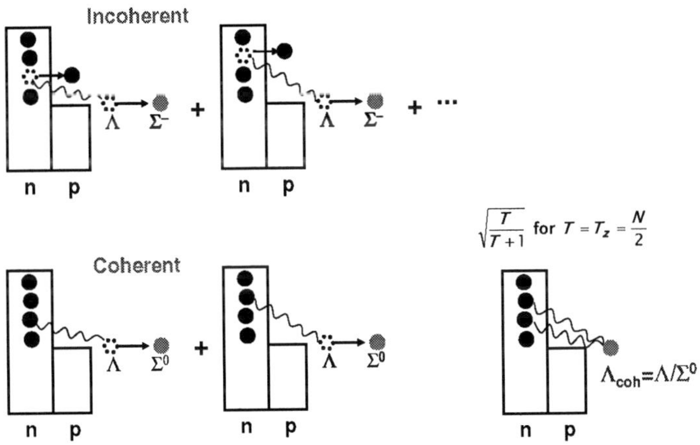

FIGURE 4. Λ-Σ coupling in a neutron-excess nucleus. In the case of Λ-Σ^0 the transition *amplitudes* are coherently added on a common nuclear-core state, bringing an effect with $(N-Z)^2$ collectivity.

isospin-selection rule. In the case of a $T \neq 0$ nuclear core, however, the $(N-Z)$ excess neutrons can collectively contribute to the Λ-Σ mixing. The collectivity is proportional to $(N-Z)$ for Σ^- mixing and to $(N-Z)^2$ for Σ^0 mixing. In the latter case we can factorize out the nuclear core as a common factor and regard the Λ/Σ^0 mixed particle to be a quasi-Λ particle, Λ_{coh}, since its $(N-Z)$ transition amplitudes are coherently added. Thus, a Λ particle in a $T \neq 0$ nuclear medium is no longer an I(intrinsic isospin)=0 object but an I-mixed quasi-Λ particle.

In neutron matter, all neutrons contribute coherently to the Λ/Σ^0 particle mixing, which would give a large effect especially at high densities. Such an effect, however, is neglected in the usual treatments of the Λ appearance in neutron-star matter [16, 17, 18]. The s-shell "overbinding / underbinding" problem tells us that the coherent Λ/Σ^0 mixing effect should be seriously taken into account. Shinmura *et al.* [19] investigated this Λ/Σ^0 particle mixing caused by the strong interaction with the use of the Nijmegen SC potential [6]. They found a large amount of Σ^0 admixture accompanying Λ hyperons in neutron-star matter at densities of 2∼3 times ρ_0. Then, the threshold density for Λ appearance becomes much lower than the usual. In order to draw any conclusion, however, we need more experimental data which could be obtained from neutron-rich hypernuclei. As a remark we mention that the neutral Σ^0 starts to appear in neutron-star matter at the same density as the Λ appears, and the Σ^0/Λ ratio grows rapidly toward 1 at high densities. The coherent Λ-Σ coupling takes place only in asymmetric $(N \neq Z)$ nuclear matter and it disappears in symmetric nuclear matter due to the cancellation between neutron-induced and proton-induced $\Lambda \to \Sigma$ transitions.

Now, let's discuss light neutron-rich hypernuclei. A superheavy hydrogen, ^{5}H, is populated by the ^{1}H (^{6}He, ^{2}He) ^{5}H reaction as a low-lying resonance at 1.7 MeV above the ^{3}H + 2n threshold [20]. When a Λ is added, it becomes $^4_\Lambda$H + 2n where the Λ-Σ coupling is much enhanced by the excess neutrons. Then, there could appear

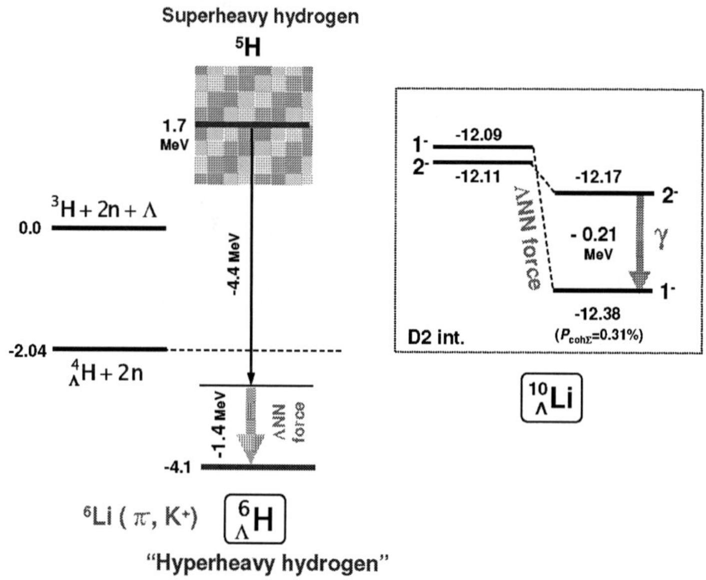

FIGURE 5. Energy levels of ${}^{6}_{\Lambda}$H(0^+) and ${}^{10}_{\Lambda}$Li(1^-,2^-) are shown together with the coherent Λ-Σ coupling effect, *i.e.* the ΛNN three-body force effect.

a particle-stable bound state. The present authors have predicted the possible existence of "hyperheavy hydrogen", ${}^{6}_{\Lambda}$H, as follows [21]. In the calculation the BHF method is used with the potential SC97f(S) shown in Fig. 2. The Λ-Σ coupling mean-field, $U_{\Lambda\Sigma}$, is obtained by folding a sum of interactions from ${}^{4}_{\Lambda}$H (we consider it as a "seed") and from two p-shell neutrons which is given as

$$\frac{1}{2}\sqrt{\frac{1}{6}}\left[3\{g_{\Sigma N,\Lambda N}(^3\mathrm{E})+g_{\Sigma N,\Lambda N}(^3\mathrm{O})\}+\{g_{\Sigma N,\Lambda N}(^1\mathrm{E})+g_{\Sigma N,\Lambda N}(^1\mathrm{O})\}\right]. \qquad (8)$$

The total Λ binding energy of the 0^+ state from the ^{5}H + Λ threshold is found to be 5.8 MeV with a 1.6% Σ admixture due to coherent Λ-Σ coupling. It is a particle stable bound state lying 2.1 MeV below the threshold of ${}^{4}_{\Lambda}$H + 2n as seen in Fig.5. When the Λ-Σ coupling term is turned off, the binding energy is reduced to 4.4 MeV and the state is only 0.7 MeV below the threshold. The Λ-Σ coupling force gives 1.4 MeV attraction for the ground state which is about two times larger than the coupling energy of the 0^+ state of ${}^{4}_{\Lambda}$H . In the case of the 1^+ state, the binding energy is 3.4 MeV and the 0^+-1^+ splitting becomes about two times larger than that of ${}^{4}_{\Lambda}$H. A possible reaction to populate ${}^{6}_{\Lambda}$H is the double-charge exchange reaction, ^{6}Li (K^-, π^+). Its observation would provide experimental evidence for much larger effects in neutron-star matter.

The ${}^{10}_{\Lambda}$Li nucleus can be modeled as ${}^{6}_{\Lambda}$H + α where an α cluster stabilizes the system. The BHF calculation predicts that the ground state becomes 1^- due to coherent Λ-Σ coupling. The first pioneering experiment of ^{10}B (π^-, K^+) ${}^{10}_{\Lambda}$Li has been already conducted by Saha *et al.* at KEK, obtaining a hopeful result [22].

CONCLUDING REMARKS

We have derived the ΛNN three-body force due to coherent Λ-Σ coupling by using the D2 potential which simulates the realistic Nijmegen D potential, and presented a reconcilable account of the repulsive and attractive nature of the ΛNN three-body force. In the case of ${}^5_\Lambda$He the ΛNN force is largely repulsive and its origin comes from Pauli forbidden terms in the truncated Σ space. There is no Pauli allowed term in the ground-state configuration of ${}^5_\Lambda$He due to the isospin selection rule. This is the reason why the ΛNN force is repulsive. On the other hand, in the 0^+ ground state of ${}^4_\Lambda$He the attractive three-body force effect is collectively enhanced due to the coherence of the Λp pair contributions [4].

A large attractive effect of the ΛNN force due to coherent Λ-Σ coupling is expected in neutron-rich hypernuclei where contributions from excess neutrons are added coherently, while in $N = Z$ symmetric-core nuclei the contributions from protons and neutrons cancel. In high-density neutron-star matter the attractive coherent Λ-Σ coupling has a chance to be strongly enhanced [19]. The attractive ΛNN force is the important result of Λ-Σ coupling dynamics in an $N \neq Z$ asymmetric nuclear medium, while the repulsive three-body force is of kinematic origin, arising from Pauli forbidden diagrams in a truncated Σ space.

Experimental observation of a bound state of ${}^6_\Lambda$H(0^+) with a Σ component of the order of 1.6% could provide additional information on the significance of the ΛNN three-body force due to coherently enhanced Λ-Σ coupling. A possible means to produce ${}^6_\Lambda$H is ^{6}He(K^-,π^0) single-charge exchange or ^{6}Li(K^-,π^+) double-charge exchange reactions. Such an observation would be a promising new development in the study of the three-body force.

ACKNOWLEDGMENTS

We thank Prof. S. Shinmura, Prof. T. Harada, Prof. M. Wada and Prof. T. Yamazaki for helpful discussions and encouragements. Y.A. acknowledges the receipt of a Grant-in-Aid for Scientific Research of Monbu-Kagakusho of Japan.

REFERENCES

1. R.H. Dalitz, R.C. Herndon and Y.C. Tang, Nucl. Phys. **B47** (1972) 109.
2. Y. Nogami, Proc. Int. Conf. on Hypernuclear Physics, Argonne Nat. Lab., 1969.
   Y. Nogami *et al.*, Nucl. Phys. **B19** (1970) 93.
3. J. Fujita and H. Miyazawa, Prog. Theor. Phys. **17** (1957) 360.
4. Y. Akaishi, T. Harada, S. Shinmura and Khin Swe Myint, Phys. Rev. Lett. **84** (2000) 3539.
5. M.M. Nagels, T.A. Rijken and J.J. de Swart, Phys. Rev. D **12** (1975) 744.
6. Th.A. Rijken, V.G.J. Stoks and Y. Yamamoto, Phys. Rev. C **59** (1999) 21.
7. B.F. Gibson, A. Goldberg and M.S. Weiss, Phys. Rev. C **6** (1972) 741.
8. H. Nemura, Y. Akaishi and Y. Suzuki, Phys. Rev. Lett. **89** (2002) 142504.

9. T. Harada, S. Shinmura, Y. Akaishi and H. Tanaka, Nucl. Phys. **A507** (1990) 715.
10. T. Nagae *et al.*, Phys. Rev. Lett. **84** (2000) 3539.
11. T. Harada, Phys. Rev. Lett. **81** (1998) 5287.
12. A.R. Bodmer, Phys. Rev. **141** (1966) 1387.
13. J . Dabrowski and E. Fedorynska, Nucl. Phys. **A210** (1973) 509.
14. A.R. Bodmer and Q.N. Usmani, Nucl. Phys. **A477** (1988) 621.
R. Sinha and Q.N. Usmani, Nucl. Phys. **A684** (2001) 586c.
15. B.F. Gibson, Proc. U.S.-Japan Seminar on Properties and Interactions of Hyperons, Maui 1993, (World Scientific, 1994) p.55.
16. N.K. Glendenning, Phys. Lett. **B114** (1982) 392.
17. J. Schaffner and I.N. Mishustin, Phys. Rev. C **53** (1996) 1416.
18. S. Balberg and A. Gal, Nucl. Phys. **A625** (1997) 435.
19. S. Shinmura, Khin Swe Myint, T. Harada and Y. Akaishi, J. Phys. G: Nucl. Part. Phys. **28** (2002) 1.
20. A.A. Korsheninnikov *et al.*, Phys. Rev. Lett. **87** (2001) 092581.
21. Khin Swe Myint and Y. Akaishi, Prog. Theor. Phys. Suppl. **146** (2002) 599.
22. P.K. Saha, T. Fukuda *et al.*, Phys. Rev. Lett. **94** (2005) 952502.

Equation of State of Nuclear Matter and the Nuclear Three-Body Force

H. Kanzawa*, K. Oyamatsu†, K. Sumiyoshi** and M. Takano‡

*Department of Pure and Applied Physics, Science and Engineering, Waseda University, 3-4-1 Okubo, Shinjuku-ku, Tokyo 169-8555, Japan
†Department of Media Production and Theories, Aichi Shukutoku University, Nagakute-cho, Aichi 480-1197, Japan
**Numazu College of Technology, Ooka 3600, Numazu, Shizuoka 410-8501, Japan
‡Research Institute for Science and Engineering, Waseda University, 3-4-1 Okubo, Shinjuku-ku, Tokyo 169-8555, Japan

Abstract. An equation of state (EOS) for uniform nuclear matter is constructed at zero and finite temperatures with the variational method starting from a realistic nuclear Hamiltonian composed of the AV18 two-body potential and the UIX three-nucleon interaction (TNI). The two-body energy is calculated with a Jastrow wave function, and the TNI energy with the Fermi-gas wave function. In this calculation, the Fujita-Miyazawa (FM) 2π-exchange term of the TNI energy is positive for symmetric nuclear matter. By taking into account the correlation between nucleons perturbatively, the expectation value of the FM term is modified to be negative in the low density region.

Keywords: Nuclear matter, Variational method, Three-nucleon forces
PACS: 21.65.-f, 21.45.Ff, 26.60.-c

INTRODUCTION

The nuclear equation of state (EOS) is an important ingredient in the studies of supernova (SN) explosions and related phenomena. At present, there are few nuclear EOSs available for SN simulations. One was constructed by Lattimer and Swesty[1], and another is by Shen et al.[2]. Recently, the latter was extended to include hyperon mixing [3] and the hadron-quark phase transition [4]. Since these EOSs are based on phenomenological calculations, nuclear EOSs constructed with microscopic many-body theories are desirable.

In this situation, we undertake to construct a nuclear EOS for SNe using the variational method. First, we construct the EOS for uniform nuclear matter at zero temperature. The two-body energy is evaluated in the two-body cluster approximation and the contribution of the three-nucleon interaction (TNI) is taken into account phenomenologically. Then, we extend the calculation to finite temperatures [5]. In this calculation of the EOS, the Fujita-Miyazawa (FM) 2π–exchange part of the TNI energy is found to be positive. Since this part is actually expected to be negative for symmetric nuclear matter, we refine the calculation of the FM term by taking into account the correlation between nucleons.

CP1011, *New Facet of Three Nucleon Force—50 Years of Fujita Miyazawa Three Nucleon Force (FM50), The International Symposium,* edited by H. Sakai, K. Sekiguchi, and B. F. Gibson

THE EQUATION OF STATE OF NUCLEAR MATTER

We start from the nuclear Hamiltonian H with the AV18 two-body potential and the UIX three-body potential. The Hamiltonian H is decomposed into H_2 and H_3: H_2 is the two-body Hamiltonian without TNI, and H_3 is the contribution from TNI.

At zero temperature, the expectation value of H_2 is evaluated using the Jastrow-type wave function:

$$\Psi = \mathrm{Sym}\left[\prod_{i>j} f_{ij}\right]\Phi_{\mathrm{F}}. \tag{1}$$

Here, Φ_{F} is the Fermi-gas wave function at zero temperature, and the correlation function f_{ij} can be expressed as

$$f_{ij} = \sum_{t=0}^{1}\sum_{s=0}^{1}[f_{\mathrm{C}ts}(r_{ij}) + sf_{\mathrm{T}t}(r_{ij})S_{\mathrm{T}} + sf_{\mathrm{LS}t}(r_{ij})(L\cdot s)]P_{tsij}. \tag{2}$$

where $f_{\mathrm{C}ts}(r)$, $f_{\mathrm{T}t}(r)$ and $f_{\mathrm{LS}t}(r)$ are the spin-isospin-dependent central, tensor and spin-orbit correlation functions, respectively. S_{T} is the tensor operator, $(L\cdot s)$ is the spin-orbit operator and P_{tsij} is the spin-isospin projection operator. Then, we minimize E_2/N, which is the expectation value of H_2 per nucleon expressed in the two-body cluster approximation, with respect to variational functions, $f_{\mathrm{C}ts}(r)$, $f_{\mathrm{T}t}(r)$ and $f_{\mathrm{LS}t}(r)$, by solving the Euler-Lagrange equations.

In the minimization of E_2/N, two constraints are imposed on the variational functions: the extended Mayer's condition and the healing distance condition. The latter implies that $f_{\mathrm{C}ts}(r) = 1$ and $f_{\mathrm{T}t}(r) = f_{\mathrm{LS}t}(r) = 0$ at r larger than the healing distance.

Next, we consider the contribution from TNI. Since the UIX potential is composed of the Fujita-Miyazawa (FM) 2π-exchange part and the repulsive part, we decompose H_3 into the corresponding two parts: the FM part, H_3^{FM}, and the repulsive part, H_3^{R}. Then, the contribution from TNI per nucleon can be written as

$$\frac{E_3}{N} = \alpha\frac{\langle H_3^{\mathrm{R}}\rangle_{\mathrm{F}}}{N} + \beta\frac{\langle H_3^{\mathrm{FM}}\rangle_{\mathrm{F}}}{N} + \gamma\rho^2 e^{-\delta\rho}. \tag{3}$$

Here, the brackets represent the expectation values using the Fermi-gas wave function, and ρ is the nucleon number density. E_3/N includes four parameters α, β, γ and δ; α and β are common to symmetric nuclear matter and neutron matter, while $\gamma = 0$ for neutron matter. The values of these parameters are chosen so that the total energy $E/N = E_2/N + E_3/N$ reproduces the empirical saturation density $\rho_0 = 0.16\mathrm{fm}^{-3}$, saturation energy $E_0/N = -15.8$MeV, incompressibility $K = 250$ MeV and symmetry energy $E_{\mathrm{sym}}/N = 30$ MeV. Corresponding values of the parameters are $\alpha = 0.41$, $\beta = -0.2185$, $\gamma = 1604\ \mathrm{MeVfm}^6$ and $\delta = 13.93\mathrm{MeVfm}^3$.

Figure 1 shows E/N for symmetric nuclear matter and neutron matter. The results obtained by Akmal, Pandharipande and Ravenhall (APR) using the Fermi Hypernetted Chain (FHNC) method[6] are also shown. It is noted that the nuclear Hamiltonian and the Jastow wave function used in the calculation by APR are the same as used in this study. It is seen in Fig. 1 that E/N are in fair agreement with those by APR.

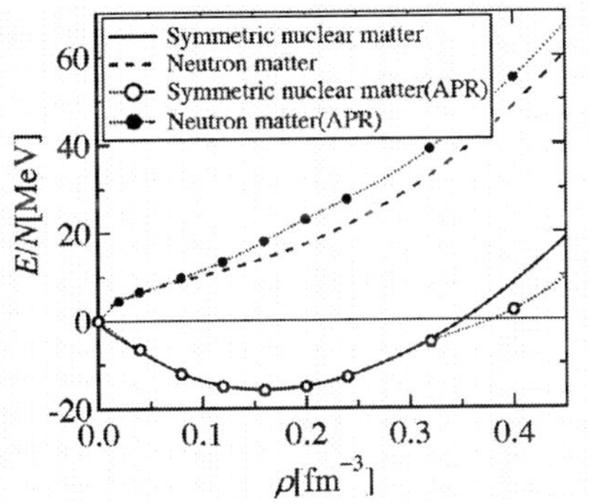

FIGURE 1. Energies per nucleon E/N for symmetric nuclear matter and neutron matter as functions of the density. Corresponding energies obtained by APR are also shown.

We apply the obtained nuclear EOS to neutron stars. The maximum mass of the neutron star is about $2.22M_\odot$. We note that causality is violated at the central mass density $\rho_{m0} \geq \rho_{mC} \sim 1.91\times10^{15}$ g/cm^3. At the critical density ρ_{mC}, the mass of the neutron star is $2.16M_\odot$.

At finite temperatures, we calculate the free energy per nucleon F/N employing the variational method proposed by Schmidt and Pandharipande[7]. In this method, F/N can be expressed as

$$\frac{F}{N} = \frac{E_{T0}}{N} - T\frac{S_0}{N}. \tag{4}$$

Here, E_{T0}/N is the approximate internal energy and S_0/N is the approximate entropy at temperature T. E_{T0}/N is expressed as the sum of E_{T2}/N and E_3/N. E_{T2}/N is the two-body cluster approximation of the expectation value of H_2 obtained with use of the Jastrow-type wave function at finite temperature, which is specified by the averaged occupation probability of the quasi-nucleon states $n(k)$. Here, $n(k)$ includes the effective mass of the quasi-nucleon m^*. The TNI contribution E_3/N is assumed to be the same as at zero temperature. The approximate entropy S_0/N is expressed in terms of $n(k)$ as in the case of the Fermi gas. Then, F/N is minimized with respect to m^*.

The obtained F/N is shown in Fig. 2. Also shown in Fig. 2 is the result by Friedman and Pandharipande (FP)[8]. The obtained F/N is higher than that by FP at high densities, which implies that the present EOS is stiffer than that by FP.

REFINEMENT OF THE TNI CONTRIBUTION

In the preceding section, we constructed a reasonable nuclear EOS. However, in the TNI contribution E_3/N, the obtained expectation value of the FM part is inadequate for the following reason.

It is known that E_3/N should be negative in the low density region and positive in the high density region so as to reproduce the empirical saturation point of symmetric

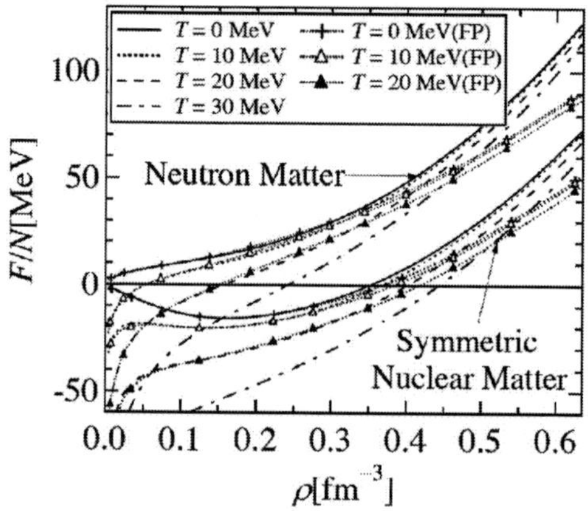

FIGURE 2. Free energies per nucleon, F/N, at $T = 0$, 10, 20 and 30 MeV for symmetric nuclear matter and neutron matter as functions of the density. The free energies obtained by FP are also shown.

nuclear matter. In the case of the UIX potential, the expectation value of the FM term, $\langle H_3^{\rm FM}\rangle/N$, is expected to be negative[9, 10].

In the present calculation, however, the expectation value $\langle H_3^{\rm FM}\rangle_{\rm F}/N$ with the Fermi-gas wave function is positive. As shown schematically in Fig. 3, $\langle H_3^{\rm FM}\rangle_{\rm F}/N$ is composed of (a) the direct term, (b) the 2-particle exchange term and (c) the circular exchange term. In the figure, a solid line labeled "π" represents the 1π-exchange potential, and an oriented curve represents the exchange correlation (the Slater function). Because of the spin-isospin dependence of the FM potential, the direct term vanishes. For this reason, the 2-particle exchange term is dominant and positive due to the exchange effect; the circular exchange term is negative but its contribution is small. Therefore, in the previous section, the factor β in Eq. (3) is chosen to be negative to obtain the negative value of $\beta\langle H_3^{\rm FM}\rangle_{\rm F}/N$.

According to Refs. [9, 10], $\langle H_3^{\rm FM}\rangle/N$ becomes negative for symmetric nuclear matter by taking into account correlations between nucleons; especially, the tensor correlation plays an important role. In this section, therefore, we introduce the central and tensor correlations $f_{{\rm C}ts}(r)$ and $f_{{\rm T}t}(r)$ in the calculation of $\langle H_3^{\rm FM}\rangle/N$. It is worth noting that

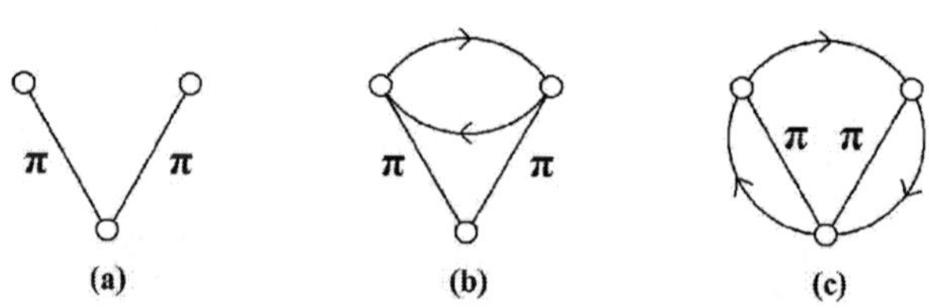

FIGURE 3. Diagrams of the components in $\langle H_3^{\rm FM}\rangle_{\rm F}/N$: (a) the direct term, (b) the 2-particle exchange term and (c) the circular exchange term.

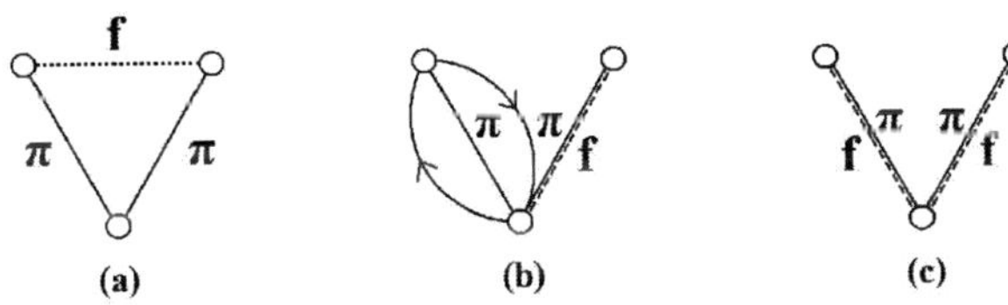

FIGURE 4. Diagrams of the correction terms for $\langle H_3^{\mathrm{FM}}\rangle/N$.

$\langle H_3^{\mathrm{FM}}\rangle_{\mathrm{F}}/N$ is regarded as the part of the three-body cluster (TBC) terms of $\langle H_3^{\mathrm{FM}}\rangle/N$ with the Jastow wave function that is the lowest order in the correlations $h_{\mathrm{C}ts}(r)$ (= $f_{\mathrm{C}ts}(r)-1$) and $f_{\mathrm{T}t}(r)$. Thus, from the standpoint of the cluster expansion, we take into account a part of higher-order TBC terms as the correction terms. In the calculation of these terms, we use $h_{\mathrm{C}ts}(r)$ and $f_{\mathrm{T}t}(r)$ which are determined in the minimizarion of E_2/N, i.e., the correlations are treated perturbatively. Hereafter, we consider symmetric nuclear matter.

First, we pick up the first-order TBC direct term in $\langle H_3^{\mathrm{FM}}\rangle/N$, as shown in Fig. 4(a). In the figure, the correlations $h_{\mathrm{C}ts}(r)$ and $f_{\mathrm{T}t}(r)$ are represented by dashed lines labeled "f". The first-order TBC direct term $\langle H_3^{\mathrm{FM}}\rangle_{\mathrm{1d}}/N$ is decomposed into the central part including $h_{\mathrm{C}ts}(r)$ and the tensor part including $f_{\mathrm{T}t}(r)$. As shown in Fig. 5, $\langle H_3^{\mathrm{FM}}\rangle_{\mathrm{1d}}/N$ is negative and its tensor component is dominant. However, the total energy $\langle H_3^{\mathrm{FM}}\rangle_{\mathrm{F}}/N+\langle H_3^{\mathrm{FM}}\rangle_{\mathrm{1d}}/N$ is still positive at $\rho \geq 0.04\ \mathrm{fm}^{-3}$.

Second, we consider a part of the first-order TBC 2-particle-exchange terms and the second-order TBC direct terms, as shown in Figs. 4(b) and 4(c). Here, for simplicity, we only take into account separable diagrams. The results including these correction terms are shown in Fig. 6. Taking into account these three types of correction terms, $\langle H_3^{\mathrm{FM}}\rangle/N$ becomes negative at low densities; the contributions of higher-order terms improve the

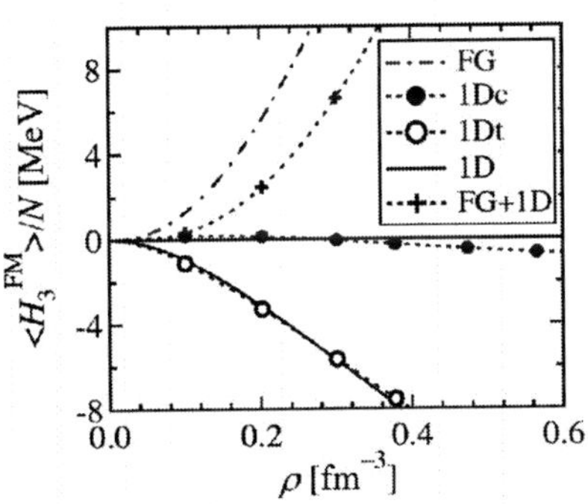

FIGURE 5. The first-order TBC direct term $\langle H_3^{\mathrm{FM}}\rangle_{\mathrm{1d}}/N$ (the curve '1D') for symmetric nuclear matter as a function of the density. The central and tensor parts of $\langle H_3^{\mathrm{FM}}\rangle_{\mathrm{1d}}/N$ (the curves '1Dc' and '1Dt') are also shown. The curve 'FG' represents $\langle H_3^{\mathrm{FM}}\rangle_{\mathrm{F}}/N$.

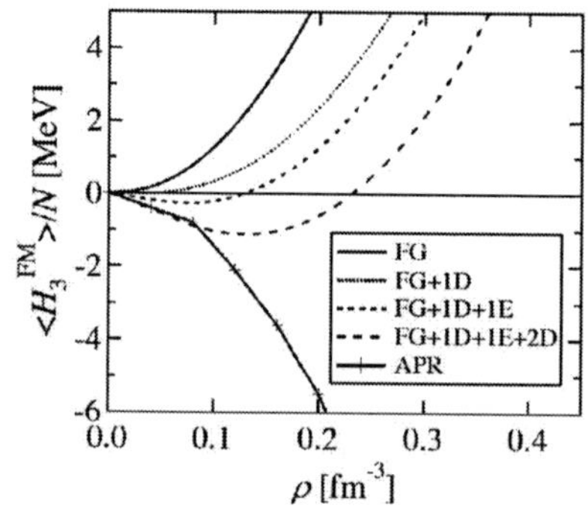

FIGURE 6. The TBC correction terms of $\langle H_3^{\mathrm{FM}}\rangle/N$ for symmetric nuclear matter as functions of the density. The contributions from the diagrams in Figs. 6(a), 6(b) and 6(c) are expressed as '1D', '1E' and '2D', respectively.

result. However, at high densities, $\langle H_3^{\mathrm{FM}}\rangle/N$ is still positive. On the contrary, $\langle H_3^{\mathrm{FM}}\rangle/N$ using the FHNC calculation by APR is negative in the density region considered here: the correction examined in this paper may not be satisfactory.

For further improvement, we are planning to take into account other higher-order TBC terms in $\langle H_3^{\mathrm{FM}}\rangle/N$ that are not included in this study. The full minimization of the total energy including the correction terms, rather than the perturbative treatment, may also be necessary. We will also proceed to calculate the nuclear EOS for nonuniform matter toward the EOS for supernova simulations.

ACKNOWLEDGMENTS

We would like to thank Prof. Y. Akaishi for valuable suggestions. This study is supported by a Grant-in-Aid for the 21st century COE program "Holistic Research and Education Center for Physics of Self-organizing Systems" at Waseda University, and Grants-in-Aid from the Scientific Research Fund of the JSPS (Nos. 18540291, 18540295 and 19540252).

REFERENCES

1. J. M. Lattimer and F. D. Swesty, *Nucl. Phys. A* **535**, 331-376 (1991).
2. H. Shen, H. Toki, K. Oyamatsu and K. Sumiyoshi, *Prog. Theor. Phys.* **100**, 1013-1031 (1998).
3. C. Ishizuka, A. Ohnishi, K. Sumiyoshi and S. Yamada, *Prog. Theor. Phys. Suppl.* **156**, 152-153 (2004).
4. K. Nakazato, K. Sumiyoshi and S. Yamada, *Phys. Rev. D* submitted (2007).
5. H. Kanzawa, K. Oyamatsu, K. Sumiyoshi and M. Takano, *Nucl. Phys. A* **791**, 232-250 (2007).
6. A. Akmal, V. R. Pandharipande and D. G. Ravenhal, *Phys. Rev. C* **58**, 1804-1828 (1998).
7. K. E. Schmidt and V. R. Pandharipande, *Phys. Lett. B* **87**, 11-14 (1979).
8. B. Friedman and V. R. Pandharipande, *Nucl. Phys. A* **361**, 502-520 (1981).
9. T. Kasahara, Y. Akaishi and H. Tanaka, *Prog. Theor. Phys. Suppl.* **56**, 96-107 (1974).
10. H. A. Bethe, *Ann. Rev. Nucl. Sci.* **21**, 93-244 (1971).

Direct Reactions of Borromean Nuclei

S. Shimoura

Center for Nuclear Study (CNS), University of Tokyo, Wako branch at RIKEN, Wako, Saitama 351-0198, Japan

Abstract. There are so-called Borromean nuclei on the drip-line in the nuclear chart, where any two-body subsystem has no bound state. The exotic structure and responses of Borromean nuclei have been studied by measurements of three-body (final-state) channels produced via direct reactions. Measurement of the three-body decay of the isospin partner of light drip-line nuclei populated by the charge exchange reaction is discussed as a means for studying the neutron halo structure. Experimental efforts to study nuclei beyond the drip-line, where many-body resonances may appear, are also presented.

Keywords: Neutron Halo, Charge Exchange Reaction, Isobaric Analog State
PACS: 21.10.Gv, 21.10.Sf, 21.45.-v, 25.40.Kv, 25.60.Lg, 27.20.+n

INTRODUCTION

The properties of neutron drip-line nuclei have attracted much attention since their exotic nature, such as neutron halos and soft collective responses, were first observed. The drip-line nuclei such as ^{6}He, ^{11}Li, and ^{14}Be have the structure of a Core + n + n three-body system and are characterized as "Borromean" when nonc of the two-body sub-system pairs have a bound state. Experimentally, the dynamics of these Borromean systems has been studied by direct reactions such as breakup[1, 2], charge exchange[3, 4], and so on. In this report, I focus on the charge exchange reaction populating the isobaric analog state (IAS), since the isospin multiplets have essentially the same nuclear structure besides a small isospin mixture due to the Coulomb interaction and a small violation of isospin symmetry in the nuclear interactions. I also present an experimental plan to reach few-body systems beyond the drip-line with a new type of charge exchange reaction at the RI beam factory (RIBF) in RIKEN, where multi-neutron correlations may be studied.

ISOBARIC ANALOG STATES OF ^{11}LI AND ^{14}BE

The neutron drip-line nuclei ^{11}Li and ^{14}Be are considered to have two-neutron halo structure because of their Borromean nature. Assuming the same nuclear structure for the isospin partners, their IAS's have a two-nucleon halo structure. The difference between the mother nucleus and its IAS appears in their Coulomb energy. The Coulomb displacement energy (ΔE_C) provides information on the difference in charge distribution between the partners. The kinematical conditions are also different due to ΔE_C: the mothers are bound whereas their IAS's are resonances. The resonance energies can be determined experimentally by analyzing the decay energy spectra. It is noted that isospin conservation restricts the decay channels from the IAS; namely, only the ^{9}Li (^{12}Be) + p

CP1011, *New Facet of Three Nucleon Force—50 Years of Fujita Miyazawa Three Nucleon Force (FM50), The International Symposium,* edited by H. Sakai, K. Sekiguchi, and B. F. Gibson

+ n channel is allowed for IAS's of ^{11}Li (^{14}Be).

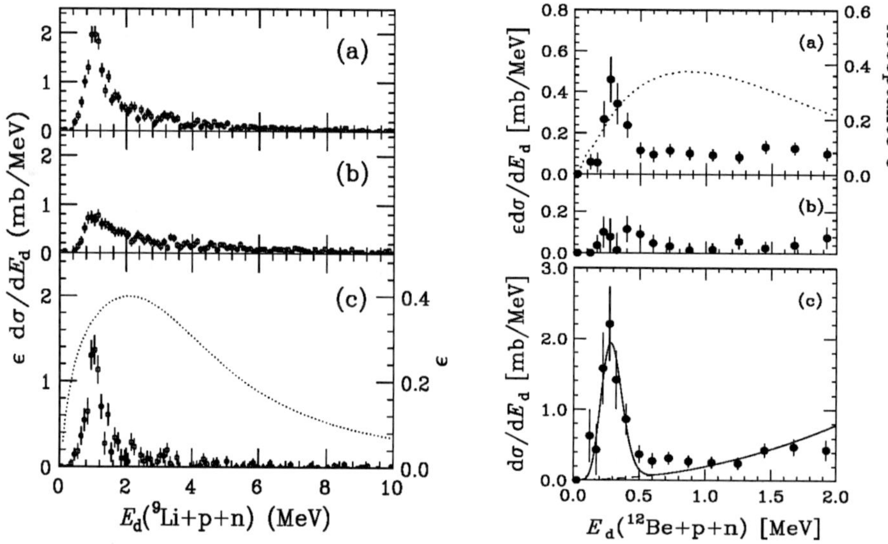

FIGURE 1. Left: Decay energy spectra of the ^{9}Li+p+n channel for (a) the ^{11}Li(p,n)^{11}Be* and (b) the ^{11}Li(d,2n)^{11}Be* reactions. (c) Fermi transition spectrum estimated by subtracting (b) from (a)[3]. Right: Decay energy spectra of the ^{12}Be+p+n channel for (a) the ^{14}Be(p,n)^{14}B* and (b) the ^{14}Be(d,2n)^{14}B* reactions. (c) Decay energy spectrum of the same channel for the (p,n) reaction after correcting for the detector acceptance[4].

In order to populate the IAS's of the ^{11}Li and ^{14}Be, the (p,n) reaction has been measured as well as (d,2n) reaction at 64 *A* and 74 *A* MeV, respectively. Both the measurements were performed using inverse kinematics, where a hydrogen and a deuterium target were irradiated by the ^{11}Li and ^{14}Be (secondary) beam. It is noted that the (d,2n) reaction excludes the Fermi transition populating the IAS.

Experimental results are shown in Fig. 1, where the decay energy spectra of the isospin allowed channels for the (p,n) and the (d,2n) reactions are depicted. Sharp resonance structures appear clearly just above the threshold for three-body channels only for the (p,n) reaction, which are identified as the IAS's. From the resonance energies , ΔE_C is deduced as shown in the left panel of Fig. 2. Rapid decreases of ΔE_C in ^{11}Li and ^{14}Be strongly indicate their halo structure, extending more than twice the radii of the respective cores[3, 4].

The correlation in the ^{9}Li+p+n channel from the IAS of ^{11}Li has been studied by examining the relative energies in the two-body subsystems (^{9}Li+n, p+n, and ^{9}Li+p). The relative energy spectrum of ^{9}Li+n shows a correlation in the low energy region, whereas that of the p+n is consistent with phase space (right panel of Fig. 2). The correlation in the ^{9}Li+n system is consistent with a prediction of effective range theory with a scattering length of -16^{+4}_{-7} fm that is comparable with that of the n-n system.

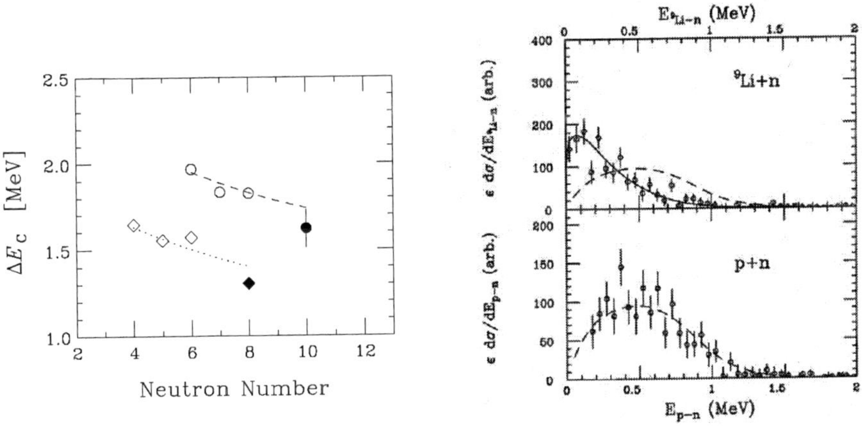

FIGURE 2. Left: Coulomb displacement energy (ΔE_C) as a function of neutron number. The circles and diamonds denote Be and Li isotopes, respectively[4]. Right: Relative energy spectra of the ^{9}Li + p (upper) and the p+n (lower) systems for the ^{9}Li + p + n channel from the IAS of the ^{11}Li. The solid line denotes a prediction taking into account the strong correlation in the ^{9}Li + n system. The dashed lines denote the corresponding phase space[5].

APPROACH TO NUCLEI BEYOND THE DRIP-LINE

The study of nuclear systems beyond the neutron drip-line is a challenge for understanding nuclear stability, nucleon correlations in multi-body scattering states, and so on. Among them, multi-neutron systems attract much attention for the understanding of few-nucleon systems and few-body forces which are to be compared with precise *ab initio* calculations based on realistic interactions. The tetra-neutron system (^{4}n) may be produced by the double charge exchange (DCX) reaction on ^{4}He. The DCX reaction using a RI beam, namely, ^{4}He(^{8}He,^{8}Be)4n, is a unique tool, since the momentum transfer of this reaction is very small (10–20 MeV/c) compared to the DCX reaction using a stable nuclear beam. Because of the small momentum transfer, the missing mass spectrum may be directly compared to theoretical predictions of the spectra by using realistic N-N interactions[6, 7]. This reaction can be used for the study of very neutron-rich "superheavy" hydrogen systems, ^{6,7}H, which are also unique multi-body scattering states.

We have a plan for measuring the (^{8}He,^{8}Be) reaction at the RIBF facility in RIKEN, since the intensity and the energy are suitable. In order to perform the measurement, we are constructing a high-resolution magnetic spectrometer SHARAQ[1] in RIBF, which will enable us to make such measurements with a momentum resolution of 1/14700 as well as an angular resolution of 1 mrad (Fig. 3)[8]. A feasibility study and required R&D for this plan are now in progress.

---

[1] Spectroscopy with **H**igh-resolution **A**nalyzer and **R**adio**A**ctive **Qu**antum beams

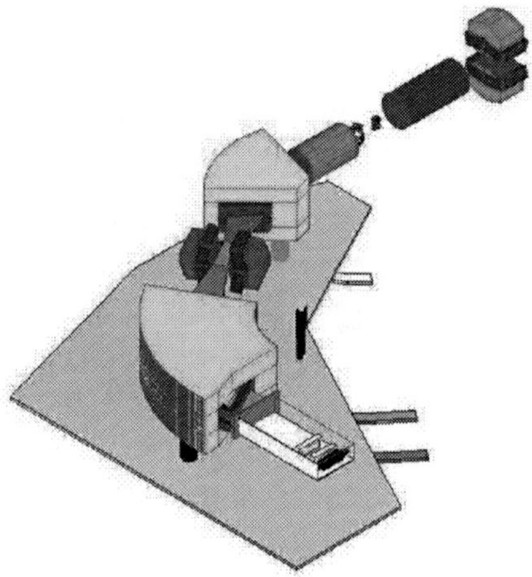

FIGURE 3. Schematic view of SHARAQ Spectrometer.

SUMMARY

Charge exchange reactions involving drip-line nuclei having a Borromean nature are discussed focusing on the IAS. The neutron halo structure extending significantly from the core reduces the Coulomb displacement energy. A large scattering length of about -16 fm for the ^{9}Li+n system is deduced, which is comparable with that of the n-n system.

An experimental plan to study the tetra-neutron and superheavy hydrogen systems using DCX of ^{8}He is presented. The high resolution SHARAQ spectrometer being constructed in RIBF will be used for the DCX experiments.

ACKNOWLEDGMENTS

A part of the present study was supported by the Grant-in-Aid for Scientific Research (No. 17002003) by the Ministry of Education, Culture, Sports, Science and Technology.

REFERENCES

1. S. Shimoura et al., *Phys. Lett.* **B348**, 29 (1995).
2. T. Nakamura et al., *Phys. Rev. Lett.* **96**, 252502 (2006).
3. T. Teranishi et al., *Phys. Lett.* **B407**, 110 (1997).
4. S. Takeuchi et al., *Phys. Lett.* **B515**, 255 (2001).
5. S. Shimoura et al., *Nucl. Phys.* **A630**, 387c (1998).
6. A. Hemmdan, W. Glockle, and H. Kamada, *Phys. Rev.* **C65**, 054001 (2002).
7. N.K. Timofeyuk, *J. Phys.* **G29**, L9 (2003).
8. S. Shimoura, *AIP Conf. Proc.* **912**, 406 (2007).

Concluding Remarks

B. F. Gibson

Theoretical Division, Los Alamos National Laboratory, Los Alamos, NM 87545, USA

Abstract. The observations collected here represent the author's personal perspective regarding the physics discussed and the ideas expressed by the participants in this International Symposium on New Facets of the Three Nucleon Force. These remarks are not intended to be a summary talk. Moreover, physics rather than the physicists who spoke are emphasized. The ideas, which were skillfully brought to life in the University of Tokyo's Koshiba Hall through erudite and colorful powerpoint presentations, will hopefully outlive the rapporteurs who gave generously of their time and talent to make it happen.

Keywords: three-body forces, nuclear, hypernuclear
PACS: 21.45.Ff, 21.45.-v, 21.80.+a

INTRODUCTION

We symposium participants owe a debt of gratitude to Hideyuki Sakai and the other organizers of this International Symposium for the effort expended to create the scientific program we have enjoyed and to make the arrangements that have made possible the meeting. Moreover we should thank Hironari Miyazawa for joining us during this symposium to explore what has happened during the fifty years since he and Fujita first published their idea[1] regarding the existence of a nuclear three-body force. Much has happened during the intervening time, and we thank Manoel Robilotta for his excellent opening review[2]. Following that we have heard much about recent developments during the course of this symposium.

The three-body force models that were built upon the dispersion theory approach of the FM beginning are several. The Tucson-Melbourne model[3], including its more recent incarnations[4], has been used by many, as has the Brazil three-body force model[5]. The more phenomenological Urbana[6] and Illinois[7] three-body force potentials have also been very successful in calculations of the properties of light nuclei. Finally, the EFT models[8, 9, 10] are coming into fruition, and we heard how, based upon that analysis, one might argue that Fujita and Miyazawa were not just lucky. One point worth emphasizing, for heavier nuclei it is important to improve upon the precision with which the three-body force potential coefficients a, b, and d are determined. The mass three- and four-body systems are relatively forgiving when one calculates the binding energies. Moreover, as we have heard oft repeated, it is important that there be consistency among the two-nucleon force model and the three-body force model used. As electromagnetic currents are included (*e.g.*, in photonuclear reactions) it is imperative that the current operators also be consistent with the Hamiltonians which generate the wave functions.

CP1011, *New Facet of Three Nucleon Force—50 Years of Fujita Miyazawa Three Nucleon Force (FM50), The International Symposium,* edited by H. Sakai, K. Sekiguchi, and B. F. Gibson

CONVENTIONAL NUCLEI

Three-body Continuum Data

The data described during the symposium were perhaps the most impressive aspect of the meeting. Even though experiments were not always in agreement, we have not seen data in such quantity nor of such quality for a number of years. Theorists have been challenged. We can only hope that they will now produce the calculations that are needed to address the puzzles and questions raised by the experiments.

The Nd scattering data at intermediate energies clearly show a discrepancy with available calculations when no three-body force is included. Adding a three-body force improves the representation of the angular distribution in pd scattering, but model spin observables still show significant defects in comparison with the data..

At low energies we have yet to solve the A_y puzzle. The calculations disagree with both pd and nd data. Likewise, in the Nd breakup arena we are faced with the Sagara anomaly as well as the space-star anomaly.

The experimentalists deserve kudos for their recent efforts. The data are much appreciated. There is obvious disagreement between experiment and theory, which is not resolved by including contemporary three-nucleon force models. However, we must remember that evidence for a three-body force effect requires comparison of a model calculation with the data and, therefore, depends in detail upon the nucleon-nucleon force used in the model calculation.

Bound State Results

The bound-state calculations show impressive agreement with the binding energy data through a major part of the p-shell, when nucleon-nucleon force models that reproduce with precision the nucleon-nucleon scattering data are combined with the best available three-nucleon force models. Failure to agree with precision data (binding energies are our most precisely measured physical observables) implies something is missing from the models, and that mantra has been utilized by adept model builders.

We build computer models of nuclei and nuclear reactions. We take the models seriously: we calculate measured observables and attempt to understand the underlying physics. Occasionally we find unexpected effects. One of those arose in introducing three-body forces into the three-nucleon system.

- Extending the Hamiltonian to include $NN - N\Delta$ coupled-channels expands the Fock space and should enhance the attraction in the Hamiltonian and the resulting triton binding energy.
- P. Sauer has often reminded us that the resulting dispersive effect in the NN interaction can considerably reduce the expected three-body force attraction[11].
- This is perhaps most obvious in hypertriton calculations such as those by I. Afnan[12].

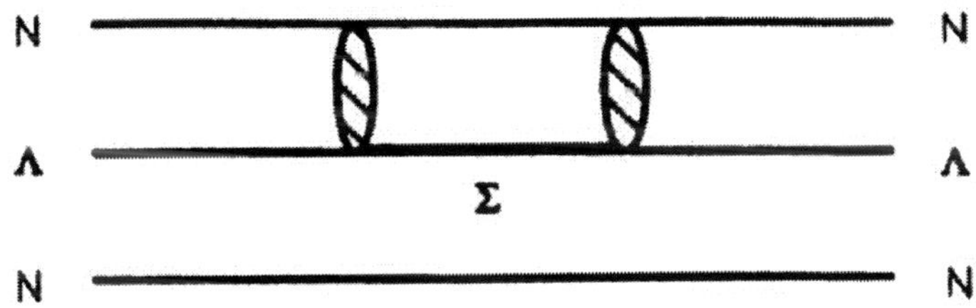

FIGURE 1. Dispersive contribution to the ${}^{3}_{\Lambda}$H binding energy due to $N\Lambda - N\Sigma$ mixing.

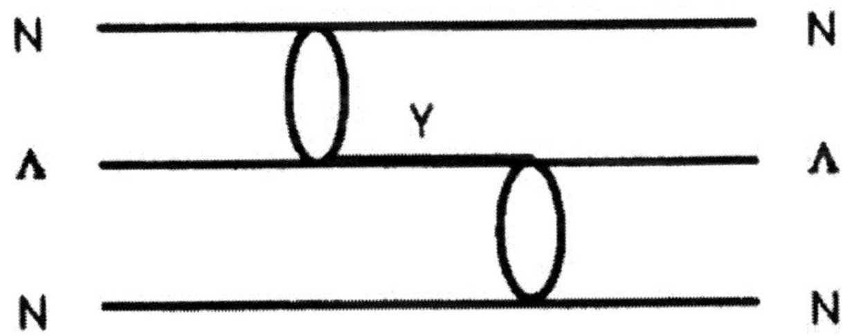

FIGURE 2. True three-body-force contribution to the ${}^{3}_{\Lambda}$H binding energy due to $N\Lambda - N\Sigma$ mixing.

A comparison of the repulsive dispersive effect in the $N\Lambda$ interaction due to its coupling to the closed $N\Sigma$ channel versus the true three-body force ($N\Lambda N$) effect in ${}^{3}_{\Lambda}$H is shown in Fig. 1 and Fig. 2. The dispersive modification of the effective $N\Lambda$ interaction weakens the two-body force, whereas the $N\Lambda - N\Sigma$ coupling generates an attractive contribution. The net effect is that ${}^{3}_{\Lambda}$H is very weakly bound.

The S Shell. The ^{3}H-^{3}He isodoublet has been thoroughly investigated. For a model Hamiltonian based upon a realistic nucleon-nucleon local potential (with strong short range repulsion and strong tensor coupling) that quantitatively reproduces the extensive NN data base of physical observables, the triton binding energy is approximately 7.6 MeV compared with the experimental value of 8.48 MeV. One is missing less than 1 MeV out of a total potential energy of some 50 MeV.

Other physical observables are off correspondingly :

- the root-mean-square radius is too large;
- the ratio of asymptotic normalization factors (C_2/C_0) is too small;
- the ^{3}He Coulomb energy is too small.

We have learned that low energy observables scale with the binding energy[13, 14, 15]. This is illustrated in Fig. 3 by the model results for the ^{3}H and ^{3}He rms radii as a function of binding energy. Introducing a small three-nucleon force adjusts most

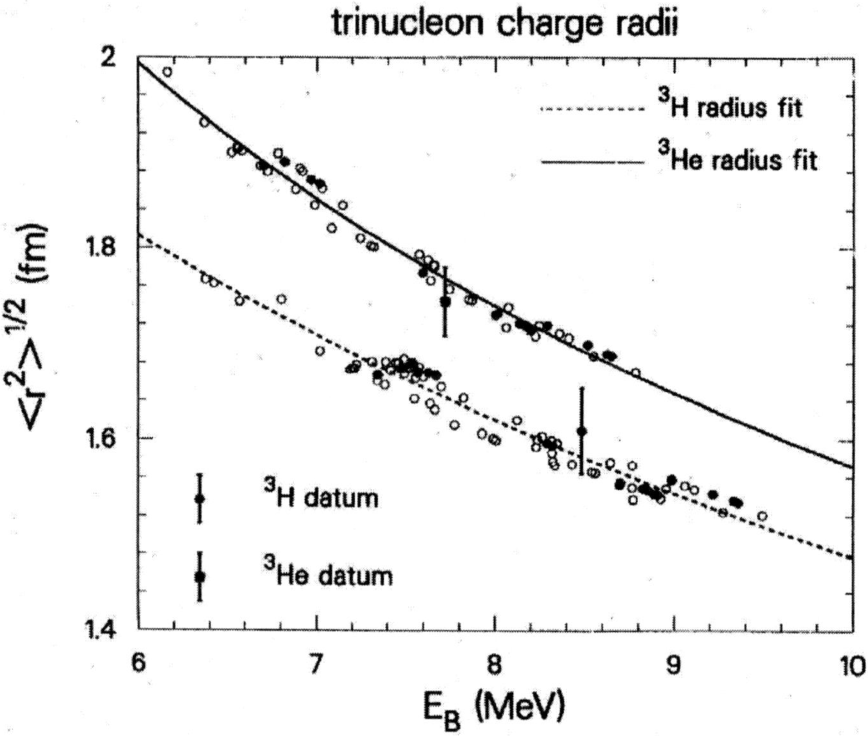

FIGURE 3. The ^{3}H and ^{3}He rms radii as a function of binding energy[13]. The symbols represent various model calculations; the solid symbols correspond to model calculations that include a three-body force.

observables to agree with experiment. For example,

$$\langle r^2(^3\text{H})\rangle^{1/2} = 1.58\,\text{fm vs. } 1.61\pm0.04,$$
$$\langle r^2(^3\text{He})\rangle^{1/2} = 1.77\,\text{fm vs. } 1.75\pm0.04.$$

Furthermore, addition of a small charge symmetry breaking force ($V_{nn} \neq V_{pp}$) can provide the 120 keV needed to increase the approximately 648 keV of Coulomb energy to match the measured 764 keV ^{3}H-^{3}He binding energy difference.

Given such a Hamiltonian that describes well the triton (realistic pairwise forces augmented by an appropriately adjusted three-nucleon force), the resulting model description of ^{4}He is comparable. One can test the model wave functions by use of an electromagnetic probe. We have seen that in α particle photodisintegration, introducing a three-nucleon force to obtain the correct ^{4}He binding energy (and therefore, the correct rms radius) produces a reasonable result for the total cross section. The low energy cross section is sensitive to the size of the α particle. Moreover, we have also seen that one

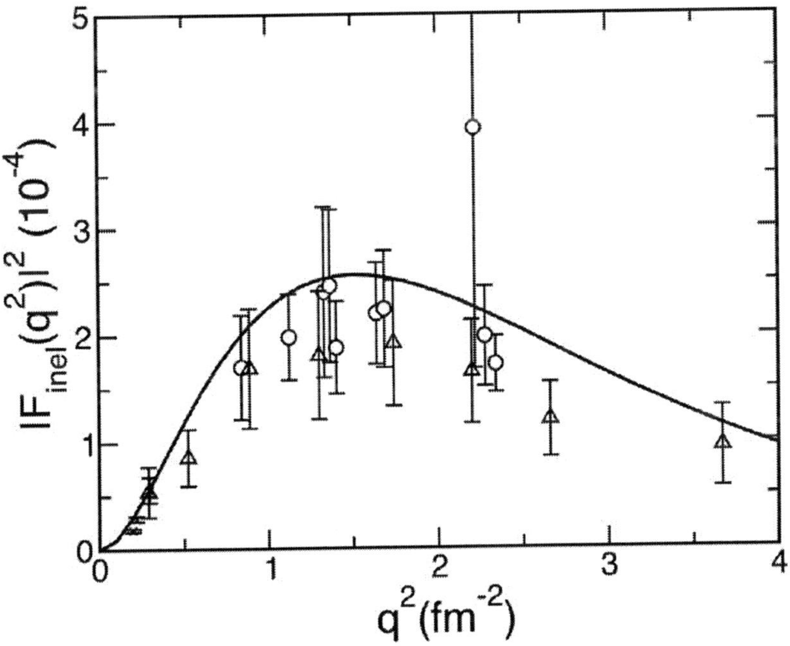

FIGURE 4. The inelastic form factor for $^4He(e,e')^4He(0_2^+)$ from Ref.[16] compared with the available data in Refs.[17, 18, 19].

can model the loosely bound ^4He 0_2^+ (excited) state with a phenomenological three-body force plus the Coulomb interaction well enough to describe the electro-excitation data[16]. (See the inelastic form factor for $^4He(e,e')^4He(0_2^+)$ depicted in Fig. 4.) The calculational methods are precise enough to describe these reactions, and three-body forces play an important role in properly describing the size of the nucleus. ^4He appears to be the smallest system with the complexity of a "real" nucleus; it has an excited state spectrum. Despite these successes, n–^3H elastic scattering is not properly described, nor are the other coupled channels[20, 21]. The A = 4 continuum suffers discrepancies similar to that seen for A = 3, and they appear to be more prominent.

The P Shell. The P Shell nuclei provide a more stringent test of the model Hamiltonian[7, 22, 23]. Green's Function Monte Carlo calculations have shown that the model Hamiltonian described above misses the ^7Li binding energy by 2 MeV compared to a total three-nucleon potential energy of approximately 18 MeV. Conclusion: a small $T = 3/2$ three-nucleon force is required to be added to the Hamiltonian. The resulting model does well through A=10 in the P shell, including the low lying excited states.

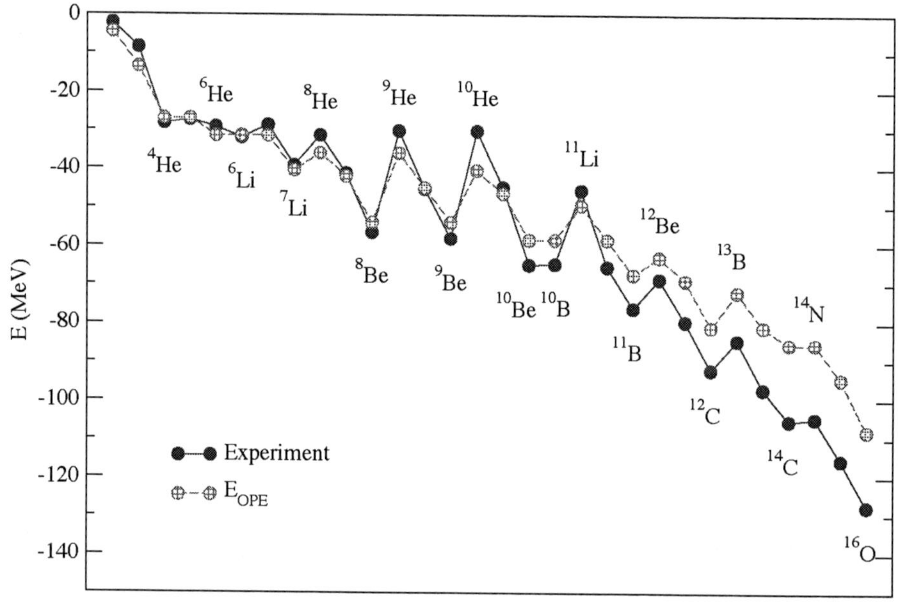

FIGURE 5. Comparison of S-shell and P-shell ground state binding energies with an OPE model calculation, omitting any three-nucleon force.

We appear to find that a proper description of the phenomenological three-body force requires at least three parameters; the few body systems need additional attraction while nuclear matter is in need of repulsion. The Illinois 2 potential that described well the light nuclei through mass 10 was too soft in nuclear matter; the Illinois 7 potential is an improvement.[7]

As an aside, it is worth noting that variational Monte Carlo calculations have demonstrated that the one-pion-exchange (OPE) potential (strong tensor force) accounts for 75%-80% of the binding energy through ^{16}O. This is illustrated in Fig. 5, where model results[24] for the OPE potential are compared with experimental binding energies. The point is that the scalar/vector model (*e.g.*, the Walecka Model) of nuclei may provide a good phenomenological description, the model does not contain the correct physics based upon what we know about the nucleon-nucleon interaction.

Heavier Nuclei. The neutron-rich nuclei is a region of interest to a large number of experimentalists and theorists, alike. Modification of closed-shell properties and the role of three-nucleon forces as well as the isospin dependence of three-nucleon forces as one approaches the neutron drip line are important questions. However, one needs to keep in mind the distinction between fundamental forces and effective interactions.

Effective Field Theory

The effective field theory (EFT) approach to modeling nuclear forces appears to be a major step forward in terms of providing consistency between the nucleon-nucleon interaction and a three-nucleon force[10]. However, one should recall that in describing nucleon-nucleon scattering by means of a partial-wave amplitude analysis one requires some 28 parameters to obtain a precision description of the *pp* and *np* scattering data[25, 26]. A similar number of parameters is needed in any formulation of a "realistic" *NN* potential model description of the data[27, 28]. Thus, one is looking at N^nLO with $n > 3$ for a precision description of the *NN* data.

Three-Body Forces and On-Shell Equivalent *NN* Forces

It should be emphasized that a three-body force cannot be measured by an experiment. Life is not fair. Experiments can constrain the elements of a three body force, but the three-body force in a given Hamiltonian depends crucially upon the nucleon-nucleon interaction on which the Hamiltonian is based.

There is no unique three-nucleon force in nature, just as there is no unique two-nucleon force of nature. If one measured the *NN* phase shifts from zero energy to infinity, one would still need to hypothesize a local interaction to determine *uniquely* a two-nucleon potential. An infinite variety of nonlocal potentials would yield the same nucleon-nucleon phase shifts.

Recall the paper by W. Polyzou and W. Gloeckle[29] which makes this quite clear. Given two-body Hamiltonians H and $\bar{H}$ such that

$$H_{ij} = H_i + H_j + V_{ij}$$

and

$$\bar{H}_{ij} = H_i + H_j + \bar{V}_{ij} ,$$

which support the same bound states and have the same scattering matrices, then there exists a three-body interaction W such that

$$H = H_1 + H_2 + H_3 + V_{12} + V_{23} + V_{31}$$

and

$$\bar{H}' = \bar{H} + W ,$$

where

$$\bar{H} = H_1 + H_2 + H_3 + \bar{V}_{12} + \bar{V}_{23} + \bar{V}_{31} ,$$

have the same bound state binding energies and scattering matrices. Thus, a three-body force can only be defined within the context of a given two-body force model.

This is equivalent to the fact that in many-body theory, one can construct a unitary transformation [*e.g.*, $U = \exp(\sum_{i<j} \ln U_{ij})$] that transforms all two-body operators into their equivalents and generates, in addition, 3-, 4-, ... body interactions needed to retain the equivalences of the N-body dynamics. One can also include U_{ijk} in the exponent, so that the two-body interactions remain unchanged but the 3-, 4-, ... body interactions are altered. However, one should be careful **not** to interpret this analysis to mean that three-body forces can be removed, although the effect of a three-body force can be minimized in such a scheme.

Although a nonlocal potential may be chosen so as to reduce the magnitude of the three-body force required to reproduce the data (or a unitary transformation may designed to help in a similar manner), the resulting electromagnetic currents are almost guaranteed to become insurmountably complex. There is no free lunch.

STRANGENESS PHYSICS

The strangeness degree of freedom adds a new dimension to our picture of the nucleus. One has the opportunity to investigate whether our intuition, which has been carefully developed in the non strange (neutron-proton) sector of conventional nuclei, extrapolates to explain our observations in the strangeness sector **or** whether the complex models that we have developed are merely sophisticated interpolation tools that are valid within the neutron-proton space where they were developed.

$N\Lambda - N\Sigma$ mixing appears to play a much larger role in Λ hypernuclei than does $NN - N\Delta$ mixing in conventional nuclei. The former mixes baryons in the same octet; the latter couples the octet and decuplet. The effective $N\Lambda N$ three-body force seems to be a much more significant factor in binding ${}^{3}_{\Lambda}\mathrm{H}$ and ${}^{5}_{\Lambda}\mathrm{He}$ than does the NNN force in ${}^{3}\mathrm{H}$ or ${}^{3}\mathrm{He}$. The charge symmetry breaking seen in the ${}^{4}_{\Lambda}\mathrm{H}$ – ${}^{4}_{\Lambda}\mathrm{He}$ Λ separation energy difference is three times that in the ${}^{3}\mathrm{H} - {}^{3}\mathrm{He}$ binding energy difference once the Coulomb energy of ${}^{3}\mathrm{He}$ has been accounted for, most likely due to $N\Lambda - N\Sigma$ mixing. Adding a Λ to ${}^{6}\mathrm{Li}$ compresses the nuclear core significantly. $\Lambda\Lambda - \Sigma\Sigma - N\Xi$ mixing is predicted to play an important role in $S = -2$ hypernuclei.

YN Data

The *NN* data, upon which our understanding of the physics of nuclei and nuclear reactions is based, are abundant. By combining an analysis of crucial spin observables with cross section measurements, one can unravel the spin-dependent partial-wave amplitudes. The deuteron (np bound state) provides additional constraints, in particular for the the tensor (S_{12}) interaction.

In contrast the *YN* scattering data are meager. There are fewer than 600 events below 300 MeV/c. There are fewer than 250 events between 300 and 1500 MeV/c. There is no $N\Lambda$ or $N\Sigma$ bound state. *YN* spin-dependent data are few in number. One has primarily differential and total cross section measurements. The sparse nature of the $p\Lambda$ data can be seen in Fig. 6, where the elastic total cross section data are compared with the Nijmegen soft-core potential model[34] prediction. The need for precision *YN* scattering data is clear, as not even the expected cusp in the $p\Lambda$ data due to the opening of the $N\Sigma$ threshold is well defined.

Hypernuclei

Japan has played a major role in hypernuclear physics. The record of KEK is well recognized. One looks forward to significant contributions from the new J-PARC facility. We eagerly await new hypernuclear measurements that will address the role of three-

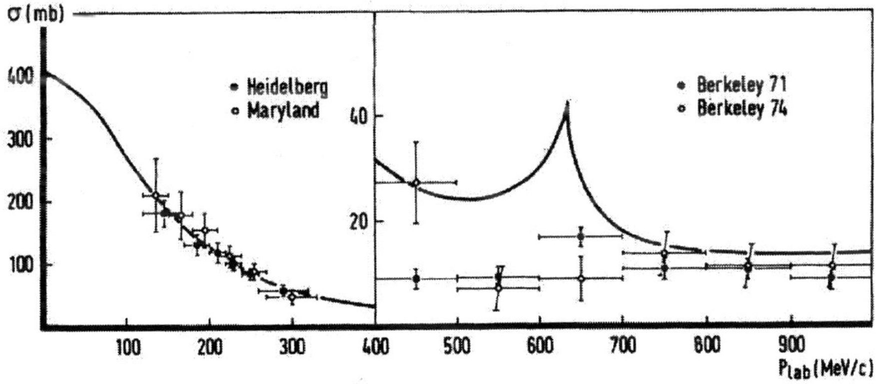

FIGURE 6. Λp total cross section data for p_{Lab} in the range from 0 to 1000 MeV/c from Refs[30, 31, 32, 33] compared with predictions from the Nijmegen soft-core potential[34].

body forces in such phenomena as the stabilization of the neutron-rich ${}^{6}_{\Lambda}$H system, the charge symmetry breaking in the excited states of the ${}^{4}_{\Lambda}$H − ${}^{4}_{\Lambda}$He isodoublet, the predicted strong binding in ${}^{5}_{\Lambda\Lambda}$H compared to the surprisingly weak binding in ${}^{6}_{\Lambda\Lambda}$He, *etc.*. However, real progress in modeling hypernuclei requires a significant improvement in the *YN* data base. J-PARC can provide that improvement.

FROM WHERE WE HAVE COME

In conclusion, let me paraphrase a great intellect: We have seen that in the past 50 years much progress has been made in our understanding of the three-nucleon force component of the nuclear interaction, **because** we have been able to stand on the shoulders of the giants who came before us. Thank you, Miyazawa sensei, for showing the way.

ACKNOWLEDGMENTS

The research of the author is supported by the U.S. Department of Energy under contract W-7405 -ENG-36. He thanks the organizers for the invitation to listen carefully to all of the presentations at this symposium, and he thanks the JUSTIPEN program at RIKEN for partial support to come to Japan.

REFERENCES

1. J. I. Fujita and H. Miyazawa, Progress of Theoretical Physics **17**, 360 (1957)
2. See M. Robilotta's contribution to this volume.

3. S. A. Coon, *et al.*, Nucl. Phys. **A317**, 242 (1979).
4. S. A. Coon and H. K. Han, Few-Body Systems **30**, 131 (2001).
5. H. T. Coelho, T. K. Das, and M. R. Robilotta, Phys. Rev. C, **28**, 1812 (1983).
6. B. S. Pudliner *et al.*, Phys. Rev. Lett. **74**, 4396 (1995).
7. S. C. Pieper, Phys. Rev. C **64**, 014001 (2001); see also S. C. Pieper's contribution to this volume.
8. U. van Kolck, Phys. Rev. C **49**, 2932 (1994).
9. E. Epelbaum, W. Gloeckle, and U.-G. Meissner, Nucl. Phys. **A747**, 362 (2005).
10. V. Bernard *et al.*, arXiv:0712.1967v1[nucl-th]; see also U.-G. Meissner's contribution to this volume.
11. P. U. Sauer, Prog. Part. Nucl. Phys. **16**, 35 (1986).
12. I. R. Afnan and B. F. Gibson, Phys. Rev. C **41**, 2787 (1990).
13. J. L. Friar, B. F. Gibson, and G. L. Payne, Phys. Rev. C **35**, 1502 (1987).
14. J. L. Friar *et al.*, Phys. Rev. C **37**, 2852 (1988).
15. J. L. Friar *et al.*, Phys. Rev. C **37**, 2859 (1988).
16. E. Hiyama, B. F. Gibson, and M. Kamimura, Phys. Rev. C **70**, 031001 (2004).
17. R. F. Frosch *et al.*, Nucl. Phys. **A110**, 657 (1968).
18. Th. Walcher, Phys. Lett. **31B**, 442 (1970); Z. Phys. Chem. (Leipzig) **237**, 368 (1970).
19. G. Koebschall, *et al.*, Nucl. Phys. **A405**, 648 (1983).
20. A. Deltuva and A. C. Fonseca, Phys. Rev. C **75**, 014005 (2007).
21. A. Deltuva and A. C. Fonseca, Phys. Rev. C **76**, 021001(R) (2007).
22. S. C. Pieper, K. Varga, and R. B. Wiringa, Phys. Rev. C **66**, 044310, (2002).
23. S. C. Pieper, R. B. Wiringa, and J. A. Carlson, Phys. Rev. C **70**, 054235 (2004).
24. R. B. Wiringa, Phys. Rev. C **73**, 034317 (2006).
25. J. R. Bergervoet *et al.*, Phys. Rev. C **41**, 1435 (1990).
26. V. G. J. Stoks *et al.*, Phys. Rev. C **48**, 792 (1993).
27. V. Stoks and J. J. de Swart, Phys. Rev. C **47**, 761 (1993).
28. V. Stoks and J. J. de Swart, Phys. Rev. C **52**, 1698 (1995).
29. W. Polyzou and W. Gloeckle, Few-Body Systems **9**, 97 (1990).
30. G. Alexander, *et al.*, Phys. Rev. **173**, 1452 (1968).
31. B. Sechi-Zorn, *et al.*, Phys. Rev. **175**, 1735 (1968).
32. J. A. Kadyk, *et al.*, Nucl. Phys. **B27**, 13 (1971).
33. J. M. Hauptman, *et al.*, Nucl. Phys. **B125**, 29 (1977).
34. P. M. M. Maessen, Th. A. Rijken, and J. J. de Swart, Phys. Rev. C **20**, 2226 (1989).

List of Participants

Abe, Daisuke	University of Tokyo, Japan	abo@nt.phys s.u-tokyo.ac.jp
Akaishi, Yoshinori	Nihon University and RIKEN, Japan	akaishi@post.kek.jp
Catto, Sultan	The Graduate Center, the City University of New York, U.S.A.	SCatto@gc.cuny.edu
Coon, Sidney A.	Department of Energy, U.S.A.	Sidney.A.Coon@science.doe.gov
Fonseca, Antonio C.	University of Lisbon, Protrugal	fonseca@cii.fc.ul.pt
Fujiwara, Yoshikazu	Kyoto University, Japan	yfujiwar@scphys.kyoto-u.ac.jp
Gibson, Benjamin F.	Los Alamos National Laboratory, U.S.A.	bfgibson@tmail.lanl.gov
Guo, Lu	University of Tokyo, Japan	lu@nt.phys.s.u-tokyo.ac.jp
Hatanaka, Kichiji	RCNP, Osaka University, Japan	hatanaka@rcnp.osaka-u.ac.jp
Hiyama, Emiko	Nara Women's University, Japan	hiyama@cc.nara-wu.ac.jp
Ichikawa, Yuichi	University of Tokyo, Japan	yuichi@nucl.phys.s.u-tokyo.ac.jp
Ikeda, Kiyomi	RIKEN Nishina Center, Japan	k-ikeda@apricot.ocn.ne.jp
Ishihara, Masayasu	RIKEN Nishina Center, Japan	ishihara@ribf.riken.jp
Ishikawa, Souichi	Hosei University, Japan	ishikawa@hosei.ac.jp
Ishikawa, Daiki	RCNP, Osaka University, Japan	disikawa@rcnp.osaka-u.ac.jp
Itagaki, Naoyuki	University of Tokyo, Japan	itagaki@phys.s.u-tokyo.ac.jp
Ito, Makoto	RIKEN Nishina Center, Japan	itomk@riken.jp
Itoh, Satoshi	University of Tokyo, Japan	itoh@nucl.phys.s.u-tokyo.ac.jp
Kalantar-Nayestanaki, Nasser	KVI, University of Groningen, The Netherlands	nasser@kvi.nl
Kamada, Hiroyuki	Kyushu Institute of Technology, Japan	kamada@mns.kyutech.ac.jp
Kamimura, Masayasu	Kyushu University, Japan	kami2scp@mbox.nc.kyushu-u.ac.jp
Kanazawa, Takuya	University of Tokyo, Japan	tkanazawa@nt.phys.s.u-tokyo.ac.jp
Kanzawa, Hiroaki	Waseda University, Japan	kann-sein@toki.waseda.jp
Kawabata, Takahiro	CNS, University of Tokyo, Japan	kawabata@cns.s.u-tokyo.ac.jp
Kimura, Hitomi	University of Tokyo, Japan	kimura@nucl.phys.s.u-tokyo.ac.jp
Kistryn, Stanislaw	Jagiellonian University, Poland	stanislaw.kistryn@uj.edu.pl
Kuboki, Hironori	University of Tokyo, Japan	kuboki@nucl.phys.s.u-tokyo.ac.jp
Kuroita, Sho	Kyushu University, Japan	kuroita@kutl.kyushu-u.ac.jp
Ladygin, Vladimir	Joint Institute for Nuclear Research, Russia	ladygin@he.jinr.ru
Lang, Liu	University of Tokyo, Japan	LLiu@nt.phys.s.u-tokyo.ac.jp
Maeda, Yukie	Kyushu University, Japan	yukie@kutl.kyushu-u.ac.jp
Makino, Junichiro	National Astronomical Observatory of Japan, Japan	makino@th.nao.ac.jp
Marcucci, Laura Elisa	University of Pisa, Italy	laura.marcucci@df.unipi.it
Meißner, Ulf-G.	HISKP, Univ. Bonn and IKP. FZ Jülich, Germany	meissner@itkp.uni-bonn.de
Miki, Kenjiro	University of Tokyo, Japan	miki@nucl.phys.s.u-tokyo.ac.jp
Miyashita, Seiji	University of Tokyo, Japan	miya@spin.phys.s.u-tokyo.ac.jp
Miyazawa, Hironari	University of Tokyo, Japan	miyazawa@phys.s.u-tokyo.ac.jp
Motobayashi, Tohru	RIKEN Nishina Center, Japan	motobaya@riken.jp
Nemura, Hidekatsu	RIKEN Nishina Center, Japan	nemura@riken.jp
Nogami, Yuki	McMaster University, Canada	nogami@mcmaster.ca

Noji, Shumpei	University of Tokyo, Japan	noji@nucl.phys.s.u-tokyo.ac.jp
Nomura, Kosuke	University of Tokyo, Japan	nomura@nt.phys.s.u-tokyo.ac.jp
Ohnishi, Tetsuya	RIKEN Nishina Center, Japan	oonishi@ribf.riken.jp
Oka, Makoto	Tokyo Institute of Technology, Japan	oka@th.phys.titech.ac.jp
Okada, Shinji	RIKEN Nishina Center, Japan	sokada@riken.jp
Okamura, Hiroyuki	RCNP, Osaka University, Japan	okamura@rcnp.osaka-u.ac.jp
Orabi, Momen Ahmad	Niigata University, Japan	orabi@nt.sc.niigata-u.ac.jp
Oryu, Shinsho	Tokyo University of Science, Japan	oryu@ph.noda.tus.ac.jp
Otsu, Hideaki	RIKEN Nishina Center, Japan	otsu@ribf.riken.jp
Otsuka, Takaharu	CNS & Department of Physics, University of Tokyo, Japan	otsuka@phys.s.u-tokyo.ac.jp
Pieper, Steven C.	Argonne National Laboratory, U.S.A.	pieper@phy.anl.gov
Quaglioni, Sofia	Lawrence Livermore National Laboratory, U.S.A.	quaglioni1@llnl.gov
Robilotta, Manoel R.	Universidade de São Paulo, Brazil	mane@axpfep1.if.usp.br
Sagara, Kenshi	Kyushu University, Japan	sagara@kutl.kyushu-u.ac.jp
Sakaguchi, Satoshi	CNS, University of Tokyo, Japan	satoshi@cns.s.u-tokyo.ac.jp
Sakai, Hideyuki	University of Tokyo, Japan	sakai@phys.s.u-tokyo.ac.jp
Sasaki, Shoichi	University of Tokyo, Japan	ssasaki@phys.s.u-tokyo.ac.jp
Sasamoto, Yoshiko	CNS, University of Tokyo, Japan	yoshiko@cns.s.u-tokyo.ac.jp
Sasano, Masaki	University of Tokyo, Japan	sasano@nucl.phys.s.u-tokyo.ac.jp
Sawada, Tetsuo	Nihon University, Japan	t-sawada@fureai.or.jp
Schwenk, Achim	TRIUMF, Canada	schwenk@triumf.ca
Sekiguchi, Kimiko	RIKEN Nishina Center, Japan	kimiko@ribf.riken.jp
Shimizu, Noritaka	University of Tokyo, Japan	shimizu@nt.phys.s.u-tokyo.ac.jp
Shimoura, Susumu	CNS, University of Tokyo, Japan	shimoura@cns.s.u-tokyo.ac.jp
Stephan, Elzbieta	Institute of Physics, University of Silesia, Poland	elzbieta.stephan@us.edu.pl
Suda, Kenji	RCNP, Osaka University, Japan	suda@rcnp.osaka-u.ac.jp
Sugimoto, Satoru	Kyoto University, Japan	satoru@ruby.scphys.kyoto-u.ac.jp
Suzuki, Yasuyuki	Niigata University, Japan	suzuki@nt.sc.niigata-u.ac.jp
Takano, Masatoshi	Waseda University, Japan	takanom@waseda.jp
Takatsuka, Tatsuyuki	Iwate University, Japan	takatuka@iwate-u.ac.jp
Tamura, Hirokazu	Tohoku University, Japan	tamura@lambda.phys.tohoku.ac.jp
Tanifuji, Makoto	Hosei University, Japan	
Tanihata, Isao	RCNP, Osaka University, Japan	tanihata@rcnp.osaka-u.ac.jp
Tornow, Werner	Duke University and TUNL, U.S.A.	tornow@tunl.duke.edu
Uchino, Shun	University of Tokyo, Japan	uchino@nt.phys.s.u-tokyo.ac.jp
Uesaka, Tomohiro	CNS, University of Tokyo, Japan	uesaka@cns.s.u-tokyo.ac.jp
Uzu, Eizo	Tokyo University of Science, Japan	jph301@ed.noda.tus.ac.jp
Witała, Henryk	Jagiellonian University, Poland	witala@if.uj.edu.pl
Yako, Kentaro	University of Tokyo, Japan	yakou@phys.s.u-tokyo.ac.jp
Yano, Yasushige	RIKEN Nishina Center, Japan	yyano@riken.jp
Yonemura, Tetsuji	Fusion Energy Forum, Japan	

International Symposium on New Facet of Three Nucleon Force
—50 years of Fujita-Miyazawa Three Nucleon Force—

Scientific Program

Oct. 29 (Mon)	Oct. 30 (Tue)	Oct. 31 (Wed)
1 Morning I 9:40-10:30 Chair: W. Tornow	5 Morning I 9:00-10:40 Chair: Y. Nogami	10 Morning I 9:00-10:35 Chair: M. Kamimura
Break	*Break*	*Break*
2 Morning II 10:50-12:40 Chair: Y. Akaishi	6 Morning II 11:00-12:45 Chair: U.-G. Meißner	11 Morning II 10:55-12:35 Chair: H. Tamura
Lunch	*Conference Photo and Lunch*	*Lunch*
3 Afternoon I 14:00-15:45 Chair: M. Oka	7 Afternoon I 14:15-15:25 Chair: Y. Suzuki	12 Afternoon I 14:00-15:20 Chair: T. Motobayashi
Break	*Break*	*Break*
4 Afternoon II 16:05-17:35 Chair: N. Kalantar	8 Afternoon II 15:45-17:15 Chair: H. Sakai	13 Afternoon II 15:40- Chair: A. C. Fonseca
Welcome Drink 17:45-	*Break*	
	9 Afternoon III 17:30-18:30 Chair: H. Sakai	
	Banquet 19:00-	

Koshiba Hall, University of Tokyo
October 29-31, 2007

http://nucl.phys.s.u-tokyo.ac.jp/FM50/

Session 1: Monday Morning I, 9:40-10:30 (Chair: W. Tornow (Duke U. & TUNL))

Time	Title	Speaker (Affiliation)
9:40 (10)	Opening	
9:50 (30+10)		M. R. Robilotta (U. Saõ Paulo)

Session 2: Monday Morning II, 10:50-12:40 (Chair: Y. Akaishi (RIKEN & Nihon U.))

Time	Title	Speaker (Affiliation)
10:50 (25+5)	Three-nucleon force effects in 3N hadronic and photonic reactions	H. Witała (Jagiellonian U.)
11:20 (25+5)	*Nd* scattering at intermediate energies: Epoch-making experiment of three nucleon force study	K. Sekiguchi (RIKEN Nishina Center)
11:50 (25+5)	What have we learned about three-nucleon systems at intermediate energies?	N. Kalantar-Nayestanaki (KVI, U. Groningen)
12:20 (15+5)	Model study of short-range three-nucleon forces	S. Ishikawa (Hosei U.)

Session 3: Monday Afternoon I, 14:00-15:45

(Chair: M. Oka (TITech))

Time	Title	Speaker (Affiliation)
14:00 (30+5)	Why Fujita and Miyazawa were not just lucky	U.-G. Meißner (U. Bonn & FZ Jülich)
14:35 (25+5)	Partial wave decomposition of 2π-1π 3NF in the chiral perturbation theory	H. Kamada (Kyushu Institute of Technology)
15:05 (15+5)	Cross sections of the deuteron-proton breakup as a probe of three-nucleon system dynamics	St. Kistryn (Jagiellonian U.)
15:25 (15+5)	Large, precise set of polarization observables of the deuteron-proton breakup at 130 MeV	E. Stephan (U. Silesia)

Session 4: Monday Afternoon II, 16:05-17:35

Time	Title	Speaker (Affiliation)
16:05 (25+5)	The three-nucleon analyzing power puzzle: The past twenty years	W. Tornow (Duke U. & TUNL)
16:35 (15+5)	Remaining problems in low-energy *pd* breakup reaction	K. Sagara (Kyushu U.)
16:55 (15+5)	Measurements of inclusive *pd*-breakup cross section at low and intermediate energies	S. Kuroita (Kyushu U.)
17:15 (15+5)	*N-d* elastic scattering at low energy with realistic interactions	L. E. Marcucci (U. Pisa)

Session 5: Tuesday Morning I, 9:00-10:40 (Chair: Y. Nogami (McMaster U.))

Time	Title	Speaker (Affiliation)
9:00 (25+5)	Structure of four-baryon systems	E. Hiyama (Nara Women's U.)
9:30 (25+5)	Experimental studies of hypernuclear structure and the ΛN and ΛNN interactions	H. Tamura (Tohoku U.)
10:00 (15+5)	A new three-baryon-force in $\Lambda\Lambda$ hypernuclei	H. Nemura (RIKEN Nishina Center)
10:20 (15+5)	Non-mesonic weak decays of ${}^{5}_{\Lambda}$He and ${}^{12}_{\Lambda}$C hypernuclei in (π^+, K^+) reactions	S. Okada (RIKEN Nishina Center)

Session 6: Tuesday Morning II, 11:00-12:45 (Chair: U.-G. Meißner (Bonn & Jülich))

Time	Title	Speaker (Affiliation)
11:00 (30+5)	The Illinois extension to the Fujita-Miyazawa three nucleon force	S. C. Pieper (Argonne National Lab.)
11:35 (25+5)	Fujita-Miyazawa 3N mechanism and neutron-rich exotic nuclei	T. Otsuka (Dept. of Phys. & CNS, U. Tokyo)
12:05 (15+5)	Three-nucleon interactions: a frontier in nuclear structure	A. Schwenk (TRIUMF)
12:25 (15+5)	3α description of ^{12}C with microscopic non-local potential	M. Orabi (Niigata U.)

Session 7: Tuesday Afternoon I, 14:15-15:25 (Chair: Y. Suzuki (Niigata U.))

Time	Title	Speaker (Affiliation)
14:15 (25+5)	Study of the few-nucleon system at intermediate energies	K. Hatanaka (RCNP, Osaka U.)
14:45 (15+5)	Application of chiral *NN* and *NNN* interactions to the ^{4}He photo-disintegration	S. Quaglioni (Lawrence Livermore National Lab.)
15:05 (15+5)	Measurements of the ^{2}H($\vec{p}$, 2p)n breakup reaction at 250 MeV	Y. Maeda (Kyushu U.)

Session 8: Tuesday Afternoon II, 15:45-17:15 (Chair: H. Sakai (U. Tokyo))

Time	Title	Speaker (Affiliation)
15:45 (25+5)	Many-body interactions in condensed-matter	S. Miyashita (U. Tokyo)
16:15 (25+5)	Gravitational many-body problem	J. Makino (National Astronomical Observatory of Japan)
16:45 (25+5)	Universal three-body repulsion suggested from neutron stars	T. Takatsuka (Iwate U.)

Session 9: Tuesday Afternoon III, 17:30-18:30 (Chair: H. Sakai (U. Tokyo))

Time	Title	Speaker (Affiliation)
17:30 (25+5)	Three nucleon force, fifty years ago	H. Miyazawa (U. Tokyo)
18:00 (25+5)	The Tucson-Melbourne family of forces and how they grew	S. A. Coon (Office of Science, DOE)

Session 10: Wednesday Morning I, 9:00-10:35 (Chair: M. Kamimura (Kyushu U.))

Time	Title	Speaker (Affiliation)
9:00 (30+5)	Coulomb effects in three- and four-nucleon scattering: a mask for the 3N force	A. C. Fonseca (U. Lisboa)
9:35 (15+5)	Four-body calculations for low-energy $3N + N$ reactions	E. Uzu (Tokyo Univ. Science)
9:55 (15+5)	Analyzing powers in the $dd \to {}^3\mathrm{He}n$ (${}^3\mathrm{H}p$) reactions at intermediate energies	V. P. Ladygin (Joint Institute for Nuclear Research)
10:15 (15+5)	Measurement of analyzing powers for *d-p* elastic scattering at Internal Target Station of Nuclotron	K. Suda (RCNP, Osaka U.)

Session 11: Wednesday Morning II, 10:55-12:35 (Chair: H. Tamura (Tohoku U.))

Time	Title	Speaker (Affiliation)
10:55 (15+5)	Triton and hypertriton binding energies with SU_6 quark-model baryon-baryon interactions	Y. Fujiwara (Kyoto U.)
11:15 (15+5)	Miyazawa supersymmetry	S. Catto (CUNY & Rockfeller U.)
11:35 (15+5)	Study of the tensor correlation in a neutron-rich sd-shell region with mean-field and beyond-mean-field methods	S. Sugimoto (Kyoto U.)
11:55 (15+5)	Three-body and Coulomb forces in few-body systems	S. Oryu (Tokyo Univ. Science)
12:15 (15+5)	Proposal to observe the strong Van der Waals force in the low energy proton-proton scattering	T. Sawada (Nihon U.)

Session 12: Wednesday Afternoon I, 14:00-15:20 (Chair: T. Motobayashi (RIKEN))

Time	Title	Speaker (Affiliation)
14:00 (25+5)	ΛNN three-body force due to coherent Λ-Σ coupling	Y. Akaishi (RIKEN Nishina Center & Nihon U.)
14:30 (15+5)	Equation of state of nuclear matter and the nuclear three-body force	H. Kanzawa (Waseda U.)
14:50 (25+5)	Direct reactions of Borromean nuclei	S. Shimoura (CNS, U. Tokyo)

Session 13: Wednesday Afternoon II, 15:40- (Chair: A. C. Fonseca (U. Lisboa))

Time	Title	Speaker (Affiliation)
15:40 (30)	Concluding remarks	B. Gibson (Los Alamos National Lab.)
16:10	Closing	

Author Index

T

U

V

W

Y

Z